International Conference on Advances in Materials and Manufacturing Engineering (ICAMME 2019)

IOP Conference Series: Materials Science and Engineering
Volume 653

Bhubaneswar, India
15 - 19 March 2019

ISBN: 978-1-7138-0911-1
ISSN: 1757-8981

Printed from e-media with permission by:

Curran Associates, Inc.
57 Morehouse Lane
Red Hook, NY 12571

CURRAN ASSOCIATES INC. proceedings.com

Some format issues inherent in the e-media version may also appear in this print version.

Institute of Physics
Dirac House, Temple Back
Bristol BS1 6BE UK

Phone: 44 1 17 929 7481
Fax: 44 1 17 920 0979

techtracking@iop.org

Additional copies of this publication are available from:

Curran Associates, Inc.
57 Morehouse Lane
Red Hook, NY 12571 USA
Phone: 845-758-0400
Fax: 845-758-2633
Email: curran@proceedings.com
Web: www.proceedings.com

TABLE OF CONTENTS

IOP Conf. Series: Materials Science and Engineering **653** (2019) 011001 doi:10.1088/1757-899X/653/1/011001

Preface

It is my great pleasure to present the proceedings of the "**International Conference on Advances in Materials and Manufacturing Engineering (ICAMME-2019)**". This conference is organized by the School of Mechanical Engineering of **Kalinga Institute of Industrial Technology** (KIIT), Deemed to be University, Bhubaneswar. We were honored to have **Prof. Mustafizur Rahman** from National University of Singapore, **Prof. D K Pratihar** from IIT Kharagpur, **Prof. Md. Mamun Habib** from BRAC University, Bangladesh, **Prof. Brij Kumar Dhindaw** from IIT Bhubaneswar, **Dr. Debasish Deb** (Executive Director) from Hindustan Aeronautical Limited (HAL) and **Prof. Biswanath Doloi** from Jadavpur University as Key Note Speakers on this special occasion.

We have received good number of research papers and after peer review 47 papers were accepted for IOP Conference Series: Materials Science and Engineering. These papers were presented in different categories of oral sessions. ICAMME-2019 covered four major areas namely Smart and Advanced Materials, Advanced Manufacturing, Design and Industrial Engineering, Thermal and fluid Engineering of Mechanical Engineering which are directly linked to socially useful research. The emerging areas like Advanced machining methods, automation, 3-D printing, robotics, waste management, biofuels, Nano-materials, PCM & smart materials, Supply chain management, which directly meets the needs of society in the form of design, planning and processing of various components.

As the convener of the conference, I extend my gratitude to our honorable Founder **Prof. A. Samanta**, KIIT and KISS, **Prof. Ved Prakash**, Chancellor, KIIT, **Prof. S. K. Acharya**, Pro-Chancellor, KIIT, **Prof. H. K. Mohanty**, Vice-Chancellor, KIIT, **Prof. SasmitaSamanta,** Registrar, KIIT, **Prof. R. Kodali,** Pro-Vice Chancellor, KIIT, **Prof. A. K. Sahoo**, Dean, SME, KIIT, **Prof. P. C. Mishra**, Assoc. Dean, SME, KIIT, for their continuous support and guidance throughout the planning and execution stages of ICAMME. I would like to thank technical program committee, organizing committee, faculty, staff and student volunteers of KIIT, SME for their dedicated support. Special gratitude is paid here to the Springer, for the Technical Sponsorship, Silver sponsor: GAIL, BBSR.

Finally I would like to thank all the authors, volunteers and persons who directly or indirectly contributed to the conference. Without their cooperation and full support, this conference would not have been possible.

IOP Conf. Series: Materials Science and Engineering **653** (2019) 011002 doi:10.1088/1757-899X/653/1/011002

Peer review statement

All papers published in this volume of *IOP Conference Series: Materials Science and Engineering* have been peer reviewed through processes administered by the proceedings Editors. Reviews were conducted by expert referees to the professional and scientific standards expected of a proceedings journal published by IOP Publishing.

Parametric Optimization of Electroless Ni-P coatings using Taguchi method

Subhasish Sarkar[1], Rishav Kumar Baranwal[1*] (0000-0001-6199-1459), Rajat Subhra Sen[1], Buddhadeb Oraon[1], Siddhartha Mukherjee[2], Gautam Majumdar[1]

[1]Department of Mechanical Engineering, Jadavpur University, Kolkata-700032, India
[2]Department of Metallurgical and Materials Science, Jadavpur University, Kolkata-700032, India

Abstract. The present work deals with surface roughness characteristics of electroless Ni-P coatings. The optimised process parameters have been determined based on L27 Taguchi Orthogonal array. Bath concentration of nickel sulphate, concentration of sodium hypophosphite and temperature of bath has been considered as variable in this study. The optimum combinations of process parameters and surface roughness have been considered as response variable of the study. It has been found that 24 g/L of Nickel Sulfate, 25 g/L of Sodium Hypophosphite and 80°C temperature gives the minimum surface roughness of the Ni-P coated sample which was 0.23μm. Characterisation of the substrate and the as deposited samples were done by scanning electron microscope (SEM) and optical microscopy for surface morphology, energy-dispersive X- ray analysis (EDAX) for element analysis and X-ray diffraction for phase analysis.

Keywords: Electroless coating: Taguchi Method: Orthogoanl Array

1 Introduction

Electroless Nickel plating is a process of depositing Nickel-Phosphorus coating over a substrate in the absence of electric current [1]. Coatings are deposited over the substrate by chemical reactions taking place in the aqueous solution. Electroless coatings are categorized into pure metal (copper, nickel, etc), polly alloy, nano and composite coatings [2]. The dissimilarity between electroless deposition process and the conventional electroplating processes is that the latter depends on an exterior source of dc in order to reduce nickel ions in the electrolyte to nickel metal on the substrate, while the former does not [3]. The plating bath conditions such as the type and concentrations of the reducing agent, stabilizer, pH and the temperature are governed by the deposition rate, properties of coated components and the structural behaviour of the deposits mainly. The characteristics and microstructures of EN coatings generally depend on the quantity of phosphorous alloyed in the deposit [4]. Due to high porosity the corrosion resistance and surface roughness of the coating is reduced. In general, phosphorus content (1–5%P) EN deposits are microcrystalline, medium content phosphorus (6–9%P) coatings has mixed crystalline and amorphous microstructures [5].
Surface roughness is an important property where the quality of the resulting surface plays an important role in determining the cost of the product. Like in a textile industry machine, the glazing surfaces of some automobiles [26-27]. Surface roughness has great impact on the mechanical properties like fatigue behaviour, corrosion resistance and creep life [6-9]. As a result, there is a need for development in modelling surface roughness and optimization of the controlling parameters to obtain a surface finish of desired level [10-12]. Thus it is important to optimize the *Centre line average roughness* (R_a). Several optimization techniques have been used in the past to optimize the input characteristics in order to obtain an optimum response [13-17].We have applied the Taguchi method for optimization. It uses the concept of orthogonal arrays and Signal to Noise ratio for optimization [18-22]. ANOVA analysis was done to find significant parameters and the significant interactions. It works on the principle of null hypothesis i.e. p value [23-25]. SEM and Optical Microscopy were done to find the surface morphology XRD and EDX was done to find the different phases and weight percentage of each element in the coating.

2 EXPERIMENTAL DETAILS

2.1. Substrate preparation and coating deposition

Copper substrate of size 30×25×.1 mm^3 is cut from a copper foil (99.0% purity, lobachemie) for the deposition of electroless nickel coatings. The rectangular substrate is cleaned for removing oxide layer and foreign particles.Firstly the samples are rinsed with25% HCl solution in 3:1 ratio for 10 min for removing foreign particles and oxide layer. Then, after washing with deionized water, the substrates are dipped into palladium chloride solution ($55^{0}C$) to activate the surface by overcoming the initial energy barrier and start the deposition process. Then dipping into distilled waterfor a few seconds, the substrate is ready to be put into electroless bath. This also introduces a significant amount of deposit thickness, higher deposition rates and ensures good adhesion of the coatings to the substrate. Nickel sulphate is used as the source of nickel with sodium hypophosphite as the reducing agent and sodium citrate is added as complexing agent. The coating is deposited over substrate by various steps of chemical reaction. Regarding the Electrochemical reaction, catalytic oxidation of the hypophosphite yields electrons at the catalytic surface, which in turn reduces nickel and hydrogen ions. The chemical concentration of the bath is pre-determined by trial and error method. The pH value of the solution is maintained around 4.5 by continuous monitoring with a pH meter and the temperature of bath is also maintained by constant supervision with a thermometer. A bath of volume 250 ml is prepared and deposition is carried out for a period of 1 hour. After that, the substrates coated with Ni-P are taken out of the bath and rinsed in distilled water. Then all samples are mounted with the help of epoxy resin .Finally it is ready for testing of surface roughness. Thus it is essential that all samples are of uniform roughness.

2.2 SURFACE ROUGH NESS MEASUREMENT

Surface roughness measurement is done using a portable stylus and skid type profilometer, Talysurf (Taylor Hobson, Surtronic 3+). The profilometer is set to cut-off length of 0.8 mm, gaussian filter, and traversed speed 1mm/s with 4mm evaluation length. The measured profile is digitized and processed through the advanced surface finish analysis software Talyprofile for evaluation of the roughness parameters. Talysurf measures the Centre line average roughness. Surface roughness was determined by taking the average of surface roughness recorded over four different places of the coating. Centre line average roughness (R_a) is defined as arithmetic mean deviation of the surface height from the mean line through the profile while the mean line is defined in such a way that it has identical area on either side of it.

2.3 Design of Experiment

Design of experiments (DOE) method is used to reduce the experimental time, number of experiments and computational cost and other resources with the help of conclusive information.Taguchi Optimization method, where orthogonal array (OA) is employed to reduce the number of experiments, is used to determine the optimal process parameters. An OA requires minimum number of experimental trials and provides the shortest possible matrix combination in which parameters are varied to determine the mean data and S/N ratio data through which optimized levels of each factor is determined. In the present investigation, an L_{27} orthogonal array is used, which has 27 rows and 13 columns where each column represent the factors and their interactions while each row represent the levels of factors and the interacting levels of different factors.

2.4. Taguchi method

Taguchi Modelling is a very effective way to optimize the input parameters in accordance with the output of a signal/experiment with the use of Signal to Noise ratio. The S/N ratio is a ratio of the significant data to the error in an experiment (noise). Hence, a higher value of S/N ratio suggests a better optimization. Hence, the most optimum experiment will have a maximum S/N ratio. S/N ratio works in three different ways according to the response. If the response needs to be optimized in a maxima region then we shall use the 'Larger the better' concept of the S/N ratio and if the response needs to be optimized in such a way so that the optimum response is minimum, then we use the S/N ratio which will give 'smaller is better'. In order to get a nominal value of the response, we shall use the 'Nominal is best' formula to get the optimized data. In case of surface roughness we use smaller is the best. Higher S/N ratio designates lesser unrequired noise and more required signal. Therefore a higher S/N ratio suggests a higher optimized results.

IOP Conf. Series: Materials Science and Engineering **653** (2019) 012001 doi:10.1088/1757-899X/653/1/012001

3. RESULT AND DISCUSSION

3.1. Surface Roughness

The surface roughness is measured using Talysurf instrument mentioned in the section 2.2. The result obtained from the experiment in given in table 1.

Table1: Surface roughness for each coating

Experiment No	Surface roughness
1	0.74
2	0.34
3	0.23
4	0.45
5	0.87
6	0.46
7	0.98
8	0.76
9	0.79
10	0.84
11	0.71
12	0.32
13	0.5
14	0.67
15	0.44
16	0.81
17	0.73
18	0.94
19	0.84
20	0.54
21	0.39
22	0.75
23	0.69
24	0.96
25	1.03
26	0.83
27	0.61

The surface roughness measured of substrate is calculated at 1.19 μm.

3.2 Taguchi analysis

It works on the basis of the Signal to Noise ratio discussed in section 2.4. The S/N ratio follows a decreasing trend in case of the 3 different levels in the factors Temperature and Nickel Sulphate. From figure 1, it is clearly seen that level 1 of temperature, level 1 of Nickel Sulphate and level 3 of Sodium Hypophosphite are the optimized parameters in order to optimize the surface roughness of the coating, as they have the maximum S/N ratio for that level. The optimized surface roughness is obtained at 0.23 μm.

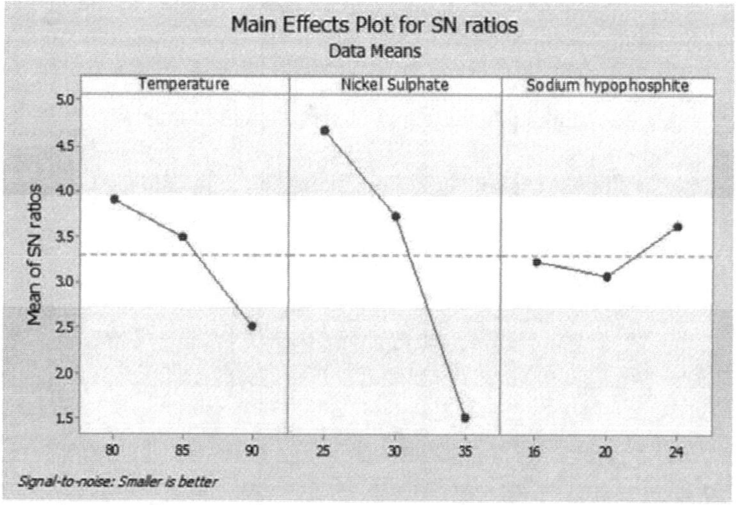

Figure 1: Main Effect Plot for SN Ratios

3.3 Characterization

The experimental sample before and after coating are characterised using SEM and Optical Microscopy, XRD and EDX.

Figure 2: Scanning Electron Microscope Image of Copper Substrate without etching

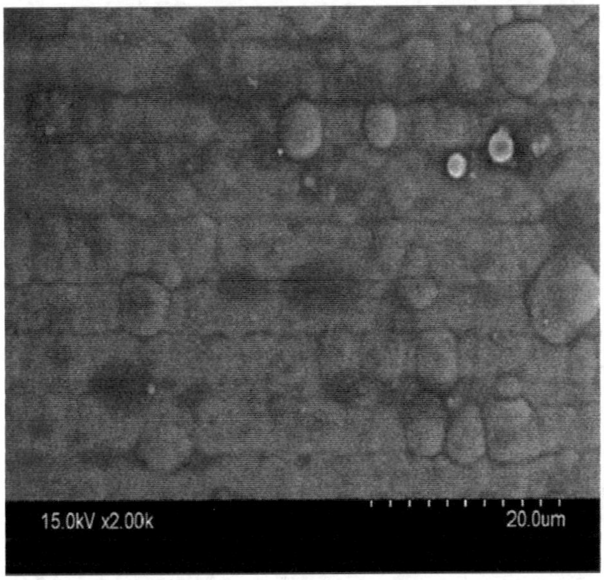

Figure 3: Scanning Electron Microscopy micrograph of optimized as-deposited coated sample

Figure 2 shows the micrograph of substrate without etching. The micrographs revealed mechanically worked rolled elongated structure. The deposited particles are spherical in nature dispersed over the surface. Figure 3 provide the nature of deposits and its morphology and alignments. The over lapping of particles and its interconnection is clearly visible. Type of deposit is quite dense and firmly attached. The corresponding EDX analysis is indicates the presence of elemental Ni and P in the optimized coated sample. The weight percentage of the elements are Nickel 87.88 and Phosphorus 12.12.

XRD analysis of Optimize Coated Sample:

XRD was done for phase determination of the optimised coated sample using X-ray diffraction in Rigaku Ultima- III machine. The range of 2 theta was from 20° to 85° with a scan speed of 2° min−1) in Cu Kα radiation. It can be concluded that compare between copper substrate and as deposite optimised coated sample cu phase converted to micro-crystalline Ni_3P phase which ware shows in below.

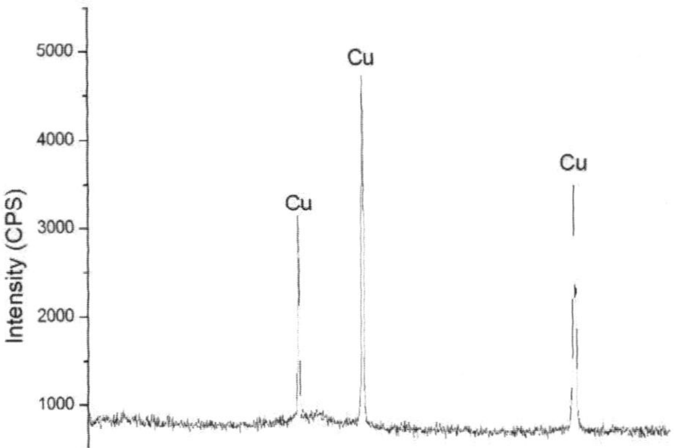

Figure 4. X-ray diffraction pattern of Copper substrate

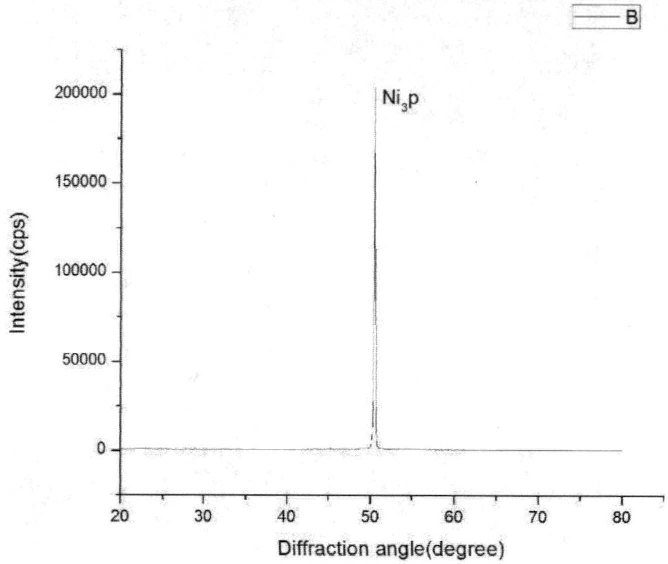

Figure 5. X-ray diffraction pattern of optimised as deposited Ni-P coated sample

4. Conclusion

Uniform coating is successfully deposited on the substrates' surface and Taguchi analysis is successfully applied to find the optimized surface roughness at 0.23 µm with 25g/L Nickel Sulphate , temperature 80 C and 24 g/Lof Sodium Hypophosphite. The ANOVA analysis shows that all the 2- way interactions between the three factors, including the Nickel Sulphate factor, are very important in minimising the surface roughness to such low value. EDAX analysis shows the presence of high percentage of Phosphorus which is thus significant in diminishing the surface roughness to a minimal value. The surface roughness of substrate is found to be of 1.19 µm. We have found that decrease in percentage of optimised coated sample to copper substrate is 75.63%.

References

1. Majumdar G, Biswas N, Pramanik A, Sen RS, Chakraborty M, 2017, Advances in Materials and Processing Technologies, 3(2):186-95.
2. Sudagar J, Lian J, Sha W, 2013, Journal of Alloys and Compounds, 571:183-204.
3. Biswas N, Baranwal RK, Majumdar G, Brabazon D, 2018 , Advances in Materials and Processing Technologies. 28:1-8.
4. Kucernak AR, Sundaram VN, 2014, Journal of Materials Chemistry A, 2(41):17435-45.

5. Narayanan TS, Baskaran I, Krishnaveni K, Parthiban S, 2006, Surface and Coatings Technology. 200(11):3438- 45.

6. Uysal M, Akbulut H, Tokur M, Algül H, Çetinkaya T, 2016, Journal of Alloys and Compounds, 654:185-95.

7. Sahin Y, Motorcu AR, 2005, Materials & design, 26(4):321-6.

8. Podgornik B, Hogmark S, Sandberg O, 2004, Surface and Coatings Technology, 184(2-3):338-48.

9. Svahn F, Kassman-Rudolphi Å, Wallen E, 2003, Wear. 254(11):1092-8.

10. Park IC, Kim SJ, 2018, Applied Surface Science.

11. Singh A, Datta S, Mahapatra SS, Singha T, Majumdar G, 2013,Journal of Intelligent Manufacturing, 24(1):35-44.

12. Oraon B, Majumdar G, Ghosh B, 2008,Materials & Design, 29(7):1412-8.

13. Tong S, Hou S, Zheng Z, Zhou J, Bao G, 2010, Coating optimization of superparamagnetic iron oxide nanoparticles for high T2 relaxivity. Nano letters, 10(11):4607-13.

14. Pesque JJ, Bouche DP, Mittra R, 1992, Optimization of multilayer antireflection coatings using an optimal control method. IEEE transactions on microwave theory and techniques. 40(9):1789-96.

15. Oksa M, Turunen E, Suhonen T, Varis T, Hannula SP, 2011, Coatings, 1(1):17-52.

16. Li JF, Liao H, Normand B, Cordier C, Maurin G, Foct J, Coddet C. 2003, Surface and Coatings Technology, 176(1):1-3.

17. Oraon B, Majumdar G, Ghosh B, 2007, Materials & design,28(7):2138-47.

18. Bharat KS, Mukund SA, Ramrao TB, 2018, International Journal of Current Trends in Science and Technology, 8(1):72-5.

19. Selvaraj DP, Chandramohan P, Mohanraj M, 2014, Measurement, 49:205-15.

20. Kumar RS, Sureshkumar K, Velraj R. 2015, Fuel, 15;140:90-6.

21. Chen PC, Yang MW, Wei CH, Lin SZ. 2016, International Journal of Greenhouse Gas Control, 45:245-52.

22. Chen WC, Nguyen MH, Chiu WH, Chen TN, Tai PH, 2016, The International Journal of Advanced Manufacturing Technology, 83(9-12):1873-86.

23. Kucuk U, Eyuboglu M, Kucuk HO, Degirmencioglu G, 2016, International journal of cardiology. 209:346.

24. Ramakrishnan R, Mao K, 2017, International Journal of Mechanical and Industrial Technology, 4(2348-7593):72- 9.

25. Ashton JC,2015, Nature methods, 12(12):1108.

26. Feldstein N, inventor; Surface Tech Inc, assignee. Metallic articles having dual layers of electroless metal coatings incorporating particulate matter. United States patent US 4,358,922. 1982 Nov 16.

27. Akafuah NK, Poozesh S, Salaimeh A, Patrick G, Lawler K, Saito K, 2016, Coatings, 6(2):24.

IOP Conf. Series: Materials Science and Engineering **653** (2019) 012002 doi:10.1088/1757-899X/653/1/012002

A State of the Art Review on Modelling Work Carried Out During Friction Stir Welding

Manish Kumar Singh[1], Rajesh Kumar Porwal[2], Sanjay Mishra[3]

[1,2]Faculty of Mechanical Engineering, Shri Ramswaroop Memorial University, Lucknow, India.

[3]Department of Mechanical Engineering, Madan Mohan Malaviya University of Technology, Gorakhpur, India.

E.mail: porwal.me@srmu.ac.in

Abstract. Welding of soft materials like aluminium, copper by fusion welding process is not achievable due to the generation of defects. These materials have a wide area of industrial applications. Therefore, their welding is one of the most challenging tasks in the joining industries. Since welding is used to join two materials or parts of materials of similar or dissimilar nature, therefore, the repudiation of the welded product due to poor weld quality is not acceptable in the modern joining industries. Friction stir welding can minimize these defects because in this process there is no melting of workpiece occurs. In the past, many researchers have examined a number of approaches to enhance the weld joint properties. The observational and notional analysis shows weld joint quality can be enhanced remarkably with the appropriate choice of process parameters. This article analyses the research work accomplished by various researchers in the field of modelling of the friction stir welding operation.

1. Introduction

FSW is a rapid growing, modernized and environmental friendly solid-state welding technique. W. Thomas et al. at The Welding Institute, Cambridge, United Kingdom, first recommended it in 1991. A schematic view is shown in figure 1 explains the process of FSW operation. The working of FSW is based on the principle of friction welding in which parts are joined by the application of heat and pressure. In the process of FSW, a non-depleting rotating tool vertically dunks into two abutting surfaces of the workpiece and move forth on the line of the joint in order to create a weld. The workpiece has to be rigidly clamped onto a fixture in such a way that prevent workpiece faces from being apart by force. Shoulder, as well as a pin, is the main components of a rotating tool. The function of tool shoulder is to generate heat and then preclude the ejection of material, whereas tool pin deforms materials around the tool and heat generation as its secondary function during operation [1-4]. FSW mostly used to join soft and lightweight materials and its alloys such as aluminium, copper, magnesium, titanium. It provides excellent mechanical properties of weld joint and has a lot of advantages over the fusion welding process like, able to weld thick plate, no solidification cracking, no shielding gas, no requirement of filler materials, no environmental hazards, low distortion, low shrinkage, low residual stress, and also able to join dissimilar metal and their alloys. Thus, this method has remarkable effectuation in marine, aerospace, rail and transportation, military and nuclear industries for making of panels for decks, marine and transport structures, fuselages, container bodies, wings, and fuel tanks. The microstructure of friction stir welded specimen can be categorized into four zones. Plastic deformation arises in the weld nugget zone and described by equiaxed, fine and recrystallized grain arrangement. The width of the weld nugget zone (WNZ) is marginally greater than the diameter of the pin. The Heat affected zone is an outermost region of the weld joint were affected by the thermal field during the welding process and yet not have any deformation. HAZ of the welded specimen is narrower during FSW as compared to fusion welding [5-6].

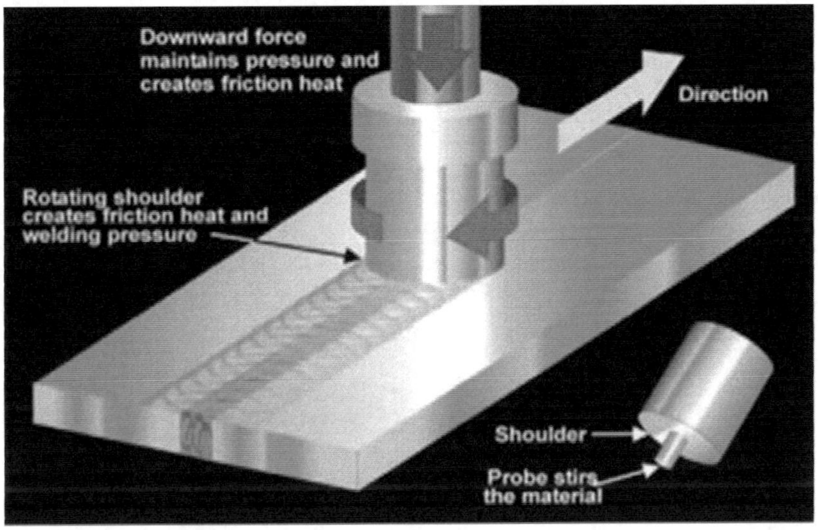

Figure 1. Schematic details of FSW technique [4].

2. Review on Modelling techniques of FSW

FSW process includes phase change, materials flow, sticking, temperature distributions, heat exchange between the workpiece and tool. To study these parameters many process techniques are used known as modelling techniques. Many industries used modelling techniques because they have benefits such as reducing the number of experimental trials in developing a new or modified process, optimizing the process, improve quality by reducing process variability and help in understanding the process. It also supports the development work through assisting the selection of process parameters and tool design [7-8].

2.1. Analytical and Numerical Modelling

The heat induced to deform materials is due to friction produced between the tool and work part, then conducted into backplate or convected from the top surface. To analyse this heat flow analytical and numerical modelling techniques are used. Schmidt et al. developed an analytical modelling technique for heat generation, which was established on various contact conditions between workpiece and tool surface, namely known as sticking, partial sticking and sliding condition during FSW of AA 2024-T3 alloy. On the basis of experimental data, they revealed that sticking condition yields an equal relationship between the rate of heat generation and plunge force [9]. An analytical model was suggested by Jinwen Qian et al. to estimate the peak temperature, using continuum mechanics and the inverse hyperbolic sine function. They concluded that the calculated value of the peak temperature was coherent with the literature value for various aluminium alloys [10]. Furthermore, numerical modelling provides a quantitative understanding of material flow and heat generations. To study the heat generation during FSW of AA2014-T6 Al alloy Mehta et al. presented a numerical model. They

IOP Conf. Series: Materials Science and Engineering **653** (2019) 012002 doi:10.1088/1757-899X/653/1/012002

used EN40 as a tool material along with four tool pin profile for their experimental investigation. They found that the resultant maximum shear stress decreased from the triangular to the hexagonal pin profile can be associated with a decrease in traverse force, whereas peak temperature accomplished by an increase in the number of tool pin side [11]. Heurtier et al. presented an investigation on the weld joint of AA2024-T351 Al alloy during FSW and developed a mechanical and thermal model. They used their model to study the path of weld material element, strain, strain rates, temperature estimation and microhardness at different weld zones. They found that increasing the velocity of the tool, the hardness profile becomes smooth near the WNZ because of the decrease in the average temperature of the weld. They also observed that the proposed model was suitable for prediction of oxide distribution in the welded zone to identify the weak zone after welding [12]. A thermal model was used for simulation and prediction of grain size as well as mechanical properties of 2 mm thick niche copper during FSW by Heidarzadeh et al. The tool rotational speed (TRS) of 400, 700, 900 rpm and welding speed (WS) of 100, 200, 300, 400 mm/min was considered for the experiment. Their modelling and experimental value for grain size at different WS and for peak temperature at different TRS are shown in figure 2 and 3, which revealed that higher TRS and lower WS results higher grain size due to the generation of higher peak temperature [13].

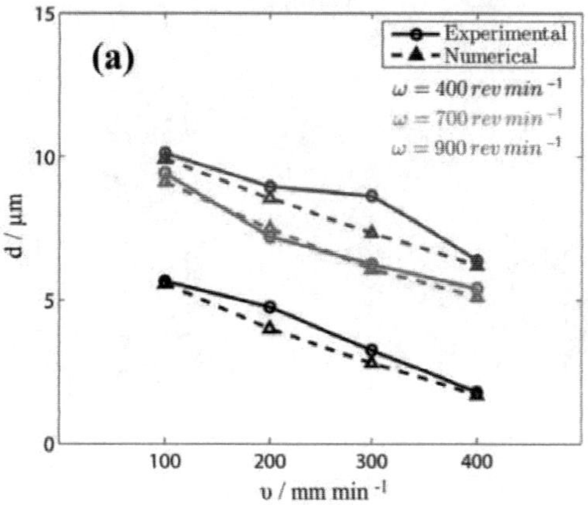

Figure 2. Predicted and measured grain size [13].

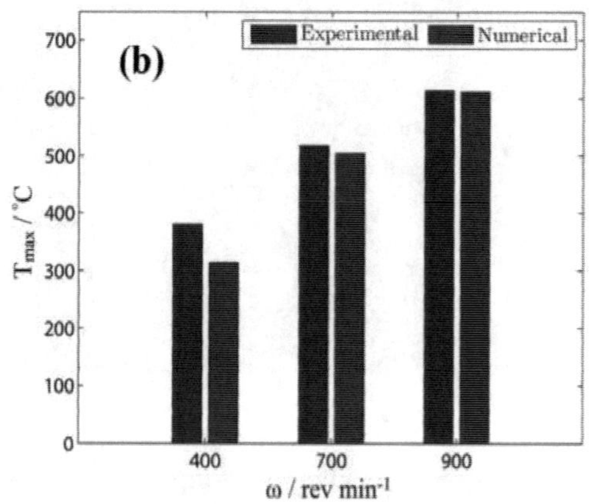

Figure 3. Predicted and measured pick temperature [13].

Three-dimensional FE model was developed by Kiral et al. with ANSYS and HyperXtrude software to model the FSW process of 6061-T6 Al alloy. With the purpose of predict temperature distribution, the Transient thermal FEA was performed. They used moving heat source with heat distribution, which simulates the generation of heat from friction between shoulder and work specimen for their analysis. They found that the used heat moving source technique was a reliable method to simulate FSW process and revealed that maximum temperature near the joint enhanced with an increase in time of tool holding and TRS, while temperature decreased with increased in WS [14]. A fully coupled 3D thermomechanical FEM has been developed by Aziz et al. to investigate the thermal heat generation and distribution during the FSW of AA219-T87. They considered both heat dissipation from friction and plastic deformation. Two weld schedule with constant TRS of 350 rpm was used. In the first schedule, welding speed was 1.27 mm/sec while in the second it was 2.54mm/sec. They found temperature distribution field obtained from finite element simulation was similar to the experimental value for both cases and concluded that the temperature around the pin shoulder was higher than adjoining location as a consequence of friction and plastic deformation. Furthermore, the maximum-modelled temperature was observed 406.74 and 450.6 ºC [15]. El-Sayed et al. reported a FEM to predict the temperature distribution and residual thermal stresses in AA5083-O Al alloy during FSW. They used infrared thermal image camera for measuring temperature experimentally. The joint was fabricated by using threaded and tapered smooth pin profile at different TRS and WS. They observed that the peak temperature was higher with threaded pin profile as compared to a tapered smooth pin profile at same TRS and WS. Furthermore, 1.22% and 3.31% errors obtained in peak temperature for measured and model value at 400 rpm and 630 rpm respectively [16]. Yang et al. created a two dimensional finite element model to calculate the linear friction welding process of GH4069 super alloy using ABAQUS software. They investigated the temperature fields of work material, stress and equivalent plastic strain fields at different welding speed during the welding process. They reported that with increase in welding time heat conducted from weld interface to specimen was more but the heat conduction was not evident when the friction time was more than 1.75 sec because there was gradual extrusion of flash. Furthermore, in the welding time range from 0.75 to 1.65 sec friction, shear stress decreased while extrusion force increased [17]. Fadi-Al-Badour et al. proposed a 3-dimensional localized finite element model considering the fact that work piece and tool was modeled by using Eulerian and Lagrangian formulation respectively for prediction of frictional coefficient, welding speed and plugging control method (position and force) of the Al6061-T6 Al alloy during FSW with H13 Steel as a tool material. They concluded that the void size was directly affected by frictional contact value and tool features. Moreover, the higher coefficient of friction value produced smaller void size [18]. A three-dimensional thermo-mechanical model including the mechanical action of the tool shoulder and thermo-mechanical effect of weld material was developed by Chen et al. for friction stir butt welding of 6 mm thick 6061-T6 Al alloy, in order to understand the thermo-mechanical process. They observed maximum temperature gradients were located just beyond the shoulder edge in longitudinal and lateral directions and residual stress in longitudinal direction was greater than the lateral residual stress at the top surface of the weld. Furthermore, high stress was located in the region extending down from the peak to the mid thickness of the weld [19]. The thermal modeling with three dimensional heat model of underwater FSW on 2219-T6 Al alloy was conducted by Zhang et al. They considered temperature dependent properties for their analysis and found that the surface heat flux of tool shoulder and maximum peak temperature of underwater weld joint were higher and lower respectively as compared to normal weld joint. They also observed that the underwater weld joint exhibited a narrow area of high temperature distribution and lower welding thermal cycles in different weld zones [20]. Fratini et al. studied an integrated numerical and experimental exploration by effects of in-process water cooling using FSW for AA7075-T6 in butt joint configurations. The temperature distribution, tensile strength and microstructure of the welded joint were observed, along with an expanded finite element model to interpret specimens obtained under various process conditions. Despite the analysed temperature histories, in-process water cooling treatment was increased joint strength, diminished destructive effect on nugget zone, besides reduced the material softening and thermal flow adjacent to the tool [21].

2.2. *Experimental Modelling*

Micallef et al. developed a computational fluid dynamics (CFD) model and analyse the amount of heat flux during FSW of 6 mm DH36 steel plate. They assumed full sticking conditions at tool shoulder and workpiece interface and the heat was generated by plastic deformation and shearing. They studied the effect of different TRS and WS on the generation of heat and WNZ and found that the amount of total heat generation at different welding conditions was correlated with the tools radial and angular location [22]. Torres developed a CFD model for the endorsement and prediction of temperature distribution during FSW of Al 6061 and AISI SAE 1020 steel. They concluded that the principal heat generation source in steel side was friction while in aluminium side it was plastic deformation [23]. RSM is constructive to establish a suitable approximation between response and independent variable for a truly functional relationship which can describe the nature of the joint. A mathematical model was developed by Elnabi et al. using RSM for tensile strength and ductility during FSW of dissimilar AA5054 and AA7075 Al alloy. They found the high accuracy of the model for tensile strength that was 97.678 and 99.56% for mean and signal to noise (S/N) ratio respectively, while for ductility it was 93.56 and 97.320% [24]. Maleki et al. developed ANN model based on back propagation algorithm to investigate the FSW parameters during joining of AA7075-T6 alloy. They used TRS, WS, applied force (AF), shoulder diameter, tool hardness, pin diameters as the input of network whereas the output of network was a tensile strength, notch tensile strength, hardness, and yield strength. They concluded that ANN modelling technique could be used for modelling and prediction of FSW parameters [25]. ANN model has been proposed by Ghetiya et al. to predict the tensile strength of AA8014 Al alloy plate during the FSW process. The inputs of the proposed network were tool shoulder diameter, TRS, AF, and WS while the weld joint properties were the output of the network. They observed that the value of a prediction network was very close to the experimental value. Moreover, the error between predicted and experimental tensile strength was less than 3% for their predicted ANN Model [26]. Okuyuca et al. assessed the influence of ANN in predicting the parameters of FSW for Al plates. In their study, they took the parameters of WS and TRS as network inputs, and for network output, the parameters were yield strength, ultimate strength and ultimate length variation of samples of the weld. Finally, the results of their work showed an appropriate correspondence between the experimental results and the neural network results [27].

3. Conclusion

FSW is the most successfully joining process for soft materials as compared to fusion welding processes. Earlier it was used only for aluminium and its alloy but regularly many researches have been done to investigate the effect of input parameters and improve its working area. Nowadays this method is also used to join hard materials like steel, nickel, metal matrix composites and its alloys. The above review gives us brief knowledge about the different modelling techniques carried out by the researchers during the FSW process. These modelling techniques provide the visualization of the behaviour of welded material and assist to analyse the effect of parameters on the weld joint properties without performing any experiment. The numerical and analytical modelling techniques found to be a good prediction of heat generation, material flow and temperature distribution in the stir zone during FSW. From the literature, it is concluded that ANN model is more relevant models as compared to other modelling technique for simulation and prediction.

References

[1] Gurel C, and Selcuk M 2014, *J. Mater. Eng. Perform.* **23**, 1936-53

[2] Singh M K, and Porwal R K 2018, *Proc. In National Seminar on Advances in Materials, Manufacturing and Renewable Energy Systems*, IJTRA, Lucknow

[3] Phillips, and David H 2016 *Welding Engineering: An Introduction* (New York: Wiley-Interscience)

[4] Raju G S, Balunaik B, Ravi Shankar D V, and Govardhan D 2017, *Int. J. Mech. Eng. Technol.* **8** (10), 774-83

[5] Grimm A, Schulze S, Silva A, Gobel G., Standfuss J, Brenner B, Beyer E, and Fussel U 2015, *Mater. Today Proc.* **2**, S169-78

[6] Seidel T U, and Anthony R P 2001, *Metall. Mater. Trans. A*, **32**(11) 2879-84

[7] Kundu J, and Singh H 2017, *Adv. Mater. Process. Technol.***4**(2), 183-99

[8] Hao S, Chuan S W, Marcel B, and Michael R 2015, *Mater. Des.* **77**, 114-25

[9] Schmidt H, Hattel J, and Wert J 2004, *Model. Simul. Mater. Sci. Eng.* **12**, 143-57

[10] Jinwen Q, Yan O, Jinglong L, Yifeng X, Liang W, and Yanfei X 2017, *Sci. Technol. Weld. Join.* **22**(6), 520-25

[11] Mehta M, Reddy G M, Rao A V, and De A 2015, *Defence Technol.* **11**, 229-36

[12] Heurtier P, Jones M J, Desrayaud C, Driver J H, Montheillet F, and Allehaux D 2006, *J. Mater. Process. Technol.* **171**(3), 348-57

[13] Heidarzadeh A, Jabbari M, and Esmaily M 2015, *Int. J. Adv. Manuf. Technol.* **77**, 1819-29

[14] Kiral B G, Tabanoglu M, and Serindag H T 2013, *Math. Comput. Appl.* **18**(2), 122-31

[15] Aziz S B, Mohammad W D, Daniel H J, Muhammad W A, et al. 2017, *Acta Metall. Sin. (English Lett.)* **31**(1), 1-18

[16] El-Sayed M M, Shash A Y, and Abd-Rabou M 2018, *J. Mater. Process. Technol.* **252**, 13-24
[17] Yang X, Li W, Li J, Xiao B, Ma T, Huang Z, and Guo J 2015, *Mater. Des.* **87**, 215-30
[18] Al-Badour F, Merah N, Shuaib A, and Bazoune A 2013, *J. Mater. Process. Technol.* **213**(8), 1433-9
[19] Chen C M, and Kovacevic R 2003, *Int. J. Machine. Tools. Manuf.* **43**(13), 1319-26
[20] Zhang H J, Liu H J, and Lei Y U 2013, *Trans. Nonferrous. Metals. SOC* **23**(4), 1114-22
[21] Fratini L, Buffa G, and Shivpuri R 2010, *Acta. Mater.* **58**(6), 2056-67
[22] Micallef D, Camilleri D, Toumpis A, Galloway A, and Arbaoui L 2015, *Proc. Inst. Mech. Eng. Part L J. Mater. Des. Appl.* **230**, 586-602
[23] Torres E2012, *Proc. Int. Conference on Trends in Weld. Res.*, pp-604-10, Chicago, Illinois
[24] Elnabi M M, Osman T A, El Mokadem A E, and Elshalakany A B 2018, *Adv. J. Grad. Res.*.**4**, 1-14
[25] Maleki E 2015, *IOP Conf. Ser. Mater. Sci. Eng.* 103
[26] Ghetiya N D, and Patel K M 2014, *Procedia Technol.* **14**, 274-81
[27] Okuyucu H, Kurt A, and Arcaklioglu E 2007, *Mater. Des.* **28**, 78-84

A novel approach in comparison and experimentation of Hybrid Metal Matrix Composites using advanced MCDM methods

Soutrik Bose[1, 2*], Samiran Samanta[3], Nabankur Mandal[1], Soumyajit De[1], Niladri Sekhar Mistry[1], Partha Koley[1] and Titas Nandi[2]

[1]Department of Mechanical Engineering, MCKV Institute of Engineering, 243 G.T. Road (N), Liluah, Howrah 711204, West Bengal, India
[2]Department of Mechanical Engineering, Jadavpur University, Kolkata 700032, West Bengal, India
[3]Department of Mechanical Engineering, Kalinga Institute of Industrial Technology (KIIT), Deemed to be University, Bhubaneswar, Odisha 751024, India

[1,2*]soutrikboseju@gmail.com, [3*]samirankgec@gmail.com

Abstract. Factual material selection is essential in the manufacturing arena with a superior product and minimum cost. This paper highlights on the advanced multi-criteria approaches namely additive ratio assessment (ARAS) and multi-attributive border approximation area comparison (MABAC) methods for the best material selection of a hybrid composite using aluminum (Al) as base material varying different reinforcement weights and recycling with various industrial wastes by stir casting. Al/WCE gives superior tribomechanical properties at lower cost than the other reinforcements. ARAS and MABAC methods are applied, based on the properties and attributes of the hybrid composites, to compare between the computational and experimental results. The results exactly corroborate with the previous research results which authenticate the expediency of these methods during the solving of complex hybrid material selection problems. Out of 48 different samples, the best hybrid composite material obtained by ranking is material number A30 by ARAS approach (rank 1) which is Al+Al_2O_3 at 12.5% wt. addition and material number A24 (rank 2) which is heat treated 12.5%wt. WCE. But the same A30 is obtained to be rank 5 and A24 is obtained to be rank 1 by MABAC method. A24 is obtained to be the best hybrid composite after the experimentation.

Keywords. ARAS, Hybrid composite, MABAC, MCDM, Wettability

1. Introduction

Enhanced physical and tribo-mechanical properties of AMC is the most key aspect in modern industrial belts like in automobile and aerospace engineering for fabrication of different light weighted, cost effective components made of various nano composites reinforcing with diverse waste materials. Modern ceramic composites' fabrication through stir casting route is the recent development used in modern industries, so in this paper eggshells (ES), waste carbonized egg shells (WCE), boron carbide (B_4C) and cow dung ash (CDA) by altering the different reinforcements' percentage of weights by stir casting route. Selection of the best material in an adept material selection approach depends on the manufacturing scope and to curtail the cost while achieving the product performance objectives. Inappropriate material assortment leads to immature product failure with high cost. So proper identification and selection of pertinent material with definite functionality to acquire the best product with least cost and high performance is the most important criteria. The physical, chemical, metallurgical and tribomechanical properties are the prime concern. Therefore, two novel, efficient and organized approaches, like ARAS [1, 2] and MABAC [3-5] based on some strong mathematical foundations, are hence proposed for the amalgamation of design and manufacturing scopes but the real performance may vary from the expected results which need to be optimized. Indecent wettability is formed on the surface of the molten metal by the oxide and other impurities addition. Therefore, additions of hybrid reinforcements are essential factors. Interfacial bonding must be strong enough for improved wettability enhancement. Hassan et. al. [6] obtained improved tribo-mechanical characteristics by increasing percentage composition of WCE compared to ES. Deep et. al. [7] obtained enhanced mechanical and metallurgical properties with increased weight fractions of reinforcements of Al-SiC to an optimum level after several comparisons. After going through all the literatures, the main gaps are identified that very few research works are carried on ES and WCE based green MMC. It is also identified that very few novel methods are proposed to fabricate green hybrid MMC reinforcing with SiC, SiC+SSA, CDA and B_4C. The main objective of this paper is to inspect the probability to fabricate Al+WCE composites through stir casting route by recycling these hazardous wastes. Hence all the tribo-mechanical properties and microstructures can be compared and optimized.

2. ARAS

The ARAS method is the perceptible calculations depending on the utility theory where the utility value designates the relative efficiency of various alternatives. It is unswervingly relative to the virtual outcome of the criteria values and weightage parameters of the criteria. Table 1 below shows the nomenclature and criteria of the 48 composite materials selected for ARAS and MABAC. The sequential steps of ARAS [1, 2] method are:

Step 1: A definite normalization process is proposed for the beneficial attributes and reciprocal of all the criteria is considered for non-beneficial attributes and hence the normalized decision matrix is determined with respect to all the alternatives:

$$X_{ij'} = \frac{1}{X_{ij}} \tag{1}$$

Step 2: The normalized values are calculated:

$$R=[r_{ij}]_{mxn} = \frac{X_{ij'}}{\sum_{i=1}^{m} X_{ij'}} \tag{2}$$

The weighted normalized decision matrix D is determined:

$$D=\left[Y_{ij}\right]_{mxn} = r_{ij}\, x\, w_j \left(i=1,2,...,m;\, j=1,2,..,n\right) \tag{3}$$

Step 3: The optimality function S_i for i^{th} alternative is hence determined:

$$S_i=\sum_{j=1}^{n} Y_{ij} \tag{4}$$

Depending on the higher S_i value, better alternative is obtained, which is directly proportional to the decision matrix values and criteria weights.

Step 4: The degree of utility U_i is computed for each alternative and compared with the maximum efficient alternative S_o which is given below:

$$U_i = \frac{S_i}{S_o} \quad (5)$$

Table 1. Nomenclature of the criteria

Criteria	Abbreviation	Beneficial	Non-beneficial
Hardness(BHN)	HN	HN	PO
Tensile	TS	TS	CR
Fatigue	FS	FS	DC
Porosity	PO		
Corrosion	CR		
Ductility	DC		

The degree of utility of the alternatives varies from 0-100% and the highest utility value U_{max} is considered to be the best preference (Rank 1) in the midst of all the alternatives. By following the above steps the best hybrid composite material is obtained by ranking and material number A30 is found to be rank 1 which is $Al+Al_2O_3$ at 12.5% wt. addition and material number A24 is found to be rank 2 (Fig. 1.) which is heat treated 12.5%wt. WCE. After experiments, best result material obtained is A24 material number.

3. MABAC

MABAC method [3-5] is incorporated after acquiring the weight coefficients. The key computational formulation of this method is to obtain the distance of the criteria function. Following are the required steps:

Step 1: At first the primary decision matrix $X = \left[x_{ij}\right]_{mxn}$ is formed.

Step 2: The ultimate decision matrix $N = \left[n_{ij}\right]_{mxn}$ is normalized. The rudiments are:

(a) For Beneficial type criteria where an upper value is preferable:

$$n_{ij} = \frac{x_{ij} - x_{i-}}{x_{i+} - x_{i-}} \quad (6)$$

(b) For Cost type criteria where a lesser value is preferable:

$$n_{ij} = \frac{x_{ij} - x_{i+}}{x_{i-} - x_{i+}} \quad (7)$$

where, x_{ij}, x_{i+} and x_{i-} are the rudiments from the primary decision matrix X, for which x_{i+} delineated as: $x_{i+} = \max(x_1, x_2, ..., x_m)$ and is the maximum value and $x_{i-} = \min(x_1, x_2, ... , x_m)$ and is the minimum value of the observed criterion according to the alternatives.

Step 3: Then the weighted normalize decision matrix V is calculated.

$$v_{ij} = w_i.(n_{ij} + 1) \quad (8)$$

Step 4: The border approximation area (BAA) matrix G is hence determined by following equation:

$$g_i = \left(\prod_{j=1}^{m} v_{ij}\right)^{1/m} \quad (9)$$

where, v_{ij} are the rudiments of the weighted normalized matrix v, and m is the total number of alternatives. Then $G=[g_1 \quad g_2 \quad .... \quad g_n]$ is obtained with the format (n is the total number of criteria)

Step 5: The distance of the alternative from the BAA is as follows:

$$Q=V-G \tag{10}$$

The border approximation area G is dividing the entire approximation area into two halves as upper approximation area G^+ and lower approximation area G^-. Alternative A_i may belong to the border approximation area G or either of the upper and lower approximation area based on following equation:

$$A_i \in \{G^+ \ if \ q_{ij} > 1, G^+ \ if \ q_{ij} = 0, G^- \ if \ q_{ij} < 1\} \tag{11}$$

If $q_{ij} > 0$, then alternative A_i is in the vicinity of the positive ideal alternative. If the value $q_{ij} < 0$, it replicates that A_i is in close proximity to the negative ideal alternative.

Step 6: The overall score of the alternatives is computed by:

$$S_i = \sum_{j=1}^{n} q_{ij} \tag{12}$$

Ranking is done depending on the descending value of S_i. By following the above steps the best hybrid composite material is obtained by ranking and material number A30 is found to be rank 5 which is $Al+Al_2O_3$ at 12.5% wt. addition and material number A24 is found to be rank 1 (Fig. 1.) which is heat treated 12.5%wt. WCE.

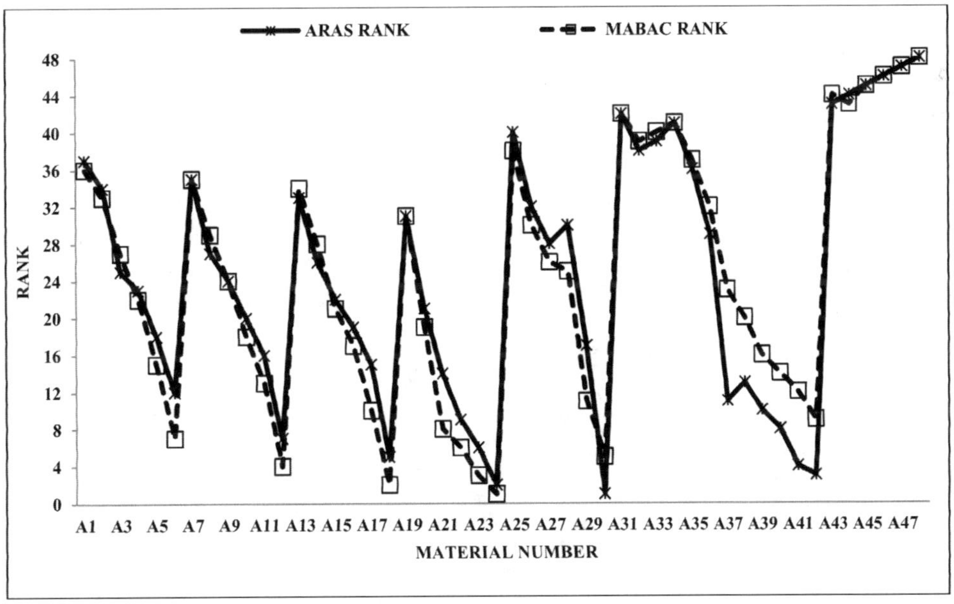

Fig. 1. Rank of hybrid composite materials

IOP Conf. Series: Materials Science and Engineering **653** (2019) 012003 doi:10.1088/1757-899X/653/1/012003

4. Experimental Investigation

4.1. Development of the material library
Table 2 depicts the material library of 48 different samples and the composition of the fabricated composite.

Table 2. Material Library and its composition

Element	Si	Fe	Cu	Mn	Mg	Zn	Ti	Ni	Cr	Al
wt.%	0.6	0.5	4.2	0.7	0.6	0.25	0.2	0.1	0.1	Balance

4.2. Development of the fabricated composite
Al matrix is heated and melted at 775 °C in a crucible and cooled in the region of 600°C and manually stirred for 5 minutes to improve wettability by removing dampness. Then the composite is superheated to 850 °C and whirled at 350 rpm (for 11 minutes @ 11 Ampere current). Different samples of the fabricated composites are taken for experiments varying the wt.% (0, 2.5, 5, 7.5, 10, 12.5 and 15) of reinforcements are then solidified in the mould and heat treated as well as shown in Fig. 2.

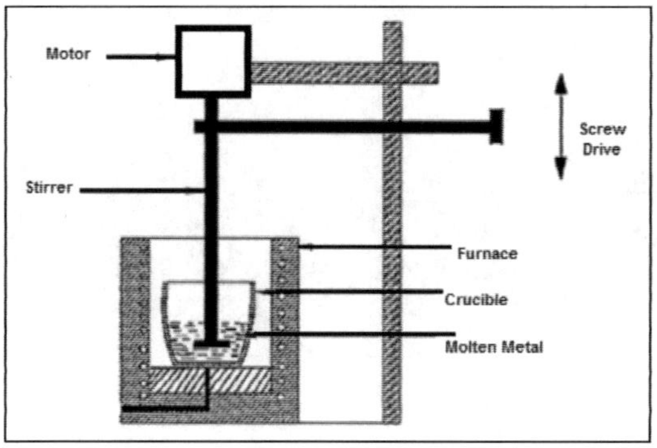

Fig. 2. Experimental setup

5. Micro structural analysis
Fig. 3 (a, b) exhibits the micro structural diagrams of Al+12.5 wt.% ES and WCE which indicate a superior performance in stipulations of porosity of WCE reinforcements because it is highly reactive to environment at elevated temperature ensuing in advanced wettability. Fig. 3 (c, d) represents the interfacial reaction layer of Al+12.5 wt.% ES and WCE, respectively which concludes superb wettability of WCE.

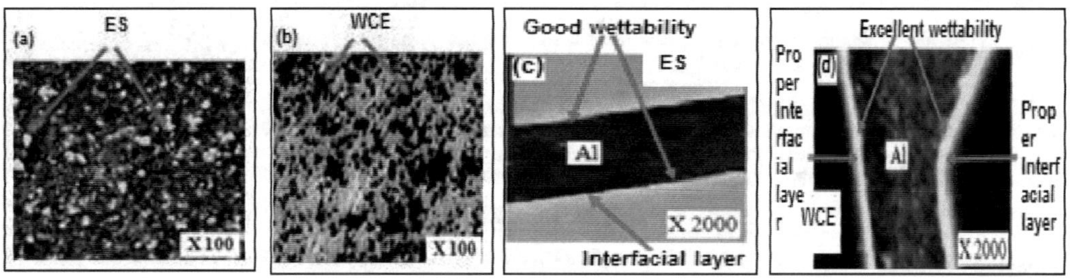

Fig. 3. Microstructure and Interfacial layer pictures of 12.5 wt.% ES and WCE

6. Discussion on mechanical properties

In Fig. 4(a). the hardness increases and gives the best value at 12.5 wt.% of ES (101 BHN) and WCE (113 BHN). Hardness increases approximately to 200 BHN with Al_2O_3+Gr by Mohanavel et al. [9] but reaches to 87 BHN after SiC addition by Deep et. al. [7] at 15 wt.%. According to Aribo et. al. [8], it reaches to the best value of 111 BHN at 7.5 wt.% of SiC+SSA and then decreases. This augmentation in hardness is due to the existence of fragile segments of ES and WCE reinforcements. However after heat treatment, the hardness values enhance to 118 BHN. Fig. 4(b). depicts the improvement of the tensile strength in the composite. It enhances to 251 MPa for ES and 272 MPa for WCE, because of the homogeneous distribution of eggshells, but then decreases because of the feeble particles-matrix interaction. Best result is obtained after the heat treatment, i.e. 264 MPa (ES) and 273 MPa (WCE) after several comparisons with various researches.

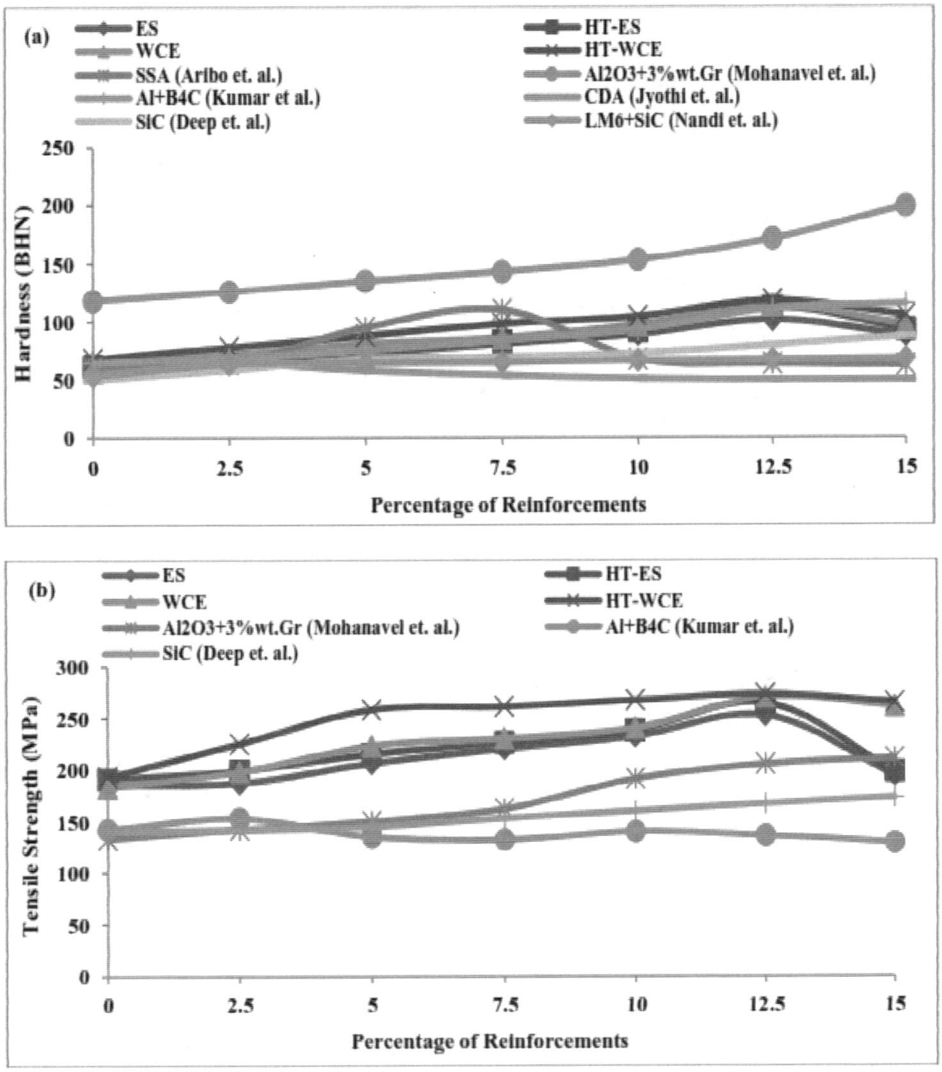

Fig.4. (a) Hardness and (b)Tensile Strength vs % of reinforcements

7. Conclusion

Recycling is the most important criterion which is the chief apprehension in this paper to fabricate novel green AMC for environmental sustainability. In this paper, ARAS and MABAC methods are used to select the best material (A24), and are compared with the experimental results. Eggshell particles are much better reinforcements than SiC+SSA at higher weights. Microstructure of Al/WCE shows excellent wettability and proper interfacial bonding resulting in the best hybrid green composite development. Best result of hardness obtained is the heat treated composites with values of 118 BHN (WCE). The most favourable value of tensile strength is 273 MPa of WCE. The fabricated composite is used in aircraft wings and automobile valves.

References

[1] Chatterjee P and Chakraborty S 2013 Gear Material Selection using Complex Proportional Assessment and Additive Ratio Assessment-based Approaches: A Comparative Study *International Journal of Materials Science and Engineering* **1** pp 104-111

[2] Zavadskas E.K and Turskis Z 2010 A new additive ratio assessment (ARAS) method in multicriteria decision-making. Technological and economic development of economy *Baltic Journal on Sustainability* **16**(2) pp 159-172

[3] Biswas T.K and Das M.C 2018 Selection of hybrid vehicle for green environment using multi-attributive border approximation area comparison method *Management Science Letters* **8** pp 121-130

[4] Pamucar D and Cirovic G 2015 The selection of transport and handling resources in logistics centers using Multi-Attributive Border Approximation area Comparison (MABAC) *Expert Systems with Applications* **42** pp 3016-3028

[5] Xue Y, You J, Lai X and Liu H 2016 An interval-valued intuitionistic fuzzy MABAC approach for material selection with incomplete weight information *Applied Soft Computing* **38** pp 703-713

[6] Hassan S.B and Aigbodion V.S 2015 Effects of Eggshell on the Microstructures and Properties of Al-Cu-Mg/Eggshell Particulate Composites *Journal of King Saud University-Engineering Sciences* **27** pp 49-56

[7] Deep G, Idrisi A.H and Siddiqui T.U 2016 A New Bio-Filler for Polypropylene Composites *International Journal of Innovative Research in Science, Engineering and Technology* **5** pp 16720-16725

[8] Aribo S, Barker R, Hu X and Neville 2013 A Erosion-corrosion behavior of lean duplex stainless steels in 3.5% NaCl solution *Wear* 302 pp 1602–1608

[9] Mohanavel V, Rajan K, Senthil P.V and Arul S 2017 Mechanical behaviour of hybrid composite (AA6351+Al_2O_3+Gr) fabricated by stir casting method *Materials Today: Proceedings.* 4 pp 3093–3101

DESIGN AND FABRICATION OF HUMAN-ELECTRIC HYBRID POWER TRI-CYCLE

Jay Prakash Srivastava[1,*], Dheeraj Joshi[1], L. Vivek[1], G. Sai Manish[1], G. Sravan[1], K. Enosh[1], E. Naresh Naik[1], K. Vamshi Krishna[1], M. Prem[1], K. Dayanand[1]

[1]Department of Mechanical Engineering, S R Engineering College, Warangal-506371, Telangana, INDIA

*Corresponding Author: jaysrvstv@gmail.com

Abstract. "EFFI-CYCLE" derived from Efficient-Cycle promotes reducing dependency on fossil fuels for domestic transport application. The objective of this paper is to conceive, design and fabricate a three-wheel configuration (tad-pole design) vehicle powered by human-electric hybrid power and capable of seating two passengers catering to the day to day mobility needs. PTC Creo 3.0 software is used for geometric modelling of the vehicle's frame. To ensure the safety of the design the model is then imported into ANSYS V16.2 for impact analyses. Results in terms of deformation, strain and von Mises stresses are plotted. After successfully completing the design and simulation of the frame and other components it is then fabricated in the college's (S R Engineering College, Warangal) workshop.

Keywords: Efficycle, Finite Element Analysis, SAE, Hybrid Vehicle, Electric Tricycle.

1. Introduction

The modern world today faces a huge crisis in fuel energy consumption and conservation. With ever increasing need for fuel and dependency of transportation systems on it, scientists predict that by 2040, the world's fuel reserves would be reduced to 20% of their current status [1]. This will lead to a price hike for fuels. So, the need of the hour is to switch over to a new form of transport that reduces the current strain on the natural resources and aims at leading in an era of sustainable development.

A large number of studies [2-5] depicts their concern over congestion and pollution associated with the use of a motor vehicle for personal transport. This paper aims to design a vehicle that could be a viable alternative to cars for short distance journeys. This designed vehicle is powered by a hybrid human-electric drive system. The combined human and electric powered vehicle is the vehicle which is driven by human muscular power as well as electric power. The working principle adopts the synergy of bicycle and electric vehicle technology. This is a three-wheeled vehicle, driven by a maximum of two members with side by side seating arrangement. The vehicle is equipped with a BLDC 1.4 HP electric motor retrofitted at the rear side, 4 sealed lead acid battery (each of 12 Volts) and chain drive used for electric power transmission. Besides this, two manual-pedaling systems also provided for the human power transmission based on the principle of bicycle. Whenever required the mode of driving (manual to electric and vice-versa) can be altered. The tadpole type configuration is used for this vehicle in which

IOP Conf. Series: Materials Science and Engineering **653** (2019) 012004 doi:10.1088/1757-899X/653/1/012004

two wheels are mounted in front side and one wheel in the back side of the vehicle. The vehicle has an Ackermann type steering system handled by one driver (Right-hand side). Disc brakes with sliding calipers in direct connection are provided in all the three wheels for proper and effective braking.
For a better understanding of the complicacies pertained in fabrication a prototype is made. From this paradigm, after doing some modification a hand-sketch is drawn proceeded by a CAD model generation. The roll cage material selection was carried out with an aim to optimize strength, weight and cost. A preliminary study for the frame's structural integrity is carried out, using finite element analysis, prior to its fabrication. The key parameters taken under consideration were safety, ergonomics, weight reduction and cost of manufacturing the chassis.

2. Design Procedure

The dimensions of the vehicle are first assumed considering the constraints and dimensions based on the rulebook of SAE. After assuming the dimensions, the focus was given to make the vehicle ergonomic for an average adult person. Then the vehicle is designed on a CAD Package PTC Creo 3.0. The structural integrity of the vehicle is tested by using Finite Element Analysis software ANSYS V16.2. To ensure the safety the design of roll cage/frame is modified and then the design is finalized. During manufacturing various parameters like serviceability, craftsmanship and cost reduction were considered. In order to make the vehicle more attractive and convincing, its aesthetic considerations like appearance were also given utmost importance. The geometric model of the vehicle is shown in figure 1 (a)-(d). Vehicle dimensions and different component specifications are given below in Table 1.

Figure 1(a): Isometric view of the vehicle

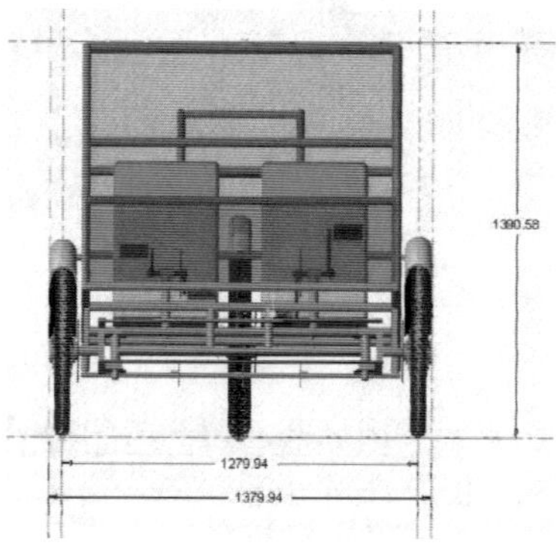

Figure 1(b): Front view of the vehicle

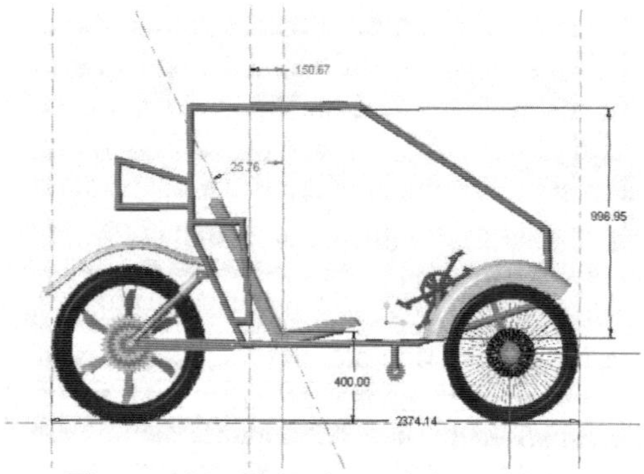

Figure 1(c): Side view of the vehicle

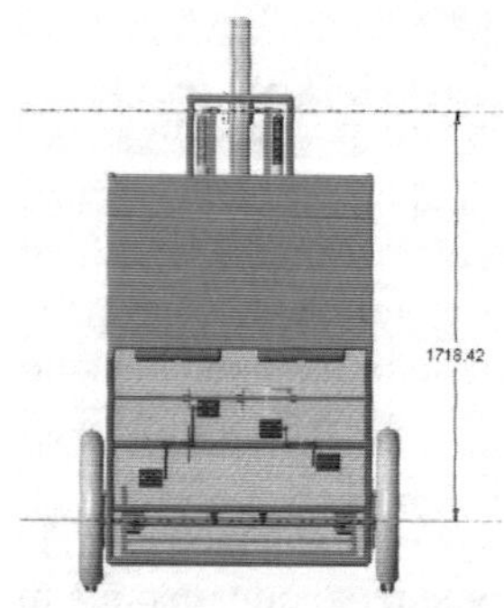

Figure 1(d): Top view of the vehicle.

Figure 1 (a)-(d): Three dimensional CAD model of the hybrid tri-cycle vehicle.

IOP Conf. Series: Materials Science and Engineering **653** (2019) 012004 doi:10.1088/1757-899X/653/1/012004

Table 1. Vehicle dimensions and component specifications

Component	Specifications / Dimensions
Motor	BLDC 1.4 HP, 500 RPM 48 Volts motor
Frame material	AISI 1018
Frame material dimension	OD- 31.75 mm, ID- 28.55 mm, Thickness- 1.6 mm
Weight of drivers (two members)	120-180 kg
Wheel configuration	Front -2, Rear - 1
Steering	Ackermann with drag link
Turning Radius	3.4 m (analytical calculation)
Turning angle	45 °
Braking	Mechanical Disc Brake on all the wheels
Dimension	Length - 2368 mm, Height - 1404.9 mm, Wheelbase - 1524 mm, Ground clearance - 152.4 mm

3. Frame

The frame was designed considering better aerodynamics and comfortable base for the riders. Additional aspects of battery mounting, suspension and seating arrangements were also designed as per the Efficycle rulebook [6], along with this arrangement for top covering is also been designed for the shelter of riders. The material used for the frame is AISI 1018 having a circular cross-section. The material composition and its properties are given in Table 2 and 3.

Table 2. Frame material (AISI 1018) composition

Element	Carbon, C	Iron, Fe	Manganese, Mn	Phosphorous, P	Sulfur, S
Content (%)	0.14 - 0.20 %	98.81 - 99.26 %	0.60 - 0.90 %	≤ 0.040 %	≤ 0.050 %

Table 3. Mechanical Properties of AISI 1018

Hardness, Brinell	126
Tensile Strength, Ultimate	440 MPa
Tensile Strength, Yield	365 MPa
Modulus of Elasticity	205 GPa
Bulk Modulus	140 GPa
Poissons Ratio	0.290
Shear Modulus	80.0 GPa
Density	7870 kg/m^3

4. Finite element analysis of vehicle-frame

The frontal, side and rollover impacts analysis were performed on ANSYS 16.2. The CAD model and the actual fabricated vehicle frame/roll cage of the hybrid tri-cycle is shown in figure 2(a) – (b).

IOP Conf. Series: Materials Science and Engineering **653** (2019) 012004 doi:10.1088/1757-899X/653/1/012004

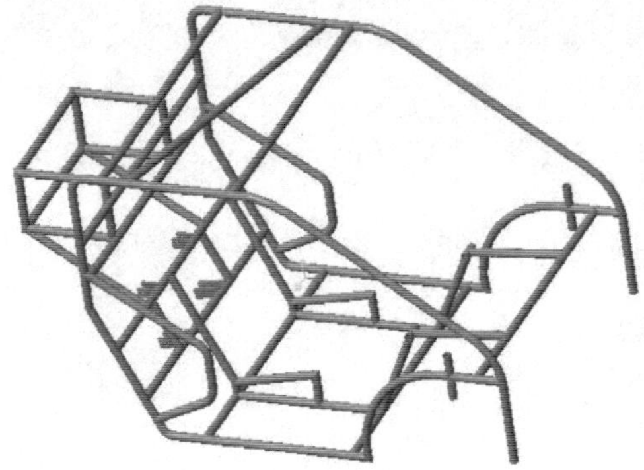

(a) CAD model

(b) Fabricated roll cage

Figure 2(a) – (b): CAD model and the actual fabricated vehicle frame/roll cage of the hybrid tri-cycle.

5. Frontal impact analysis

Assumption & Considerations:

In pretext to calculate frontal impact, a top speed of 35 kmph to final zero velocity is assumed to be attained by the vehicle in 0.6 sec. The weight of the vehicle is considered as 150 kg and two drivers each of weight 90 kg. The vehicle might be subjected to any impact, inattentively or deliberately. The vehicle is expected to sustain through the impact, so it is need of the hour to find out impact forces and, perform impact analysis.

Calculation of Impact Forces:

Total mass (m) = 150 + 90 + 90 = 330 kg, Time (T) = 0.6 sec, Initial velocity (u) = 35 km/hr = 9.723 m/s, Final velocity (v) = 0 km/hr. Also, Front Impact Force F = m dV/dT = 330 * 9.723/0.6, F = 5348.16 N. A maximum force of 5348.16 N is considered for front impact

Analysis Results:

The simulation is performed on ANSYS Workbench platform. The model is meshed with tetrahedral elements. Convergence study is also done to finalize the finite element mesh size. Result plots in terms of total deformation, von Mises stress, factor of safety and equivalent strain are shown in the figure 3 (a)-(d). Their quantification is as follows: Maximum Deformation = 4.341 mm, Maximum stress = 146.37 MPa, Safety Factor = Yield strength / Max stress = 365/146.37 = 2.49, Total Strain = 0.00088327. The above values taken from figure 3 (a) – (d) are well within the permissible range that ensure safe design.

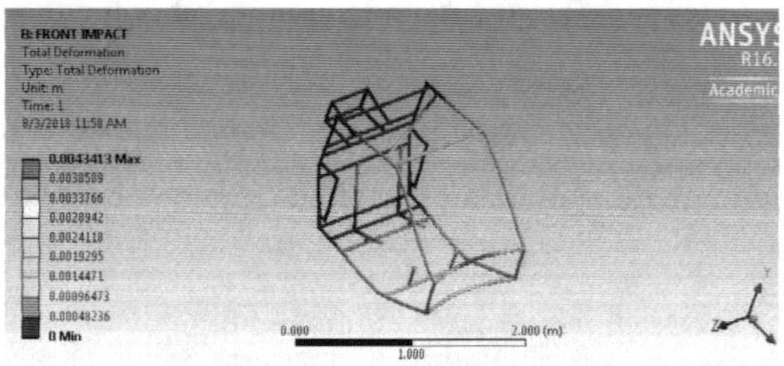

Figure 3(a): Total deformation plot.

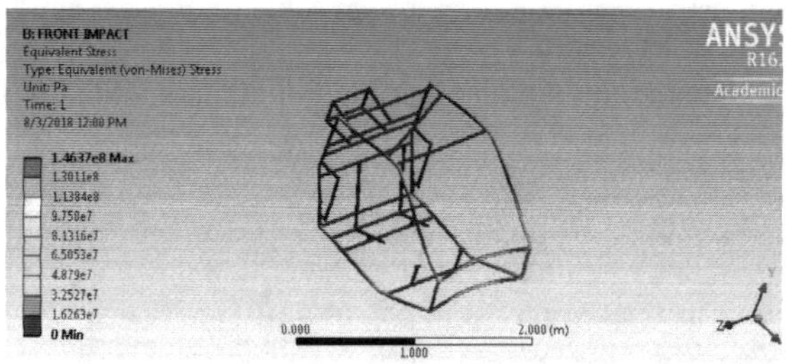

Figure 3(b): Equivalent von-Mises stress distribution.

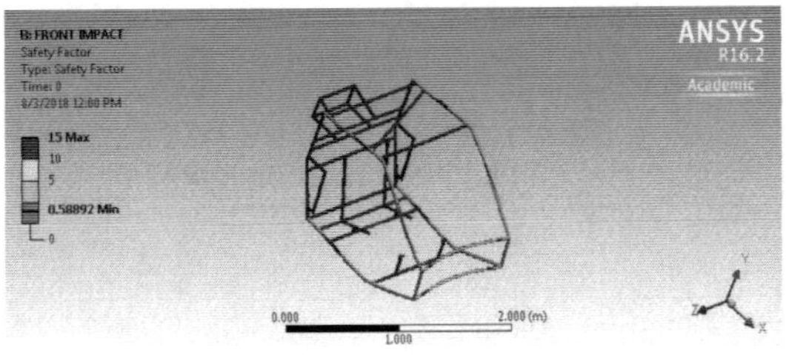

Figure 3(c): Factor of safety plot

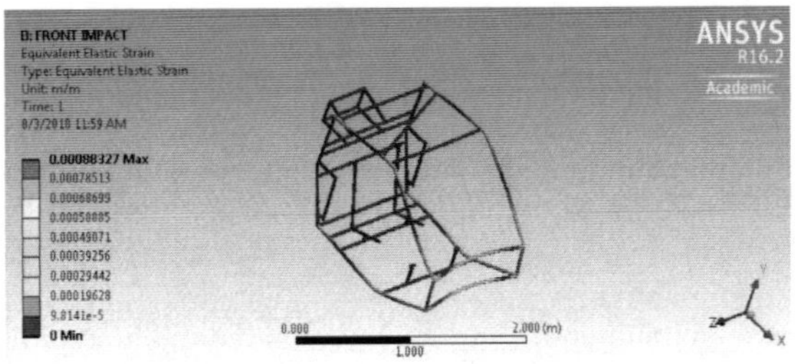

Figure 3(d): Equivalent strain plot

Figure 3: Result plots after front impact analysis.

6. Side impact analysis

Assumption & Considerations:

Assumptions are similar to that of front impact except for the time taken to stop by the vehicle after a collision with the wall is assumed to be 1s.

Calculation of Impact Forces:

Total mass (m) = 150 + 90 + 90 = 330 kg

Time (T) = 1s

Initial velocity (u) = 35 km/hr = 9.723 m/s, Final velocity (v) = 0 m/s

Also, Side Impact Force, F = m * dV/dT = 330 * 9.723/1, F = 3208.59 N

Analysis Results:

A maximum force of 3208.59 N is considered for side impact. Result plots in terms of total deformation, von Mises stress, factor of safety and equivalent strain are shown in the figure 4 (a)-(d). Their quantification is as follows: Maximum deformation = 0.5236 mm, Maximum stress = 29.207 MPa, Safety factor = Yield strength/Maximum stress, = 365/29.207 = 12.5, Total strain = 0.0001619.

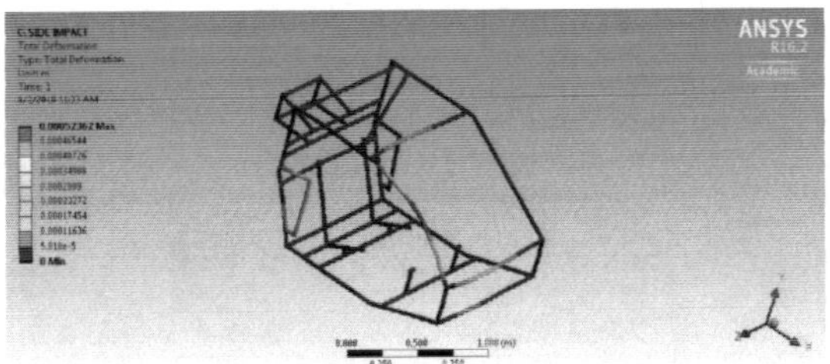

Figure 4(a): Total deformation plot

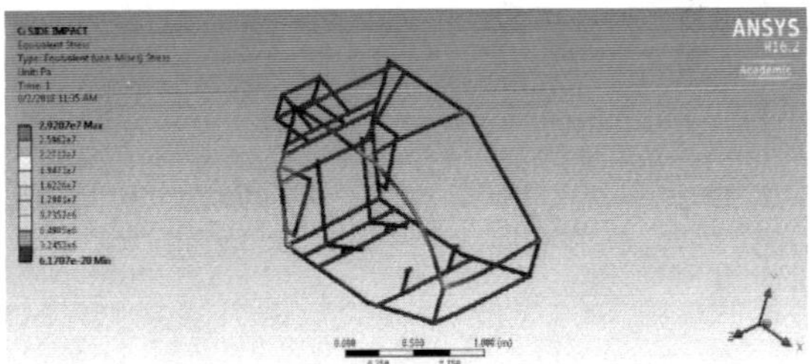

Figure 4(b): Equivalent von-Mises stress distribution.

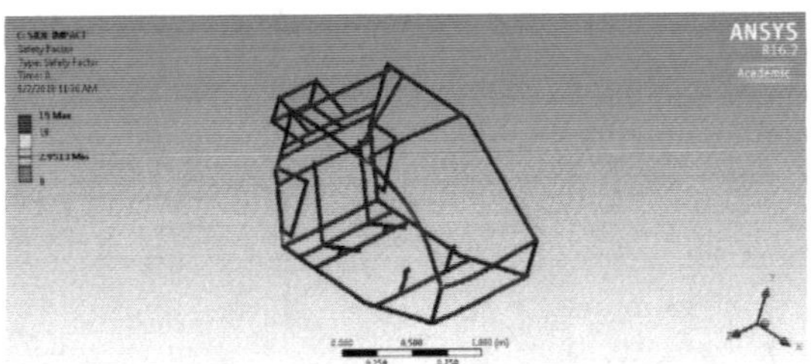

Figure 4(c): Factor of safety plot

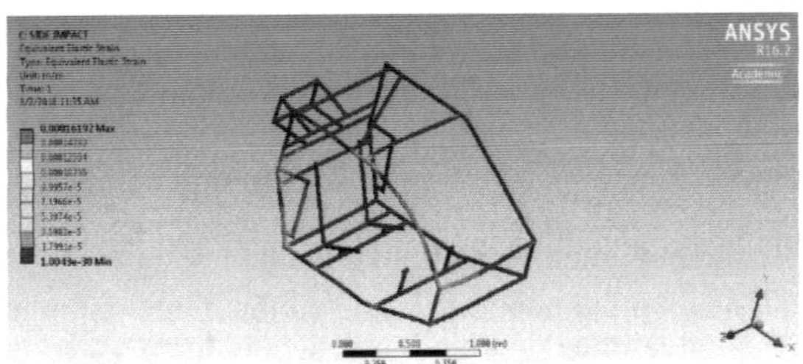

Figure 4 (d): Equivalent strain plot

Figure 4: Result plots after side-impact analysis.

IOP Conf. Series: Materials Science and Engineering **653** (2019) 012004 doi:10.1088/1757-899X/653/1/012004

7. Rollover analysis

Assumption & Considerations:

In pretext to calculate roll-over impact, a top speed of 35 kmph and final velocity as zero is assumed to be attained by vehicle in 6 sec. The weight of the vehicle is considered as 150 kg and two drivers each of weight 90 kg.

Calculation of Impact Forces:

Mass of trike = 150 kg

Mass of drivers = 90 + 90 = 180 kg

Total Mass = 330 kg

Force = F = $\left(\frac{N \times w}{2} \times h\right)$; $F = \frac{330 \times 9.81 \times 48}{2}$; F = 1438.8 N

Analysis Results:

Result plots in terms of total deformation, von Mises stress, factor of safety and equivalent strain are shown in the figure 5 (a)-(d). Their quantification is as follows: Total Deformation = 3.007 mm, Maximum Stress = 70.205 MPa, Safety factor = Yield Strength / Maximum stress = 365 / 70.205 = 5.19, Total Strain = 0.000435

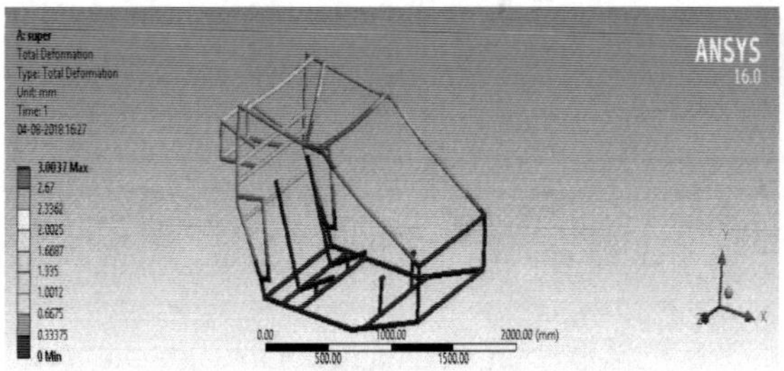

Figure 5(a): Total deformation plot

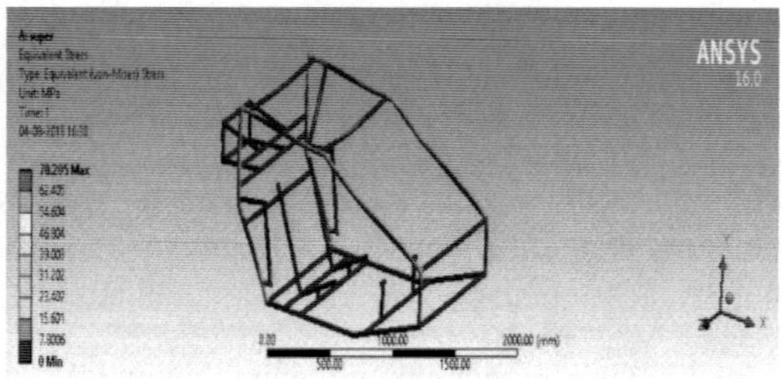

Figure 5(b): Equivalent von-Mises stress distribution.

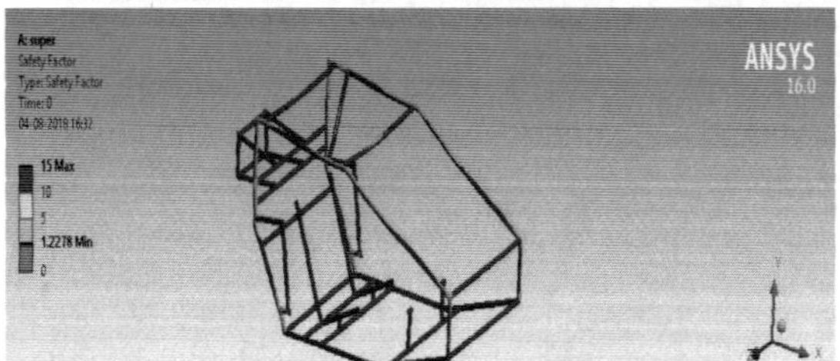

Figure 5(c): Factor of safety plot

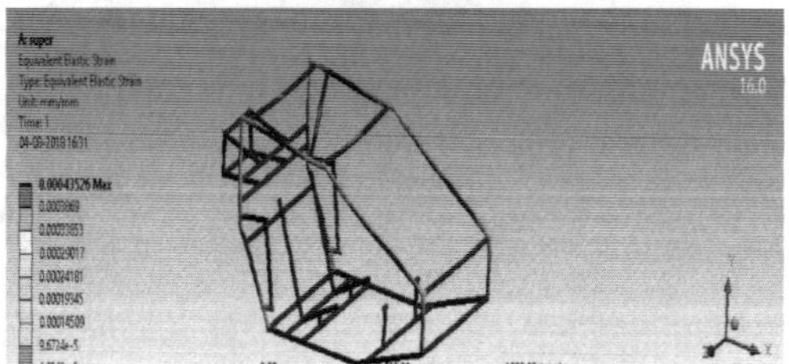

Figure 5(d): Equivalent strain plot

Figure 5: Result plots after roll-over impact analysis.

8. Safety features of the vehicle

The drivers are provided with seat belts to overcome the hazards of collision. All the electrical wires are well insulated, and the battery case is made leak proof. Drivers are also equipped with a safety helmet, knee pads, gloves, side mirrors for rear view etc. Always safety is considered over comfort. Seats are provided with synthetic foam material so that drivers could feel comfortable. Handling of the vehicle is easy as it is provided with good steering mechanism. The vehicle is made attractive using paints. The design is also unique that makes it appealing.

9. Conclusion

The initiative towards EFFICYCLE was taken with the aim to have hands-on experience in designing a Hybrid Human-Electric Tri-Cycle keeping in view the ease of manufacturing and economy. The design was made by the use of CAD software PTC Creo 3.0 and the design was analyzed by FEM analysis software ANSYS V16.2. The simplicity of design and Hybrid drive train were the key parameters in the vehicle design. This design and fabrication of Efficycle will create awareness towards environment-friendly mobility solution.

Acknowledgement

A special thanks to Prof V Mahesh (Principal, SREC), Prof P Sammaiah (HOD, ME, SREC) and college management for providing the computational laboratory and workshop facilities to carry out this work. Their encouragement and support are appreciably acknowledged.

References

[1] Zou C, Zhao Q, Zhang G and Xiong B 2016 Energy revolution: From a fossil energy era to a new energy era *Natural Gas Industry B* **3**(1) pp 1-11
[2] Zhang K and Batterman S 2013 Air pollution and health risks due to vehicle traffic *Sci Total Environ* **15** pp 307–316
[3] Solanki H.K, Ahamed F, Gupta S.K, and Nongkynrih B 2016 Road Transport in Urban India: Its Implications on Health *Indian J Community Med* **41**(1) pp 16–22
[4] Bhuyan P.J, Ahmed F 2013 Road traffic accident: An emerging public health problem in Assam *Indian J Community Med.* **38** pp 100–104
[5] Xia T, Zhang Y, Crabb S, and Shah P 2013 Co-benefits of Replacing Car Trips with Alternative Transportation: A Review of Evidence and Methodological Issues, *Journal of Environmental and Public Health*, Article ID **797312** pp 14-28
[6] EFFI-CYCLE Rule book, *SAEINDIA*, 2018
[7] Gillespie T.D 1992 Fundamentals of Vehicle Dynamics, *Published by SAE International*, ISBN 10: 1560911999 ISBN 13: 9781560911999

IOP Conf. Series: Materials Science and Engineering **653** (2019) 012005 doi:10.1088/1757-899X/653/1/012005

Mechanical characteristics of Al-Al_2O_3-SiC metal matrix composites made by stir casting technique

S V Gopala Krishna [1*], B V Subrahmanyam[2] and B Lakshmi [3]

[1&2]Department of mechanical engineering, Sir C R Reddy College of Engineering, Eluru, INDIA
[3]Mechanical engineering department, Adikavi Nannaya University, Rajahmundry, INDIA

Abstract. The present investigation focuses on Al-Al_2O_3-SiC metal matrix composites were prepared by stir casting technique. In this, Pure aluminium (varies proportionately to the reinforcement) act as a base metal and reinforced with Al_2O_3 (standard wt 3%), SiC (varies Wt 3%, 5% and7%) has taken. The samples were prepared at the temperature of 750°C Aluminium (liquid state) and to warm the reinforcement materials at 300°C mixed with it by stirring of 350rpm for equally distributing the reinforcement material while it was consolidated at one place due to density differences. Hybrid composites are potential application for aircraft and space industries because of higher resistance, lower weight to strength ratio, creep resistance. To investigated the mechanical behavior of hybrid composites like hardness, Tensile strength and density.

1. Introduction

Aluminium metal matrix composites (AMMCs) are mostly used in aerospace, automobile industries due to their strength to weight ratio, low coefficient of thermal expansion (CTE) and high hardness. Stir casting is generally accepted for reducing the fabrication cost and easily to prepare very large sized components. Pure aluminium reinforced with micron sized particles of SiC to made as aluminium matrix composite and also 3 % Al_2O_3 was added to improve the wettability and percentage of ceramic particles are incorporated.

Hashim et al was studied that there are some limitations which reveals the Non-uniform distribution and poor wettability during the process happened. The reinforcement particles are uniformly distributed in molten metal those are making float and sinking to the molten metal due to the density differences If, agglomeration and clustering is done by injecting the particles with an inert gas into the molten metal through stirring[3].

Vijaya ramnath et al were studied about the addition of reinforcement materials like al_2O_3, SiC and B_4C etc when the reinforcement of SiC with aluminium they were attained good wear resistance due to the preparation of brake drums is very useful .similarly the reinforcement of Al_2O_3 has been done to increase the volume fraction and to reduced the fracture toughness [18]. U Aybarc et al were studied aluminium metal matrix was the reinforcements of Al_2O_3, SiC and graphene however, the incorporation of reinforcement is major concern and particularly for wettability, formation of porosity have been evaluated. To understood that the type and amount of reinforcement directly reflect on raising of mechanical properties [19]

Singla et al was studied on two step stir casting technique .Aluminium (98.41%) has been chosen as matrix and SiC as reinforcement then the number of samples were made varying weight fractions of Sic, while keep up other parameters are constant. When conducted the test, results showed that an increasing trend of hardness 45.5 BHN and impact strength maximum is 36N-m were obtained at 25% of SiC [7].

Surappa et al has been investigated that an overview of AMMC material on aspects related type of processing, properties and microstructure [1].In the present days investigating the Al-SiC particulate composites for the wear resistance. The present Composites are having good wear resistance compared to Al-Si piston alloys [8].

Abdul sahib et al studied that the Al reinforced with SiC & graphite particles results were showed that raising of hardness when weight of reinforcement increased. The optimum value (maximum hardness) were obtained at 25% of silica and 4% of graphite reinforced [9]. Rabindra Behara et al studied on forge ability and machinability on AMMC's If, MMC samples were varied with the change of thickness of the casting .The results showed that the Forge ability of Cast MMCs are minimum at mid section compared to both end sections. [11].

Huang et al presents the paper on Al reinforced with SiC composites with higher volume fraction and larger particles were carried out for high speed milling experiment compared with TiC ceramic tools was also examined with the same work. The results showed that the main tools are suitable for Al –SiC composites because the large SiC particles are showed heavy abrasive nature [16]. S. Ray introduced aluminium reinforced with different ceramic particles through stir casting and powder metallurgy techniques. The stir casting process mainly do the work of stirring for distribution of ceramic particles and sophisticated up to 30% of volume fraction of ceramic particles [17].

2. Experimental Procedure

The pure aluminium has taken in the form of ingots these are heated up to 750°C when it was reached to liquid state. After, Al_2O_3 and SiC took with proportions like Al_2O_3 Standard wt 3%, SiC 3% wt. Before mixing of the reinforcements gets warm of 300°C for do not affecting the mixing process. The stirrer was rotated 450rpm while mixing the reinforcement particulates to take 10 minutes of time .the stirrer has done to distribute the particles uniformly with the span of time then, pour it into a graphite coated cylindrical shaped holes of a moulding box then getting the specimens after removing of moulding box .The same procedure was repeated for other proportions

Figure 1. Hot molten mixture in furnace

Figure 2. Bottom pouring stir casting m/c

Figure 3. Moulding box

Figure 4. Specimens prepared

Generally the following factors are affected in stir casting process those are stirring temperature, time duration, speed. Stirring temperature is one of the significant parameter which can affect the process of stirring. Viscosity of metal matrix can be reduced when the matrix temperature increased.

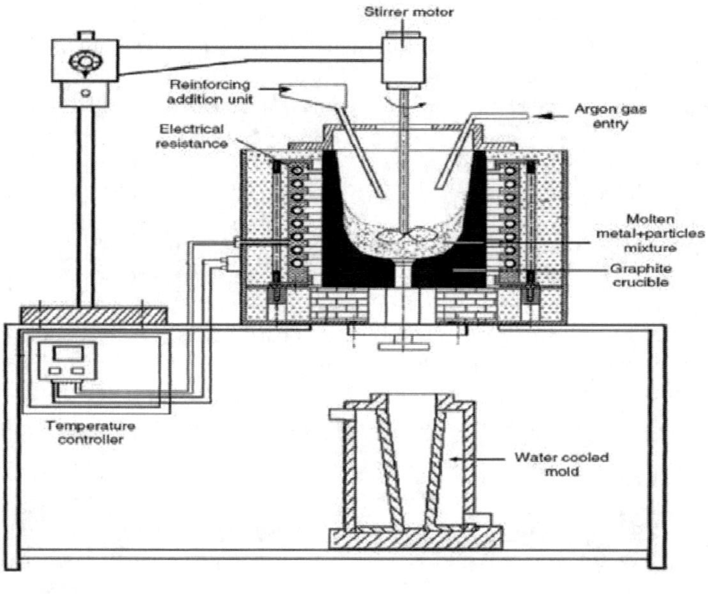

Figure 5 Sketch of Bottom pouring stir casting process

3. Results and discussions

The metal matrix composites are having good mechanical properties compared with base metals and some alloys. In this study, when varying the percentage of 3%,5% &7% SiC and standard wt of 3% Al_2O_3 gave good hardness values similarly to achieved the ultimate stress value when the reinforcements have been raised proportionately density also been raised

Table 1: Hardness, ultimate stress and density values of stir casted specimens

S.no	% of Pure aluminium	% of Al_2O_3	% of SiC	Specimens	Micro vicker's Hardness value	Ultimate stress(tenso meter in N/mm^2)	Density (gr/cm^3)
1	100	0	0	Pure Al	90	200	2.71
2	94	3	3	Composition1	105.8	250	2.81
3	92	3	5	Composition2	107.7	330	2.82
4	90	3	7	Composition3	110.9	325	2.83

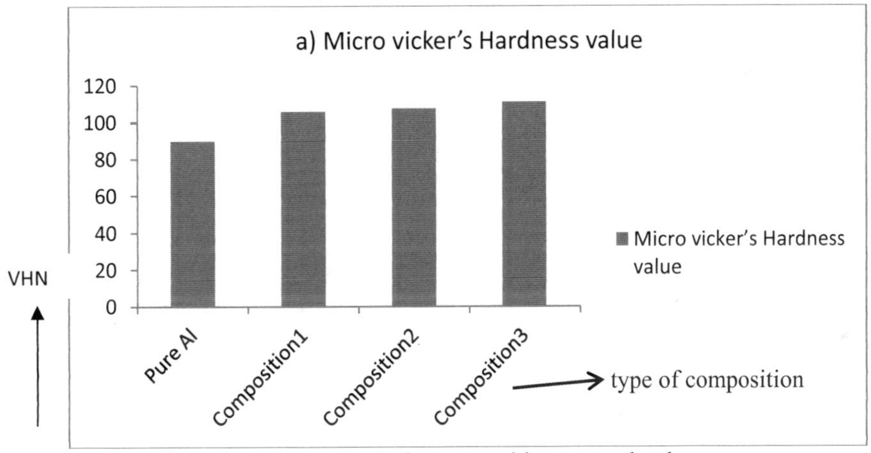

Fig. 6a Graph represents the composition verses hardness

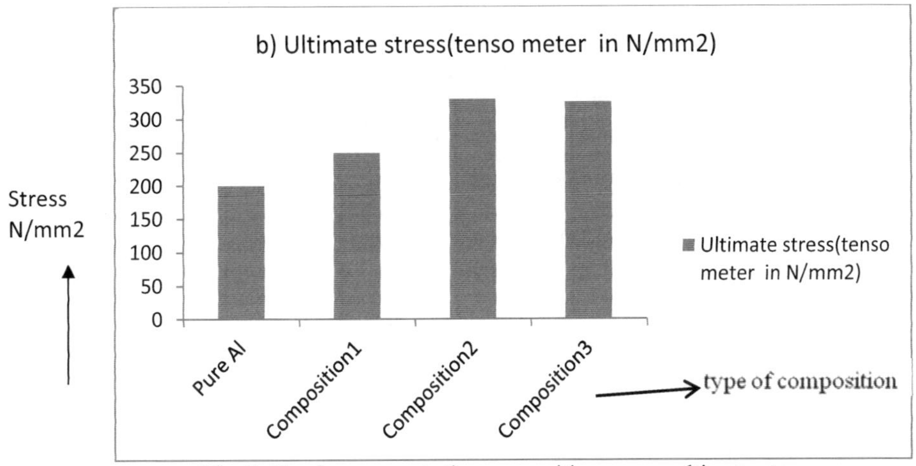

Fig.6b Graph represents the composition verses ultimate stress

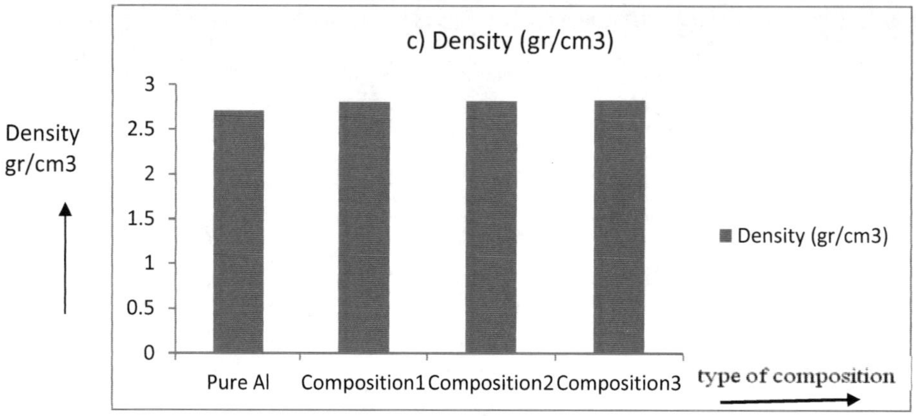

Fig. 6c Graph represents the composition verses density

Form above Figure represents the trend of hardness values when the reinforcements were increased and similarly to show the ultimate stress that was obtained good value of composition2. The density were also calculated by the experimentation those are also raised when the increase of reinforcement percentage

4. Conclusions

The work showed that the increasing of reinforcement materials like SiC (varies 3,5 &7 percentages) and standard weight of 3% Al_2O_3 with pure aluminium to attained good mechanical properties for helpful applications of aerospace, automobile industries usaged. Clustering of reinforcement particles (Al_2O_3 & SiC) and wettability of Al is the first concern through stir casting then samples have been taken minimum number of 3 and tests were conducted. Finally concluded that Aluminium metal matrix composite specimens are enhanced the hardness (90, 105, 107, 110), ultimate stresses (200,250,330,325) and density (2.71, 2.81, 2.82, 2.83)

References

[1] M K Surappa 2003 Aluminium Matrix Composites: Challenges and Opportunities, *Sadhana* Vol. 28, Parts 1 & 2, February/April, pp 319– 334

[2] S. Naher, D Brabazon and L Looney 2004 Development and Assessment of a New Quick Quench Stir Caster Design for the Production of Metal Matrix Composites, *Journal of Material Processing Technology,* Vol. 166, pp 430-439.

[3] J. Hashim, L Looney and M S J Hashmi 1999, Metal Matrix Composites: Production by the Stir Casting Method, *Journal of Material Processing and Technology*, Vol. 92,pp 1-7

[4] Y. H. Seo and C. G. Kang 1999, Effects of Hot Extrusion through a Curved Die on the Mechanical Properties of Al-SiC Composites Fabricated by Melt-Stirring, *Composites Science and Technology*, Vol.59 pp 643-654

[5] Rajan T, Pillai R, Pai B, Satyanarayana K, Rohatgi P. 2007, Fabrication and characterisation of Al–7Si–0.35 Mg/fly ash metal matrix composites processed by different stir casting routes. *Compos Sci Technol* 67(15):3369

[6] Amit Pal, Abhishek Verma, Sansar Swaroop Saxena, Bhaskar Chandra Kandpal 2015, Stir casing of metal matrix composites, *International Journal of Computer & Mathematical Sciences* Volume 4, Special Issue, February

[7] M.Singla, D.Dwivedi, L.Singh and V.Chawla 2009, Development of Aluminium Based Silicon Carbide Particulate Metal Matrix Composite, *Journal of Minerals and Materials Characterization and Engineering*, vol.8 No.6, pp.455-467

[8] Z.Hassan, R.Pandey and D.Sehgal 2011,Wear Characteristics in AL-Sic Particulate Composites, *Journal of Minerals and Materials Characterization and Engineering*, Vol. 10 No14, pp. 1329-1335

[9] Dunia Abdul Saheb 2011, Aluminium silicon carbide and Aluminium graphite, *ARPN Journal of Engineering and Applied Sciences* VOL. 6, NO.10

[10] Anand K. and Kishore 1983, the wear of aluminium-corundum composites, *Wear*, 85

[11] Rabindra Behera, S. Das, D.Chatterjee and G. Sutradhar 2011, Forgeabiliity and Machinability of Stir Casting Aluminium Alloy Metal Matrix Composites, *Journal of Minerals & Materials Characterization & Engineering*, Vol. 10, No.10, pp.923-939,

[12] S.L.Kakani and Amit Kakani, Material science, new age publishers, First edition

[13] Sydney and H.Avner, *Introduction to physical metallurgy* Tata McGraw Hill publishers, second edition

[14] Mr.Prashant Kumar Suragimath and Dr. G. K. Purohit A Study on Mechanical Properties of Aluminium Alloy (LM6) Reinforced with SiC and Fly Ash.

[15] G.V.Lai, High temperatures corrosion & materials application

[16] S.T.Huang & L.Zhou & X.L.Yu and Y.Cui 2012, Experimental study of high speed milling of Sic/Al composites with PCD tools, *Int J Adv Manuf Techno* PP: 487-493

[17] S.Ray 1998, synthesis of cast metal matrix particulate composites, *Journal of material science* Jan vol 28 issue 20 pp 5397-5413

[18] B.Vijayaramnath, C.Elanchezhian, RM.Annamalai, S.Aravind, T.Sri anand Athreya, V.Vignesh and C.Subrahmanian 2014,Aluminium Metal matrix composites-A review, *Rev.Adv.Mater.Sci* 38

[19] U.Aybarc, D.Dispinar and M.O.seydibeyoglu 2018, Aluminium metal matrix composites with SiC, Al2O3 and Graphene-review, *Archives of foundry engineering* volume18 issue2

IOP Conf. Series: Materials Science and Engineering **653** (2019) 012006 doi:10.1088/1757-899X/653/1/012006

Modeling & Analysis of B_4C Powder Mixed Wire EDM Process for Improving Performance Criteria of Ti6Al4V

Sadananda Chakraborty[1*], Souren Mitra[2] and Dipankar Bose[3]

[1,2] Production Engineering Department, Jadavpur University, Kolkata -700032, India
[3]Mechanical Engineering Department, NITTTR, Kolkata -700106, India

[1*]sadananda116@gmail.com

Abstract. The present investigation represents the effect of B_4C powder properties on performance characteristics of Ti6Al4V material. Boron carbide has good chemical resistance, low density and non-oxide properties. Due to that, it has a great impact on powder mixed wire EDM. The machining of Ti6Al4V is very difficult by other non-conventional processes when it is considered particularly through cutting operation with desired surface quality and strong economically. In this study pulse on time, pulse off time, peak current and gap voltage are considered as input parameter and analyzed the performance characteristics such MRR and SR. B4C Powder concentration and size in the dielectric are analyzed to get significant output. The optimum result is obtained from machining condition i.e. powder size 10μm and concentration 6gm/lit as for both performances like MRR and SR while machining controllable parameters are fixed i.e. Ip=2, Ton=30 and Toff=8. The machined surface is studied through microstructure image

1. Introduction

Electrically and thermally conductive materials are machined through the EDM process. The basic mechanism of metal removal Wire electric discharge machining is same as EDM process. In this process tool and workpiece always maintain a constant gap, say 0.05-0.50mm and flooded the gap area with a dielectric medium. The rate of metal removal is depending upon the workpiece material properties i.e. melting point, thermal conductivity and density as well as polarity and discharge intensity. In the modern manufacturing industry, a lot of advanced materials and superalloys are used for their enormous applications. Ti6Al4V is an alpha-beta titanium alloy which has excellent corrosion resistance and high strength to weight ratio. Ti6Al4V has numerous applications in aerospace and marine industry. Machining of this material is very difficult by other non-conventional when it is considered through cutting operation with a complex profile. WEDM is one of the best choices for perfect machining with precision dimension for its practical advantages. Alternatively, powder mixed wire EDM is the best options for high accuracy machining of advance alloys with desired product quality. In this process, powder is added with dielectric and during machining powder properties are varied as well as changing the machining parameter. Several machining inputs are involved in powder mixed technology in WEDM process.

1.1. Literature Review

Kumar et al [1] studied the multi-objective optimization through Grey Relation Analysis. The influence parameters such as peak current, pulse on time, duty factor and SI powder concentration and showed that silicon powder of 2gm/ltr in kerosene dielectric has the best effect on performance. Kumar et al [2] focused on the tool material (copper, copper-chromium and graphite) and powder particle (tungsten carbide, cobalt, boron carbide) in a dielectric on machining of Inconel 800 and developed the quadratic model of MRR and TWR. Marashi et al [3] reported a state of art in PMEDM. The authors have classified different publishing research works related to PMEDM into two major areas such as dielectric and powder properties which have a significant role in the EDM process. Machining of β-Ti; Alloy by Si powder mixed EDM, an attempt was made to fabricate submicro and nanoscale topography for enhancing biocompatibility by Prakash et al [4]. The best measured dimension of recast layer thickness (~3μm) is obtained at 4gm/ltr concentration of Si powder. Öpöz et al [5] studied the effect of SiC powder on surface topography of Ti-6lAl-4V-ELl work material and observed that high powder concentration with low pulse current enhanced the process capability. Chakraborty and Bose [6] analyzed the process parameter of the WEDM process and find out the optimal parametric setting for best MRR and surface finish as well as die corner accuracy of Inconel 718. Paul et al [7] investigated the machining parameters and showed an improvement of surface finish from 3.865 to 2.135μm in gear cutting operation. From the past research work is it evident that a few attempts have been made to enhanced performance criteria through powder mixed wire electric discharge machining. Most of the cases, researchers have analyzed the process parameter of powder mixed EDM. The present investigation is mainly focused on the improvement of MRR and SR by adding B_4C abrasive powder with deionized water in WEDM process. Finally, the surface topography is analyzed through the microscopic image to detect the microcrack generation, crater and recast layer on the machined surface.

1.2. Experimental Details

The present investigation, three significant process parameters such as peak current, pulse on time and pulse off time are chosen as input parameter based on the past research and trial run. Brass wire with 250μm diameter wire is selected as a tool electrode material. 2.5 mm thickness of Ti6Al4V workpiece material is machined at different machining condition.

Table 1. Experimental value of input and output parameter without powder

Exp No.	Input Parameter			Output Parameter	
	Ip (A)	Ton (μs)	Toff (μs)	MRR (mm^3/min)	SR (μm)
1	3	60	8	3.624	3.32
2	3	60	8	3.624	3.32
3	3	60	12	2.568	4.568
4	2	30	12	3.974	3.747
5	2	60	8	2.951	2.684
6	4	30	4	4.872	5.031
7	4	90	4	3.658	4.726
8	2	90	4	3.125	3.254
9	3	60	8	3.624	3.32
10	4	90	12	4.637	5.425
11	4	30	12	4.872	5.031
12	3	60	8	3.624	3.32
13	3	60	8	3.624	3.32
14	3	60	4	3.248	3.987
15	4	60	8	4.558	4.126
16	3	60	8	3.624	3.32
17	3	30	8	4.15	4.258
18	2	90	12	3.825	4.254
19	2	30	4	3.674	3.547
20	3	90	8	3.561	4.213

Conventional WEDM is done as per the design of experiment created by Response Surface Methodology. MRR has been measured through the weight-based method. Surface roughness is measured by Mahrsurf PS1 surface roughness tester with a cutoff value of 0.08 mm. Twenty no of experiments are conducted which is tabulated in Table 1.

1.3. Result and discussion

Regression model provided a quadratic model using Minitab statistical software. The fitted model equation for MRR and SR is given in equation 1 and 2 respectively.

MRR = 6.46 - 0.778 Ip - 0.0862 Ton + 0.062 Toff + 0.165 Ip*Ip + 0.000554 Ton*Ton- 0.00822 Toff*Toff + 0.00262 Ip*Ton + 0.0187 Ip*Toff + 0.000353 Ton*Toff 1

SR = 3.81 + 2.298 Ip - 0.0884 Ton - 0.615 Toff - 0.243 Ip*Ip + 0.000653 Ton*Ton + 0.03936 Toff*Toff - 0.00052 Ip*Ton - 0.0157 Ip*Toff + 0.001561 Ton*Toff 2

The coefficient of R^2 and Adj. R^2 value for MRR is 92.77% and 86.27% which indicates that the model is accepted for prediction. Similarly coefficient of R^2 and Adj. R^2 value for SR is 96.47% and 93.29% respectively. The most significant parameter MRR is peak current and pulse on time. Similarly, peak current and pulse of time are the significant factors for SR.

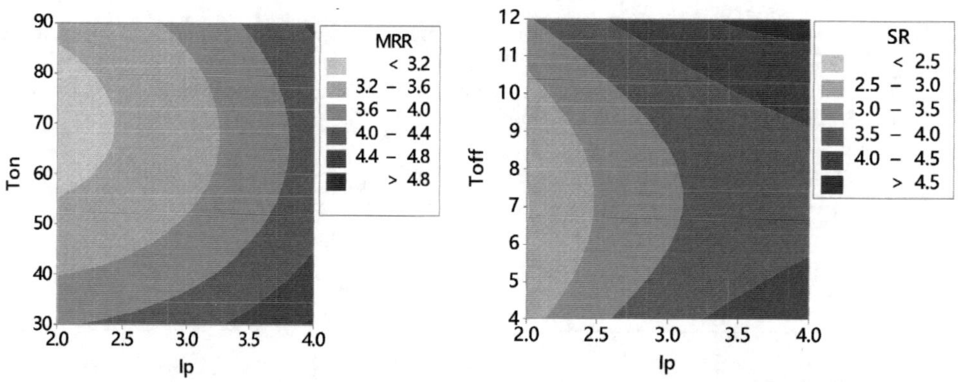

Figure 1. Contour graph for MRR and SR with significant machining parameter

The lower value of MRR (in Figure 1) is found at pulse on time = 70μs & pick current = 2 A. After that MRR is rapidly increased with increasing peak current steadily. Also, lower surface roughness(less than 2.5μm) is found at pulse off time =7μs & peak current =2A.Surface roughness is rapidly increased with increasing peak current up to a certain limit. Optimal setting for best output is Ip =2, Ton=30 & Toff =8. The predicted surface roughness is 3.0695μm and MRR is 3.9855 mm^3/min. In confirmation test, 4.052% error of MRR and 1.026% error of SR is found which is quite acceptable.

2. Powder Mixed Wire EDM

The mechanism of powder mixed electric discharge machining is different from the conventional electric discharge machining. The mixed powder particles in the dielectric are charged and act as a conductor by the very high potential energy. The charged particles are arranged themselves in chain formation which causes an early explosion. A series of spark creates on a point on the work surface and dielectric erodes debris particles from the gap zone in a quick manner, as a result, increasing MRR as well as improving surface finish.

Table 2. Experimental observation of B_4C powder mixed WEDM

EXP No.	Particle size	Concentration (gm/ltr)	MRR (mm^3/min)	SR (μm)
1	10	2	5.024	3.256
2	10	4	5.268	3.899
3	10	6	4.251	2.584
4	10	8	3.529	2.042
5	10	10	2.685	3.245
6	40	2	4.327	3.098
7	40	4	3.658	2.756
8	40	6	4.521	3.991
9	40	8	3.104	2.785
10	40	10	2.954	2.243

In the present study, Ti6Al4V is chosen as workpiece material which has low machinability due to its high hardness and toughness. In conventional WEDM, it is observed that Ti6Al4V has low MRR and poor surface finish during this process and it is required to improve performance for achieving better quality output. To overcome these issues, PMWEDM process is imported for further improvement of MRR and SF. B_4C powder of particle size 10μm and 40μm are suspended in deionized water (distilled water to supercool 100 oil ratio 20:1). Five concentrations of B_4C powder are varied i.e. 2gm/ltr, 4gm/ltr, 6gm/ltr, 8gm/ltr, and 10 gm/ltr in the dielectric by mixing a magnetic stirring system. Each experiment is performed for three times and taking the average output which is tabulated in Table no 2.

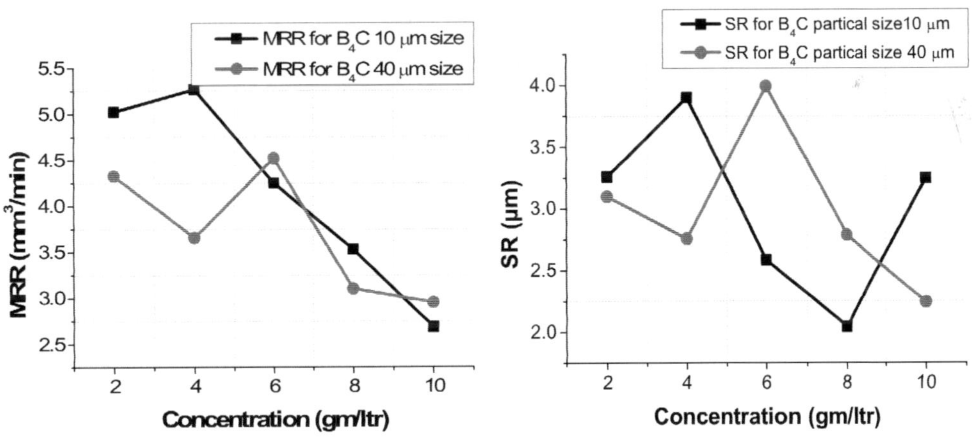

Figure 2. Graph for MRR and SR with different powder concentration

The optimum machining condition of input parameters is found in conventional WEDM of Ti6Al4V material. In this section, PMWEDM is done by varying powder concentration and size while machining controllable parameters are fixed i.e. Ip=2, Ton=30 and Toff=8. The result shows that 10 μm particle size gives a better result as compared to 40 μm particle size. If an individual response parameter is considered, then maximum MRR (5.268 mm^3/min) and minimum surface roughness (2.042 μm) are found from the experimentation. From Figure 2, it is clearly observed that concentration 6gm/ltr gives better MRR and SF. Therefore, 32.19% improvement in MRR and

33.47% improvement in SF have been achieved in PMWEDM as compare to conventional WEDM process.

2.1. Microstructure Study

PMEDM is a hybrid technique of electro-thermal spark erosion process where different powder particles are added with dielectric. Due to high intense heat and rapid cooling of the dielectric, microcracks are formed on the machined surface. These cracks are filled up to a certain level by using powder particle. In Figure (3a), many cracks are found through the microstructure image when powder size 10μm and concentration 4gm/ltr. Recast layers, as well as cracks (Figure (3b)), are more in powder size 10μm and concentration 2gm/ltr as compare to Figure (3a). The optimum result is obtained from machining condition i.e. powder size 10μm and concentration 6gm/ltr as illustrated in Figure (3c).

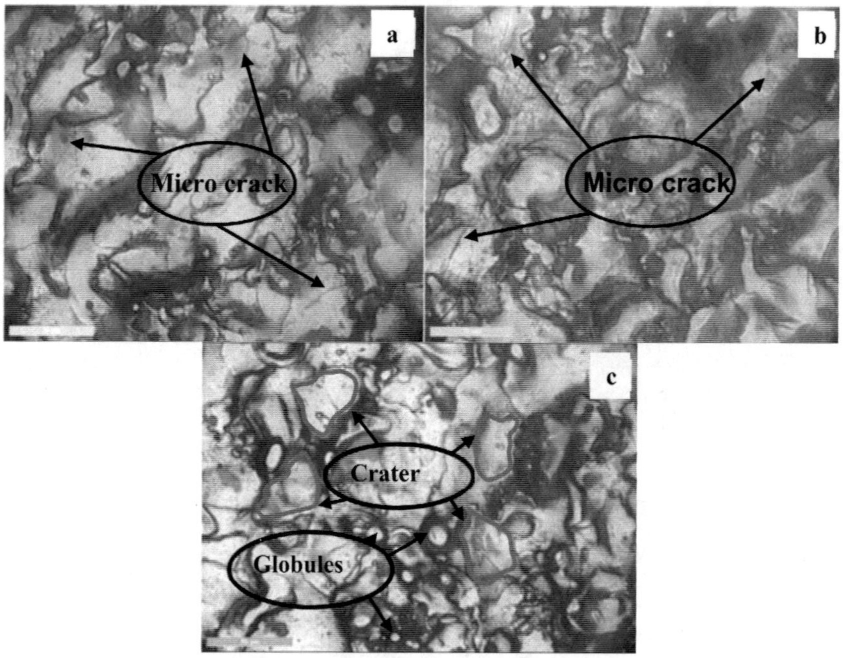

Figure 3. Microstructure image of PMWEDM sample with (a)powder size 10μm and concentration 4gm/ltr, (b) powder size 10μm and concentration 2 gm/ltr, (c) powder size 10μm and concentration 6gm/ltr

3. Conclusion

PMWEDM is a hybrid technique process where powder in the dielectric is used to improve surface finish as well as productivity. The present paper states that the use of powder enhanced performance efficiency means increasing MRR and decreasing surface roughness. The optimum machining setting for desired output is Ip=2, Ton=30 and Toff=8. From the PMWEDM experiment, it is evident that that 10μm particle size and 6gm/ltr concentration of B_4C powder in deionized water has a stronger effect on MRR and SR. Therefore, 32.19% improvement in MRR and 33.47% improvement in SF have been achieved in PMWEDM as compare to conventional WEDM process. Further, powder mixed with dielectric WEDM technology can be utilized for machining of other advance material for enhancing process stability.

IOP Conf. Series: Materials Science and Engineering **653** (2019) 012006 doi:10.1088/1757-899X/653/1/012006

4. References

[1]. Kumar A, Maheshwari S, Sharma C and Beri N, *A Study of Multiobjective Parametric Optimization of Silicon Abrasive Mixed Electrical Discharge Machining of Tool Steel.* Mat. & Manuf. Proc., 2010.**25**: p.1041–1047.

[2]. Kumar S, Dhingra A K and Kumar S, *Parametric optimization of powder mixed electrical discharge machining for nickel based superalloy inconel-800 using response surface methodology*, Mech. of Adv. Mat. & Mod. Proc., 2017. **3**(7).

[3]. Marashia H, Jafarlou D M, Sarhana A A D, and Hamdi M, *State of the art in powder mixed dielectric for EDM applications*' Prec. Engg., 2016.**46**: p.11-33.

[4]. Prakash C, Kansal H K, Pabla B S and Puri S, *Experimental Investigations in Powder Mixed Electric Discharge Machining of Ti-35Nb-7Ta-5Zr β- Titanium Alloy*, J of mat & manuf. proc., 2017. **32**(3): p.274-285.

[5]. Öpöza T T, Yasarb H, Ekmekci N, and Ekmekci B, *Particle migration and surface modification on Ti6Al4V in SiC powdermixed electrical discharge machining* J. of Manuf. Proc., 2018.**31**: p.744-758.

[6]. Chakraborty S and Bose D, *Improvement of die corner inaccuracy of inconel 718 alloy using entropy based GRA in WEDM process*, Adv. Eng.g Forum, 2017.**20**: p.29-41.

[7]. Paul T, Chakraborty S, and Bose D, *An experimental investigation on surface topography of miniature spur gear by WEDM of inconel 718*, Int. Conf. on Prec., Meso, Micro & Nano Engg., 2017: p.623-626.

Dry sliding wear behavior of LD slag filled epoxy composites using Experimental Design

Pravat Ranjan Pati[1]and Alok Satapathy[2]

[1] Faculty of Science & Technology
ICFAI Foundation for Higher Education, Hyderabad
[2]National Institute of Technology, Rourkela
pravatpati99@gmail.com

Abstract. This article reports the fabrication and investigation of sliding wear properties of Linz-Donawitz (LD) slag reinforced epoxy composites. Hand lay-up technique is used for composite fabrication. LD slag is varied from 0 to 30% by weight and sliding wear tests are performed. The sliding wear performance of these composites is evaluated at various sliding velocities and normal loads. It is found that the incorporation of increased LD slag content resulted in significant enhancement in sliding wear resistance. This work also shows that LD slag content and sliding velocity are the most influencing factors as far as the minimization of specific wear rate is concerned.

Keywords: LD slag, Epoxy composite, Sliding wear, Taguchi technique.

1. Introduction

Polymer composites find wide applications in automotive as well as aerospace industries due to their high mechanical properties. It has various properties like hardness, strength, wear resistance, conductivity and density. But among all the properties, wear resistance plays a strong role for polymer matrix composites. Due to high strength properties, industrial wastes are oftenly used as reinforcement in polymer composites. Numerous researchers have investigated and described the influence of industrial wastes on wear characteristics of polymer composites. Beside mechanical properties, improvement of the wear resistance of composites is a major challenge for researchers to satisfy the design requirements of the fabricated components when subjected to the sliding environment. Satapathy et al. [1] investigated the sliding wear response of polyester composites filled with titanium dioxide. Sliding wear characteristics of epoxy composites filled with different fillers like TiO_2, rice husk, nano-sillica, SiC/Al_2O_3, fly ash, silicon carbide and graphite has been described by different researchers [2-10]. Gyurova and Friedrich [11] reported the sliding friction and wear properties of polyphenylene sulfide composites. Umeda et al. [12] studied the characterization of wear particles. Purohit and Satapathy [13] developed an epoxy composites filled with LD sludge. Padhi et al. [14] examined the sliding wear behavior of polypropylene composites using BFS particles. Satapathy and Patnaik [15] discussed the wear response of polyester composites filled with red mud. However, the research works on the applicability of LD slag particles as filler in polymer composite yet to be studied. The purpose of the present work is to analyze LD slag as filler material in epoxy composite using hand lay-up fabrication technique and to study the sliding wear properties.

2. Experimental details

2.1. Preparation of composite materials

A sequence of composites are fabricated keeping epoxy resin constant and varying LD slag in a stable manner as shown in Table 1 through efficient increase in filler content. The epoxy resin and corresponding hardener are slowly mixed by using hand. During mixing, LD slag is poured into the resin in different weight proportions. Then, the mixed dough is stirring for some time. A silicone releasing agent is sprayed over glass molds before the mixture poured into it to easy removal of the fabricated composite from the mold after curing. The setup is then kept at room temperature for 24 hours for initial curing to obtain the required composite.

Table 1. Designation and composition of the composites.

Designation	Composition
C_1	EP + 0 wt% LD slag
C_2	EP + 7.5 wt% LD slag
C_3	EP + 15 wt% LD slag
C_4	EP + 22.5 wt% LD slag
C_5	EP + 30 wt% LD slag

Note: EP: Epoxy

2.2. Dry sliding wear test

Sliding wear tests are performed on pin-on-disc apparatus at different sliding velocities, normal loads and sliding distances. The disc of the pin-on-disc set up is made up of hardened steel. Each sample is cleaned with acetone and polished prior to test. Then, it is held fixed on the sample holder and the disc are revolved while a load is being applied through a lever machine. The experiments are performed based on Taguchi technique with L_{25} orthogonal array. The loss of material from the surface of the composite body is computed using a precision electronic balance with an accuracy of ±0.0001 g and the specific wear rate (W_s) is then obtained as follows:

$$W_s = \Delta m/(\rho t V_s F_n) \tag{1}$$

where Δm is the mass loss of the composite material (g), ρ is the density of the composite (g/mm^3), t is the duration of experimental test (s), V_s is the velocity of sliding (cm/s), and F_n is the average normal load (N).

2.3. Taguchi Experimental Design

Design-of-experiment is a technical approach to efficiently plan and execute experiments. The proper selection of the parameters is the most vital stage in experimental design; therefore, a quantity of factors is involved so that non-significant variables can be recognized at the earliest opportunity. Table 2 represents the four parameters such as sliding velocity, normal load, sliding distance and LD slag content each at five levels which is considered for the wear test and the effects of these four parameters are studied using L_{25} array design. These experiments are conducted at room temperature. Using Taguchi's L_{25} array, the number of experiments are reduced from $5^4 = 625$ runs to only 25 runs by saving lot of time and cost. The investigational observations are further transformed into signal-to-noise (S/N) ratios. The S/N ratio for minimum wear rate can be specified as "smaller is better" characteristic, which is calculated as logarithmic conversion of loss function as given below.

$$\frac{S}{N} = -10\ log \frac{1}{n}(\Sigma y^2) \tag{2}$$

where, y is the observed data and n is the number of observations.

Table 2. Control factors and their selected levels.

Control Factor	Level					
	1	2	3	4	5	Units
A: Sliding velocity	105	209	314	420	527	cm/sec
B: Normal load	5	10	15	20	25	N
C: Sliding distance	1000	1500	2000	2500	3000	m
D: LD slag content	0	7.5	15	22.5	30	wt%

3. Results and discussion

The specific wear rates and the equivalent S/N ratio are obtained for all the 25 test runs, presented in Table 3. The total mean of the wear rate for the S/N ratio is found to be -2.733 db. MINITAB 14 software is used for design-of-experiment applications. The influences of different factor are evaluated by computing the response. The conclusion of this investigation shows that factor combination of A_1, B_4, C_3 and D_5 gives the minimum wear rate, shown in Fig. 1.

Table 3. L_{25} Orthogonal array design with output and S/N ratio.

Test run	Sliding Velocity (A) (cm/sec)	Normal Load (B) (N)	Sliding Distance (C) (m)	LD slag Content (D) (wt %)	Sp. Wear Rate (10^{-9} mm^3/N-m)	S/N ratio (db)
1	105	5	1000	0	1.912	-5.62976
2	105	10	1500	7.5	1.634	-4.26504
3	105	15	2000	15	0.947	0.47300
4	105	20	2500	22.5	0.829	1.62891
5	105	25	3000	30	0.783	2.12476
6	209	5	1500	15	1.225	-1.76272
7	209	10	2000	22.5	1.017	-0.14642
8	209	15	2500	30	0.904	0.87663
9	209	20	3000	0	1.881	-5.48778
10	209	25	1000	7.5	1.598	-4.07154
11	314	5	2000	30	0.986	0.12246
12	314	10	2500	0	1.922	-5.67507
13	314	15	3000	7.5	1.637	-4.28097
14	314	20	1000	15	1.218	-1.71295
15	314	25	1500	22.5	1.133	-1.08460
16	420	5	2500	7.5	1.728	-4.75087
17	420	10	3000	15	1.421	-3.05188
18	420	15	1000	22.5	1.334	-2.50312
19	420	20	1500	30	1.012	-0.10361
20	420	25	2000	0	2.115	-6.50621
21	527	5	3000	22.5	1.548	-3.79542
22	527	10	1000	30	1.232	-1.81221
23	527	15	1500	0	2.339	-7.38060
24	527	20	2000	7.5	1.863	-5.40426
25	527	25	2500	15	1.608	-4.12572

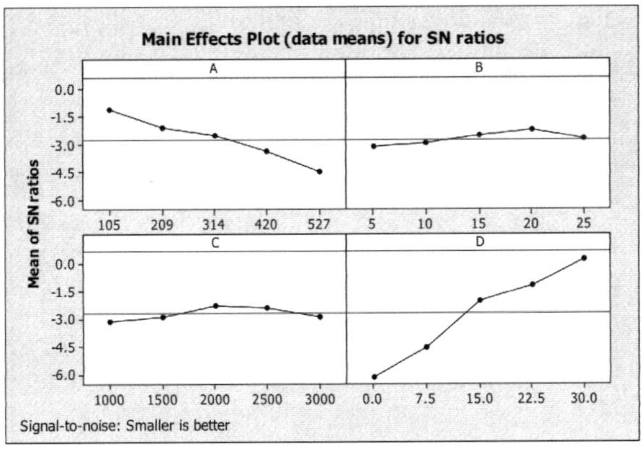

Fig. 1. Effect of control factors on erosion rate.

Table 4. Response table for signal to noise ratios.

Level	A	B	C	D
1	-1.1336	-3.1633	-3.1459	-6.1359
2	-2.1184	-2.9901	-2.9193	-4.5545
3	-2.5262	-2.5630	-2.2923	-2.0361
4	-3.3831	-2.2159	-2.4092	-1.1801
5	-4.5036	-2.7327	-2.8983	0.2416
Delta	3.3700	0.9473	0.8536	6.3775
Rank	2	3	4	1

Table 4 represents the response table for signal-to-noise ratio. The delta value of different control factors are given in this table. LD slag content having higher delta value than others and it is given as first rank. Thus, the wear rate of the composite is influenced by LD slag content followed by other factors.

4. Conclusions

This experimental and analytical investigation leads to the following conclusions:

a. LD slag is efficiently used as reinforcing filler in polymer matrix composites.

b. The influence of LD slag particles in epoxy resin resulted in enhancement of wear resistance of the composite. Sliding wear characteristics of these composites can be effectively examined using Taguchi technique.

c. Factors like LD slag content, sliding velocity, normal load and sliding distance are found to be significant for minimization of wear rate.

5. References

[1] Satapathy A, Tarkes DP and Nayak NB 2010 *Int J Plast Technol* **14** S24
[2] Siddhartha, Patnaik A and Bhatt AD 2011 *Materials and Design* **32** 615
[3] Rout A and Satapathy A 2012 *Procedia Engineering* **38** 1218
[4] Rashmi, Renukappa NM, Suresha B, Devarajaiah RM and Shivakumar KN 2011 *Materials and Design* **32** 4528
[5] Shivamurthy B, Udaya Bhat K and Anandhan S 2013 *Materials and Design* **44** 136
[6] Kang Y, Chen X, Song S, Yu L and Zhang P 2012 *Applied Surface Science* **258** 6384
[7] Ahmed KS, Khalid SS, Mallinatha V and Amith Kumar SJ 2012 *Materials and Design* **36** 306

[8] Basavarajappa S, Ellangovan S and Arun KV 2009 *Materials and Design* **30** 2670
[9] Dadkar N, Tomar BS and Satapathy BK 2009 *Materials and Design* **30** 4369
[10] Basavarajappa S and Ellangovan S 2012 *Wear* **296** 491
[11] Gyurova LA and Friedrich K 2011 *Tribology International* **44** 603
[12] Umeda A, Sugimura J and Yamamoto Y 1998 *Wear* **216** 220
[13] Purohit A and Satapathy Alok 2017 *Journal of Composite Materials* **51** 899
[14] Padhi PK, Satapathy Alok and Nakka AM 2015 *Journal of Thermoplastic Composite Materials* **28**, 656
[15] Satapathy A and Patnaik A 2010 *Journal of Reinforced Plastics and Composites* **29** 2883

Microstructure and Mechanical Behaviour of ABOw Reinforced Aluminium Matrix Composite Synthesized by Powder Metallurgy route

Neeraj Pandey[1], I Chakrabarty[2], S C Ram[2], M R Majhi[1]

[1]Department of Ceramic Engineering Indian Institute of Technology (BHU), Varanasi, U.P., India

[2]Department of Metallurgical Engineering Indian Institute of Technology (BHU), Varanasi, U.P., India

Corresponding Author:E-mail:neerajp.rs.cer16@itbhu.ac.in(N.Pandey)

Abstract. In present study, the effect of Aluminum borate whiskers (ABOw) content (5, 10 and15 wt.%) on the development of Al-ABOw composite using powder metallurgy route has been investigated. Firstly, Aluminum borate whiskers (ABOw) were produced using sol-gel method. The commercial pure aluminum powder was used as matrix material in this investigation. Amongst, the various casting technology, the powder metallurgy route is most economical and extensively used by researchers. The microstructures of composites were observed using optical microscopy (OM), scanning electron microscopy (SEM). Transmission electron microscopy (TEM) analysis was performed for the observation of ABOw powder morphologies. The XRD analysis was carried out to confirm the phases and peak intensity of developed alloy and composites. The sintering was carried out at 600°C for all samples. After that, the bending test was carried out to calculate flexural strength of alloy and composites. The flexural strength 172±5 MPa was obtained maxima in composite with 10 wt.% ABOw along with improved hardness around 40.2HV. The result shows that, the composite exhibited better interfacial bonds, uniformity in microstructure and optimum flexural strength.

Keywords: Powder metallurgy, ABOw reinforcement, flexural strength, fractography.

1. Introduction

Aluminum and its alloys have been extensively used in military industries, aerospace, auto industries and transport, marine industries, construction structures and electrical industries. Although in some industrial uses, benefits of aluminum alloys have been limited because of lower mechanical strength, wear & erosion resistance and surface hardness. Aluminum-ceramics based composites are widely used materials as MMCs due to their light weight, low density, high tensile strength and ease of fabrication [1-3]. Reinforcements used may be continuous, discontinuous or particulate in nature. Whiskers are gaining their importance as reinforcements as they provide large surface area for bonding with matrix material. The whiskers mainly used as reinforcements are SiO_2, Al_2O_3, $Mg_2B_2O_5$, Mg_2Si and ABOw [2,4-9]. Among these ABOw is gaining importance as they are low cost one dimensional material. The ABOw($Al_{18}B_4O_{33}$) may have potential oxidation résistance, the physical properties such as bulk density 1.46g/cm^3, open porosity 54.4, linear shrinkage percentage -1.8% and flexural strength 55±2 at 1500°C sintering temperature[18]. Several fabrication techniques using whiskers as reinforcements have been reported for manufacturing of such composites such as squeeze casting, stir casting and centrifugal casting [10]. However, due to lack of research work reported on powder metallurgy route for the fabrication of ABOw whiskers reinforced composite, the many researchers are in favoured to develop the composites for high performance engineering components in automotive and aircraft industries. The ABOw whisker which is used as reinforced material was generally prepared using sol-gel method [11]. X. Gao et al.[12] found that, the coating of ABOw and the addition of soybean sacrificial filler can effectively decrease the volume fraction of whiskers in the composite without changing its ultimate tensile strength. However, at the room temperature, the elongation to fracture of the composite evidently increases around 11.1%. Many researchers have studied ZrO_2, Al_2O_3 reinforced aluminum metal matrix composites but their tensile and fracture behavior needs much improvement. Mehta, Niraj Singh, et al.[13] have studied that reinforcing with zirconia (ZrO_2) particles in porcelain ceramic. Roseline, S.et al.[14] varied zirconia percentage in ceramics and found that, the impact strength and tensile strength are slightly enhanced with improved ductility. The higher porosities developed in the composites due to stir casting process causes low strength of components. The adaption of powder metallurgy route shows better results in respect of lower porosities, defect free ceramic reinforced composites fabrications.

Powder metallurgy route as shown in Figure 1 was used to prepare the final objects for characterization; this method involves blending of finely powdered materials, consequently pressing the mixture into desired form by using metal dies, followed by sintering the compacted green specimens in a controlled atmosphere condition. The desired shape component was obtained which requires minimum machining. Powder metallurgy can also be used to produce fully finished components of automotive parts like connecting rods [15-17]. The aim of this investigation was to fabricate light weight and high strength composites.

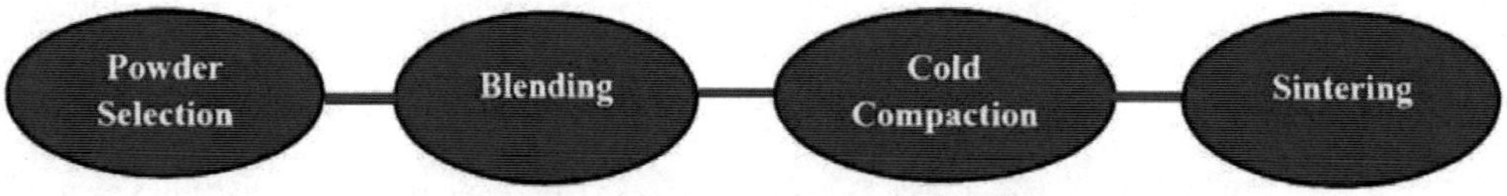

Figure 1. Powder metallurgy processes

2. Experimental Procedure

Aluminum borate whiskers reinforcement was synthesized using sol-gel method calcinized at 1300°C. The aluminum powder was used as matrix material procured from Poly-Chem private India Limited Bangalore. Pure aluminum metal powder (purity 99.5%) and particle size 20-40 μm was used as matrix. It was vigorously mixed with ethanol in magnetic stirrer for 40 min then ABOw was added in required percentage at 80°C and further mixed through magnetic stirrer for 15 min. After hydrolysis sol was dried in oven at 110°C for 4 hours, then dried sol was mixed with PVA (Polyvinyl Alcohol) solution and pellet was pressed with hydraulic press at 180 MPa. Rectangular pellets of dimension 50x10x10 mm were obtained. The pellets each were fired in Argon gas environment in a sintering furnace at heating rate 5°C/min followed by heating at 600°C and 2h soaking time then cooled to room temperature.

X Ray Diffraction by Rigaku Smart Lab at 40kV and 15mA using CuKα (λ= 0.154 nm) radiation was used to confirm the phases in sintered composite samples. Microstructures were studied using Scanning Electron Microscopy (SEM) ZEISS EVO 18 operating at 20 kV was used to identify the morphologies of sintered composites. Transmission electron microscopy (TEM TECHNAI G^2 T20) operated at 200 kV was carried out for morphology and crystal structures of ABOw reinforcement. Density of the composites was calculated using Archimedes 'immersion method. The flexural strength is measured by universal testing machine using Tinius Olsen (H10KL). Bending strength was calculated using the following expression as given below.

$$\text{Bending strength}=\frac{3FL}{2bt^2}$$

where F is the load at failure, L is the distance between the supports, b is the sample width, and t is the thickness of the sample.

3. Result and Discussion

The density of three samples of each composite with different volume fraction was calculated using immersion method, based on Archimedes' principle. Finally, the average porosities were calculated of each composite. The details of this method can be found in ASTM C20.The results obtained as shown in Table 1. From Table 1 it can be observed that, the density of the specimens decreases with increasing the volume fraction of whiskers in the aluminum matrix. It shows the micro sized whiskers decrease the density of the specimens with increased percentage in matrix. Also the increased percentage of whiskers increases the average porosities in the sintered composites. In contrast, composite with 15%ABOw shows decrease in porosity percentage due to more interlocking and rounded shape of whiskers.

The optical microscopy images for all samples with variations of ABOw percentages are shown in Figure 2.Figure 2(a) shows the microstructure without addition of ABOw reinforcement. It clearly confirms the dark flaky pattern which represents silicon eutectic phase and white phase as matrix grains. In contrast, optical microstructures of composites reinforced with 5%, 10% and 15% ABOw whiskers shown in Figure 2(b),(c) and (d) clearly confirmed the ABOw whisker particles in the form of clusters.

Table 1. Table captions should be placed above the tables.

Composites/Al Powder	Sample Number	Experimental density(g/cc)	Theoretical density ($\rho_{th}=V_f\rho_f+V_m\rho_m$)	Average porosity (%)
Al-Powder	1	2.56	2.70	5.43
	2	2.53		
	3	2.58		
Al(ABOw)$_5$	1	2.47	2.65	7.48
	2	2.48		
	3	2.45		
Al(ABOw)$_{10}$	1	2.32	2.61	11.17
	2	2.30		
	3	2.29		
Al(ABOw)$_{15}$	1	2.30	2.57	9.86
	2	2.33		
	3	2.32		

IOP Conf. Series: Materials Science and Engineering **653** (2019) 012008 doi:10.1088/1757-899X/653/1/012008

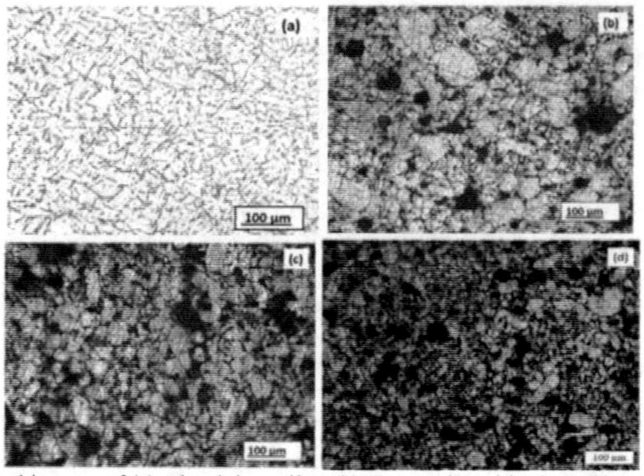

Figure 2. Optical images of (a) Aluminium alloy (b) Composite with 5% ABOw (c) Composite with 10% ABOw (d) Composite with 15% ABOw sintered at 600°C

The dark clustures represent the ABOw particles and white regions represent the metal matrix respectively. The scanning electron micrographs are shown in Figure 3 calcinized at 1300°C .The scanning electron micrograph shows the distribution of ABOw whiskers throughout the volume of cast composites. It was found that, after sintering, the uniform distribution of reinforcement in matrix materials offered excellent wear resistance and mechanical properties. However some regions show agglomeration of whiskers which may be due to the porosity incorporated in composite during sintering process.

Figure 3. SEM micrographs of ABOw used as reinforcement calcinized at 1300°C

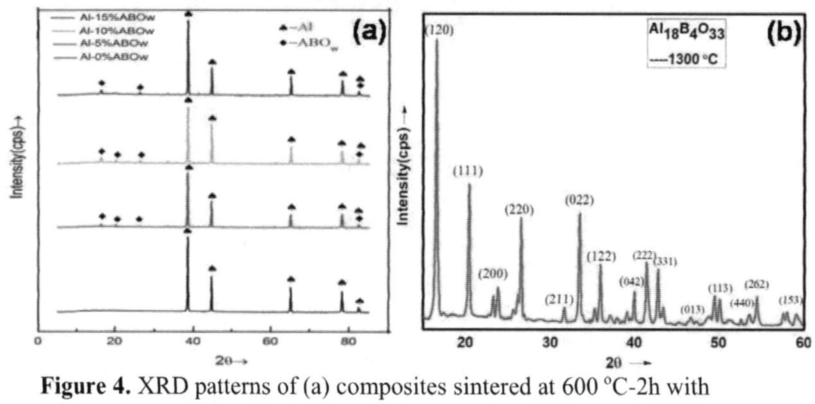

Figure 4. XRD patterns of (a) composites sintered at 600 °C-2h with different composition (b) ABOw reinforcement calcinized at 1300°C.

XRD patterns of aluminum matrix composites samples sintered in the range of 600°C is shown in figure 4.It shows that except aluminium and ABOw, no other phases were present in the composites manufactured. As the volume content of ABOw increases the height of aluminium peaks decreases whereas height of ABOw peaks increases. Moreover, From XRD results shown in Fig.4 (b), it is clear that, with higher ABOw percentage, more $Al_{18}B_4O_{33}$ (JCPDS, 32-0003), phases are present and due to their high flexural strength, high hardness (Mohs 7) [3] can be improved mechanical and physical properties of fabricated composites. The XRD pattern of ABOw as shown in Figure 4(b).

The hardness of the samples increases with increasing the percentage of whiskers in the matrix refer Table2.This may be due to inhibition of the progress of plastic deformation by whiskers present in the matrix.As agglomeration of whiskers proceed in specimens with 15% ABOw volume fraction then on uniform sintering driving force leads to localized shrinkage strains introducing more defects (dislocations and point defects, etc.) at the interfaces of ABOw and matrix during grain growth so that the interface strength decreases during sintering. This behavior of composite increases with increasing volume fraction in aluminum matrix [13] .The flexural strength of the samples were calculated using three point bending test, with a span of 30 mm and a crosshead speed of 0.5mm/min. Resultant, it was found that in Al-10wt%ABOw have more flexural strength upto 172Mpa refer Table 2.The TEM Bright field images with corresponding diffraction patterns of ABOw powder prepared by sol-gel method as shown in Figure 5.The morphological and discontinuous structure of the micro rods were characterized and the crystal parameters are determined as a=7.6874 Å, b=15.0127 Å and c=5.6643Å. The parameters confirm the orthorhombic of ABOw reinforcements [11].

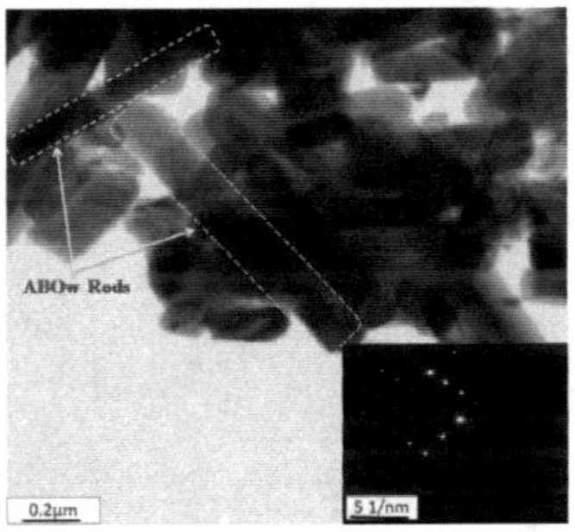

Figure 5. Bright field Image and corresponding SAED (inset) of ABOw powder calcinized at 1300 °C

Table 2.Bending Strength and Vickers Hardness of samples

Volume fraction of ABOw	Bending Strength at room temperature (MPa)	Hardness(HV)
0	125±5	27
5%	160±5	37.8
10%	172±5	40.2
15%	165±5	39

The SEM fractographs of sintered composites is as shown in Figure 6. It was observed from Figure 6(a) the fractured surface of aluminum alloy shows a ductile nature of the fracture along the micro holes and dimples on the surface. The fractographs of Al/ABOw composites in figure 6(b), (c) and (d) reflect brittle nature of fracture specifically at whisker and matrix interface. Figure 6(b) shows more agglomerates of aluminum on the fractured surface and whiskers may provide better interfacial bonding and enhanced mechanical properties were observed .From Figure 6(c) more sites of whisker pull out reflect that the interfacial bonding between matrix and whisker led to load transfer from matrix to the whisker thus giving good bending strength to the composite.

Figure 6. Fractographs of (a) Aluminium alloy (b) Composite with 5% ABOw (c) Composite with 10% ABOw (d) Composite with 15% ABOw sintered at 600°C

4. Conclusion

In this study, Al-(ABO)w metal matrix composite with the variations of ABOw whiskers reinforcements in pure commercial aluminum metal powder used as matrix materials has been synthesized using powder metallurgy route.

The sol-gel method for preparation of ABOw reinforcement was used. The OM, SEM were used for microstructures observations i.e. phases as well as grain structure and TEM analysis was done to characterized the morphological and discontinuous structure of the micro wishkers. The density of the composites is decreases with increasing percentage of whisker in the matrix due to increase in porosity with increasing content of whisker. The micro hardness of composite improved in Al-10%ABOw with reduced porosity and more interfacial bonding between matrix and reinforcement. The flexural strength around 172 MPa with 10%ABOw reinforcement was observed due to lower porosity than other sintered samples. The study clearly shows whisker content in the composite led to improvement in mechanical properties due to effect of grain refinement. This grain refinement was achieved after sintering the samples. The fractographs of fractured samples show more agglomerates of aluminum on the fractured surface along with whiskers which may provide more interfacial bonding and enhanced mechanical properties.

IOP Conf. Series: Materials Science and Engineering **653** (2019) 012008 doi:10.1088/1757-899X/653/1/012008

References

[1] Torralba, JM D., C. E. Da Costa, and F. Velasco.(2003), P/M aluminum matrix composites: an overview, *J .Mat. Pross. Tech.* 133.1-2, 203-206.

[2] Mehta, N.S., Sahu, P.K., Tripathi, P., Pyare, R. and Majhi, M.R.(2018), Influence of alumina and silica addition on the physico-mechanical and dielectric behavior of ceramic porcelain insulator at high sintering temperature.,*Boletín de la Sociedad Española de Cerámica y Vidrio*, *57*(4), 151-159.

[3] Hernández, María F., Gustavo Suárez, Mariano Cipollone, María S. Conconi, Esteban F. Aglietti, and Nicolás M. Rendtorff.(2017), Formation, microstructure and properties of aluminum borate ceramics obtained from alumina and boric acid., *Ceramics Inter.* 43(2),2188-2195(2017).

[4] Qu, Xiaoyao, Fucheng Wang, Chunsheng Shi, Naiqin Zhao, Enzuo Liu, Chunnian He, and Fang He.(2018), In situ synthesis of a gamma-Al_2O_3 whisker reinforced aluminium matrix composite by cold pressing and sintering, *Mat.Sc.Engg: A* 709, 223-231.

[5] Yu, Zhenyang, Naiqin Zhao, Enzuo Liu, Chunsheng Shi, Xiwen Du, and Jian Wang(2011)., Low-temperature synthesis of aluminum borate nanowhiskers on the surface of aluminum powder promoted by ball-milling pretreatment, *Powd. tech.*,212(2), 310-315.

[6] Jin, Pei-peng, C. H. E. N. Geng, H. A. N. Li, and Jin-hui Wang(2014). Dry sliding friction and wear behaviors of $Mg_2B_2O_5$ whisker reinforced 6061Al matrix composites. *Trans.Nonferr.Met Society of China* 24(1), 49-57.

[7] Gudlur, Pradeep, Adam Forness, Jonathan Lentz, Miladin Radovic, and Anastasia Muliana(2012)., Thermal and mechanical properties of Al/Al_2O_3 composites at elevated temperatures., *Mat. Sci. Engg: A* 531, 18-27.

[8] Ram, S. C., K. Chattopadhyay, and I. Chakrabarty(2017)., High temperature tensile properties of centrifugally cast in-situ Al-Mg_2Si functionally graded composites for automotive cylinder block liners., *J. Alloys.Compd.*, 724, 84-97(2017).

[9] Gao, X., H. Y. Yue, X. Y. Lin, L. H. Yao, W. D. Fei, and E. J. Guo(2016)., Effect of $ZnAl_2O_4$ coating on the compressive behaviors of aluminum borate whiskers-reinforced aluminum composites at elevated temperatures *Comp.Interf.*23(5), 383-394.

[10] Kala, Himanshu, K. K. S. Mer, and Sandeep Kumar(2014)., A Review on Mechanical and Tribological Behaviors of Stir Cast Aluminum Matrix Composites.,*Procedia Mat.Sci.*6,1951-1960.

[11] Wang, Jun, Jian Sha, Qing Yang, Youwen Wang, and Deren Yang(2005)., Synthesis of aluminium borate nanowires by sol–gel method, *Mat.res. Bullt.* 40(9), 1551-1557.

[12] Gao, X., H. Y. Yue, E. J. Guo, X. Y. Lin, L. H. Yao, and B. Wang(2017)., Design and tensile properties of aluminum borate whiskers reinforced aluminum composite with low whisker volume fraction,*Comp.Interf.*24(4), 371-379.

[13] Mehta, Niraj Singh, J. C. Pandey, Neeraj Pandey, Ram Pyare, and Manas Ranjan Majhi.(2018), Developing a high strength physico-mechanical and electrical properties of ceramic porcelain insulator using zirconia as an additive.,*Mat. Res. Exp.* 5 (7), 075202.

[14] Roseline, S., V. Paramasivam, and S. Jayabal.(2018), Evaluation of Structure and Mechanical Properties of Zirconia Reinforced Aluminium Matrix Composites., *J.Comput.Theor.Nanosci.* 15(3), 830-838(2018).

[15] Dileep, B. P., V. Ravikumar, and H. R. Vital(2018)., Mechanical and Corrosion Behavior of Al-Ni-Sic Metal Matrix Composites by Powder Metallurgy, *Mat.Today: Proce.* 5 (5),12257-12264.

[16] Hardy, Dale, and David J. Green.(1995), Mechanical properties of a partially sintered alumina, *Alumina J. Eur. Ceram. Soc. IS* 15, 769-715.

[17] Pickens, J. R.(1981), Aluminium powder metallurgy technology for high-strength applications, *Journal of Mat. Sci.* 16(6),1437-1457.

[18] Readey, M.J(1992)., Formation and Sintering Characteristics of Aluminum Borate Whiskers, Journal of the American Ceramic Society, 75(12), pp.3452-3456.

IOP Conf. Series: Materials Science and Engineering **653** (2019) 012009 doi:10.1088/1757-899X/653/1/012009

Study on Microstructure and Tribological Properties of Plasma Processed LM6 Alloy

J Parida[1], S C Mishra[2] and S C Pattnaik[3]

[1,3]Department of Metallurgical and Materials Engineering, I.G.I.T, Sarang, Odisha, India.

[2]Department of Metallurgical and Materials Engineering, NIT, Rourkela, Odisha, India.

E-mail: jagadish.igit@gmail.com

Abstract. In the research work, LM6 alloy prepared by plasma processed technique is studied for wear behavior. The wear behavior of the plasma processed LM6 alloy sample is study at room conditions at four different loads ,i.e. 10N, 20N, 30N, 40N and with different sliding speeds, i.e. 200, 300, 400, 600 rpm's for about 2 min,5 min,10min,15 min respectively, using a pin-on-disk wear testing instrument. From the pin-on-disk experiment, cumulative mass loss varying with the sliding speed, applied load and time was studied. The existence of eutectic silicon is sensed by the microstructure. The wear of plasma processed LM6 alloy sample increased as applied load, sliding speed and time increased.

Key words: LM6 Alloy, Dry Sliding wear, sliding speed, Wear resistance, Tribological Properties.

1. Introduction

Aluminium-silicon alloys are widely and favorably implemented in automobile engineering applications like pistons, clutch housing, and liners, in which tribological and mechanical properties of the material are significant, due to the metallurgical feature, like more strength to weight fraction and high thermal conductivity [1]. The tribological wear behavior of Al-Si alloys depend on different parameters, such as composition, size, shape, and distribution of micro components. All the while, it also relies upon service conditions and mechanical properties for example, interface condition, applied load, temperature and sliding speed [2-3]. Many research studies are worked out for investigation of analyzing the wear quality of these alloys under different testing conditions. For simulated wear testing in the laboratory, pin-on-disc testing equipment is one of the most commonly implemented. It is attached with a continuously rotating disc, in which the effect of sliding speed, normal load, alloy composition and the presence of interface film etc. can be effortlessly investigated [4]. Torabian et al. have considered the impacts of normal load, alloy composition, sliding speed and distance on the wear of Al-Si alloys. The wear rate is strictly depending upon the applied load. It increments straightly with load [5].

2. Experimental Details

2.1. Selection of Material

For organizing the wear test, cylindrical shaped samples considered here are of diameter 10mm and length 30 mm dimensions. In this present research work, LM6 alloy chemical composition is tabulated and represented in Table 1.

Table 1. Chemical composition of LM6 alloy.

Elements	Si	Cu	Mg	Pb	Ni	Fe	Mn	Zn	Sn	Ti	Al
Composition	12%	0.1%	0.1%	0.1%	0.1%	0.6%	0.5%	0.1%	0.05%	0.2%	Rest

Initially, LM6 alloy has been prepared by plasma casting approach. This casting process is one of the advance stirring techniques that stir molten metal and alloy in both conductive and non-conductive dies during the process of solidification. The some merits due to more stirring like better chemical homogeneity, finer micro structure, and much more orderly provide in both micro and macro scales. During the process of solidification, the plasma treatment casting approach implements a varying electric arc on the molten metal top and constitutes varying current flux named as changing current density per unit volume which induces altering magnetic field and thus builds an intense stirring inside the molten bath. In this paper, during casting of LM6 alloy, plasma treatment casting approach has been implemented.

2.2. Microstructure study

The LM6 alloy samples bearing the thickness of 15 mm and diameter of 20 mm have been cut and cold mounted for the study of microstructure. The polished samples are being carried out by considering normal metallographic approach and those are also etched in addition to Keller's reagent having a composition of (2.5wt% HNO3, 1.5wt% HCl, 1wt% HF and remaining water). Once this process is completed, the micro structural performance has been analyzed by stereo-microscope.

2.3. Wear Test

Dry sliding wear experiments are experimented on plasma-based approachLM6 alloy by the help of pin-on-disc wear testing equipment. Specification of cylindrical based samples are having diameter of 10 mm and length of 30 mm has been mounted vertically on a pin shaped holder. The last end of the samples has been polished with an emery paper bearing different grades such as 1/0, 2/0, 3/0, 4/0. During the conduction experiment, the pin is being suppressed against the steel disc (EN32) having the hardness of 60 HRC. Prior to every trail, the steel counter-face has been grounded with two various grit sizes such as320 and 600grits of SiC abrasive for some interval. Various testing situations constitute speed settings of 100, 200, 300 and 400 rpm with an applied load of 10, 20, 30 and 40 N for the duration of 2, 5, 10 and 15minrespectively. The initial weight of the sample is being calculated by an electronic weighing machine with an accuracy of ± 0.0001gm. For calculation and validation purpose, experimental data are being considered and calculated for wear, and represented in the form of weight loss.

3. Results and Discussion

3.1. Microstructure study

Computerized stereo-microscope is used for obtained microstructures of LM6 alloy sample under the magnification of 200X are shown in figure.

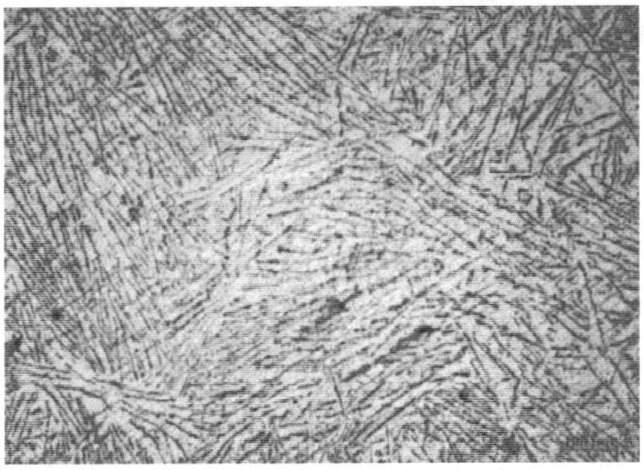

Figure 1. Microstructure of plasma processed LM6 alloy.

Above figure exhibit an optical micrograph of LM6 alloy. Here, Al – Si eutectic are light areas, the eutectic aluminium are dark dots and the PSPs (primary Silicon particles) are dark patches.

3.2. Wear test results

The wear behavior of plasma processed LM6 alloy is performed at atmospheric conditions of four distinct loads such as 10N, 20N, 30N, and 40N with different sliding speeds like 200, 300, 400,600 rpm, and with the time interval of 2 min, 5 min, 10 min, 15 min with the help of a pin-on-disk wear testing equipment. The weight loss is count with the help of weighing apparatus. The wear tests on the specimens are conducted by changing the time, sliding speed and applied load. The impacts of these considered variables have been observed. The relation between Cumulative Mass Loss (CML) i.e. (decreasing in mass of sample by successive additions of applied load and sliding speed) with respect to cumulative time (C.T) i.e. (increasing in time by successive increment) at constant load and sliding speed has been evaluated. The diameter of track is kept constant at 90mm. Following graphs are representing using the experimental data.

3.2.1. Effect of Load

The effect of normal load on wear of samples under different normal load at constant sliding speed is shown in Figures. It is clear that, all the cases increase wear of samples with normal load increases.

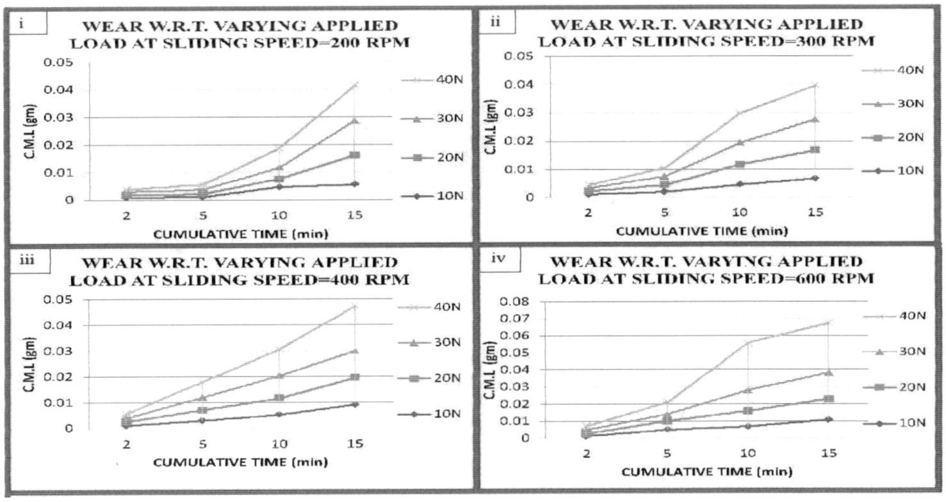

Figure 2. Plot of C.M.L Vs Time for (i) 200 rpm (ii) 300 rpm (iii) 400 rpm and (iv) 600 rpm.

All above figure show the relation between wear and time of samples at a constant sliding speed. It is seen that at a lower load, low wear and it increase slowly with normal load. As load increase, deeper grooves created due to increase pressure and temperature. So, at a higher load, wear is high.

3.2.2. Effect of Sliding Speeds

The effect of sliding speed of LM6 alloy samples on the wear under different sliding speeds, with constant applied load is shown in Figures.

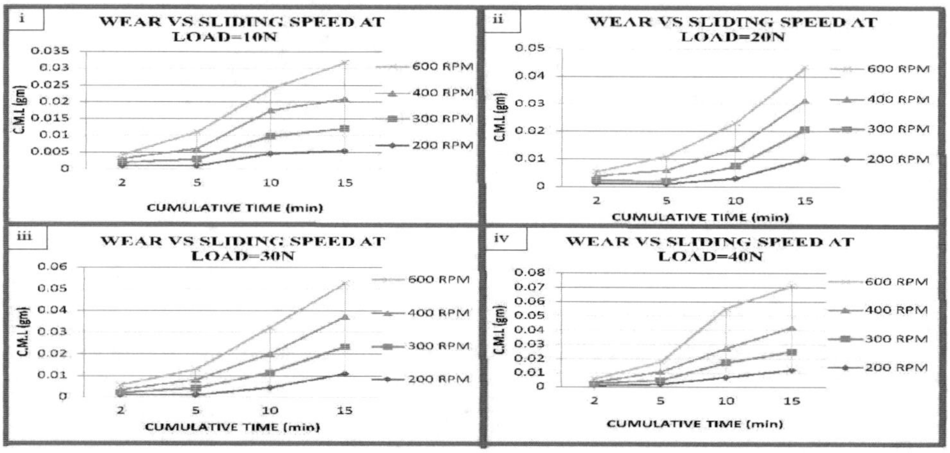

Figure 3. Plot of C.M.L Vs Time for (i) 10 N (ii) 20 N (iii) 30N and (iv) 40N.

Above all figure show the relation between wear and time of sample at a constant load. It is seen that at an initially wear is low and after a certain sliding speed the wear become high and increases with increasing sliding speed.

3.3. Wear Surface Study

To determine the chemical compositions of different inter-metallic phases in the as-cast alloy samples, Energy dispersive X-ray (EDX) analysis was used. Figure shows the SEM image and EDX spectrum taken from the as-cast specimen. With the help of SEM and EDX Analyzer, Wear surface and sub surfaces of the alloy are verified. Wear surface of the alloy shown various surface morphology in figure below.

3.3.1. Parameter (Load=40N, Time-15 minute, Sliding Speed= 200 rpm.)

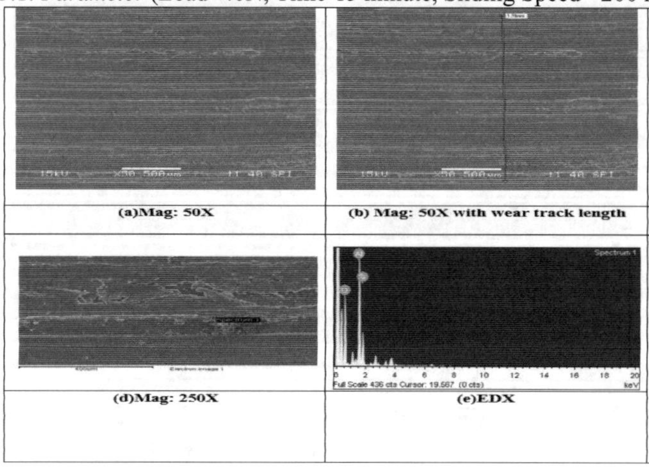

Element	Weight%	Atomic%
O K	43.30	56.56
Al K	40.97	31.74
Si K	15.73	11.71
Totals	100.00	

Figure 4. SEM micro graph and EDX micro analysis for samples.

3.3.2. Parameter (Load=10N, Time-15minute, Sliding Speed= 600 rpm.)

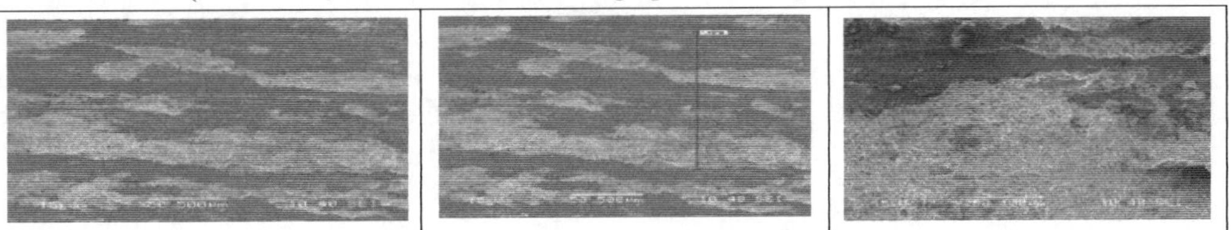

(a)Mag: 50X | (b) Mag: 50X with wear track length | (c)Mag: 250X

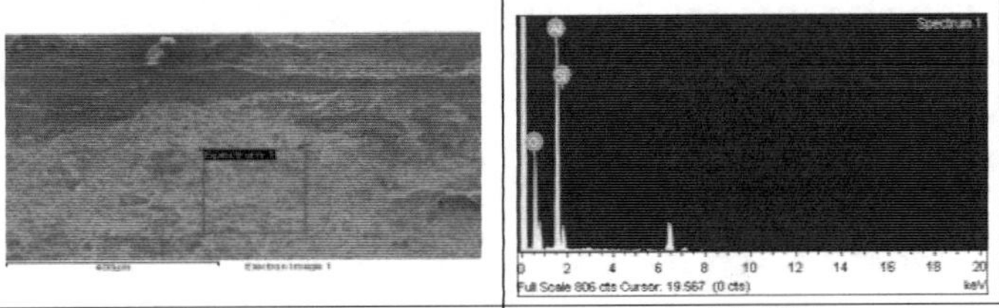

(d)Mag: 250X | (e)EDX

Element	Weight%	Atomic%
O K	38.97	51.97
Al K	53.57	42.36
Si K	7.46	5.67
Totals	100.00	

Figure 5. SEM micro graph and EDX micro analysis for samples.

All above figure show the worn surface of plasma processed LM6 alloy sample and the material loss are very high along sliding direction. All case figure micrographs reveal the formation of deeper abrasive grooves along the sliding direction leading to more material loss and also micrographs reveal the formation of deep abrasive grooves running along the sliding direction forming a delaminated surface. The EDX analysis was employed to determine the elements of worn surface of alloy and relieved that the layer is composed of O, Al, and Si element.

4. Conclusions

For the plasma casting processed LM6 alloys, due to increase in sliding speeds and applied load, the wear of LM6 alloy also increases. At higher applied load and sliding speed, it has been noticed that the severe wear regimes. The experimental data calculated and reported that the development in tribological behavior of plasma casting processed LM6 alloys and also from microstructure, presence of eutectic silicon is also observed.

References

[1] Elwin L. Roy "Introduction to Aluminum and Aluminum Alloys" ASM Handbook Vol-**24**) Hatch John E, "Aluminium: Properties and Physical Metallurgy", ASM, 1984, Metals Park, Ohio.

[2] Yasmin Tuti, Khalid Asad A, Haque M.M, "Tribological (wear) properties of aluminum–silicon eutectic base alloy under dry sliding condition", Journal of Materials Processing Technology, 153–154 (2004), pages-833-838.

[3] Kori S A, Chandra shekharaia T M, "Studies on the dry sliding wear behavior of hypoeutectic and eutectic Al–Si alloys", Wear, Volume **263** (2007), pages-745-755.

[4] DwivediDheerendra Kumar, "Wear behaviour of cast hypereutectic aluminium silicon alloys", Materials and Design **27** (2006) pages-610–616.

[5] Torabian H, Pathak JP and Tiwari SN, "Wear characteristics of Al-Si alloys", Wear, volume**172** (1994), pages 49-58.

[6] Tuti Y. Alias and M.M. Haque/ wear properties of Al-Si eutectic alloy.

IOP Conf. Series: Materials Science and Engineering **653** (2019) 012010 doi:10.1088/1757-899X/653/1/012010

Effect of nano-crystalline TiO_2 addition on reciprocating frictional behaviour of alumina ceramics

Partha Haldar[1], Tapas Kumar Bhattacharya[1] and Nipu Modak[2]

[1] Government College of Engineering and Ceramic Technology, Kolkata-10, India

[2] Mechanical Engineering Department, Jadavpur University, Kolkata-32, India

E-mail: partha.jumech@gmail.com

Abstract. We report the effect of normal load and sliding frequency on the tribological performance of alumina ceramics in relation with nano-crystalline TiO_2 addition. Tribological studies are conducted by reciprocating a silicon nitride ball on the prepared samples in dry condition in a linear reciprocating tribotester. Reciprocating friction tests are performed with sliding frequency of 15, 30, 45 and 60 Hz and at normal loads of 0.3, 0.5, 0.7 and 1.0 kg. The friction coefficient increases with increasing sliding frequency, normal load but decreases with nano titania addition. The coefficient of friction sharply increases at the level of sliding frequency from 30 Hz to 45 Hz. As the coefficient of friction gradually decreases with increase in TiO_2 addition, it can be inferred that alumina ceramics with TiO_2 addition upto 4 weight percent can be used as ceramic engine components, cutting tool inserts.

1. Introduction

Alumina ceramics is used for tribological applications due to its high hardness, high melting point, chemical inertness and high wear resistance. It is used in building tools like gas turbine engines, cutting tools, roller and bearings etc. [1-3]. One of the most important factors which impact the mechanical properties and wear resistance in alumina is grain refinement [4-6]. Fine grained ceramics improve wear resistance as compare to coarse-grained ceramics. Nano sized powders enable us to easily fabricate fine grained ceramics. Coefficient of friction (COF) of nano-crystalline TiO_2 based alumina ceramics is measured by varying sliding frequency, normal load and weight percentage of nano titania addition when reciprocating against a Si_3N_4 counter face. The main claims of this paper are as follows:

- COF is directly proportional to sliding frequency for TiO_2/Al_2O_3 system
- COF is directly proportional to normal load for TiO_2/Al_2O_3 system
- COF is inversely proportional to weight percent of nano TiO_2 addition

2. Experimental work

2.1. Materials

Reactive alumina (Al_2O_3) is purchased from M/S Alcoa. Nano-TiO_2 is synthesized with the help of chemical grade titanium (IV) isopropoxide ($Ti[OCH(CH_3)_2]_4$), citric acid ($C_6H_8O_7$) and ethylene glycol ($C_2H_6O_2$) (all of these precursors are 99.9% pure, purchased from M/S Merck), which are used as starting material without any further purification.

2.2. *Synthesis of nano-crystalline TiO_2 powder*

For the synthesis of nano TiO_2 powder, citric acid and titanium (IV) isopropoxide are mixed in stoichiometric ratio in aqueous solution under constant stirring and heating at about 60°C until complete dissolution. Ethylene glycol is then added with the same volume proportion under constant stirring and heating at about 80°C until gelation occurs. Then the gel is oven dried at 120°C for 24 hours. The dried material is ground in an agate mortar and heat treated in a muffle furnace at 800°C for 3 hours for proper crystallisation.

2.3. *Sample preparation*

Five set of samples are prepared with addition of 0, 0.5, 1, 2 and 4 weight percent of nano-TiO_2 to Al_2O_3 matrix and the samples are marked as $ATiO_2-0$, $ATiO_2-0.5$, $ATiO_2-1$, $ATiO_2-2$ and $ATiO_2-4$ to identify the samples containing alumina and respective weight percentage of TiO_2 in it. To prepare the batches the raw powders are weighed and mixed with required different weight percentages by making slurry with distilled water. The mixed suspension is dried in a drier for 24 hours followed by grinding of the dried powder in agate mortar pestle. 6% PVA (poly vinyl alcohol) is used as the binding agent. The samples are produced by powder metallurgy route. The dried mixtures are cold pressed in the form of cylindrical pellets (diameter $\approx$ 12.9 mm and thickness $\approx$ 3.8 mm) in a hydraulic press at a uniaxial pressure of 1000 kgf/cm^2. After 2 minutes of dwelling time pressure is released by operating pressure releasing valve. All the green compacts are dried in air at 110°C in a drier for 24 hours to remove the binder from the samples. The green samples are then sintered at 1600°C with a heating rate of 8°C/min up to 1000°C then 10 minutes soaking and 2°C/min upto 1600°C is maintained with a soaking time of 2 hours at the maximum peak temperature. After the definite isothermal holding, the samples are slowly cooled down to room temperature in the furnace.

2.4. *Test procedure*

Friction tester, TR-282, DUCUM is used to conduct the reciprocating friction test of the specimens under observation. Samples of ϕ10mm$\times$3mm are produced from the fabricated material. Si_3N_4 ball is used as the counter face. The normal load to the sample is applied through a pivoted lever. The test is carried out by applying different normal loads, sliding frequencies and sliding duration of 10 minutes as indicated in Table 1. The reciprocating stroke is fixed as 1 (one) mm. The COF at the interface of the ball and specimen is recorded once per second.

Table 1. Process parameters and their levels

Control parameters	Units	Levels			
		I	II	III	IV
Sliding frequency	Hz	15	30	45	60
Normal load	kg	0.3	0.5	0.7	1
Time	Min	10	10	10	10

For all the samples two set of experimental scheme is adopted. In the first set, the normal load of 500 gm, sliding time of 10 minutes are kept constant and the sliding frequency is varied as per levels. In the second set, the sliding frequency of 30 Hz, sliding time of 10 minutes are kept constant and the normal load is varied as per levels.

3. Results and discussions

The nano-crystalline TiO_2 based alumina ceramics densify more than 96% with apparent porosity less than 1.5%. Various factors affect the COF [7] but in our work we have concentrated on the impact of normal load and sliding frequency on the COF for TiO_2/Al_2O_3 system.

3.1. Effect of reciprocating frequency on coefficient of friction

The variation of COF with sliding frequency at 500 gm normal loads for all the samples has been shown in Fig. 1(a-e). The friction coefficient as obtained in all the graphs captured from the tribotester can be divided into two stages: wear in stage and stable stage [8]. In the wear in stage, contact occurs between the surface asperities of the sample with the ball. At this initial break-in period, the asperities deform, break or get ruptured. As a result the COF is unsteady at this stage. With the progress of sliding, steady level of COF is achieved due to the worn track becoming flat as well as due to polishing process during the wear test [9]. It is seen from the Fig. 1(a-e) that the COF values fluctuate around an average value which is mainly caused by non-uniform wear track or stick-slip phenomenon [10]. The friction coefficients in the stable stage are listed in Table 2 as well as in plotted in a graph as shown in Fig.2.

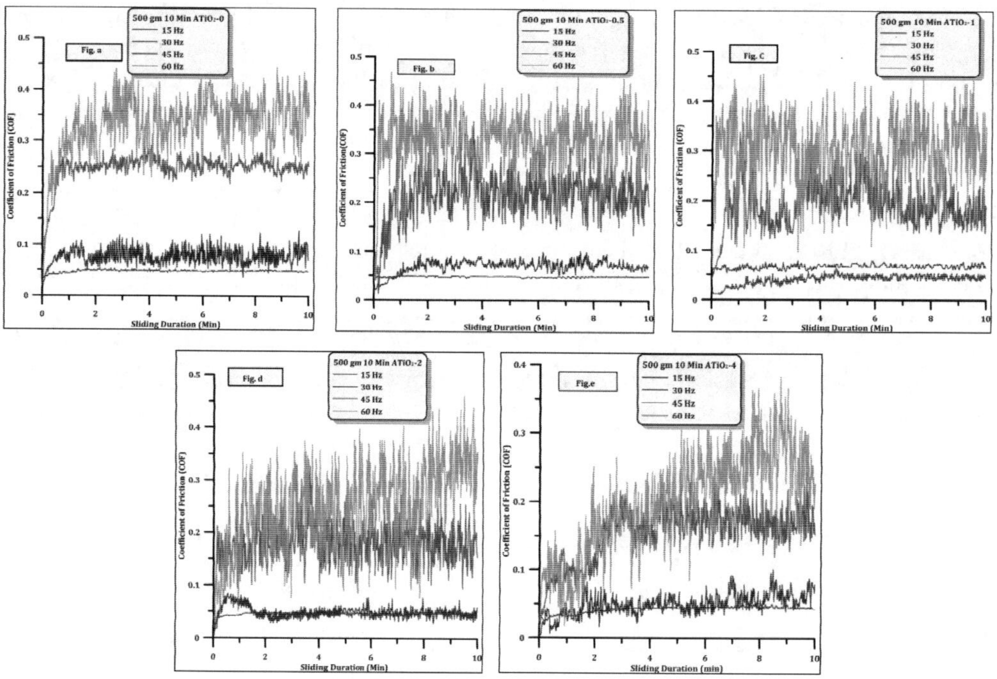

Figure 1. Variation of COF with reciprocating frequency at 500 gm normal loads for (a) $ATiO_2 - 0$ (b) $ATiO_2 - 0.5$ (c) $ATiO_2 - 1$ (d) $ATiO_2 - 2$ (e) $ATiO_2 - 4$

IOP Conf. Series: Materials Science and Engineering **653** (2019) 012010 doi:10.1088/1757-899X/653/1/012010

Table 2. COF of materials at different frequency at constant normal load of 500 gm

Sliding frequency (Hz)	$ATiO_2 - 0$	$ATiO_2 - 0.5$	$ATiO_2 - 1$	$ATiO_2 - 2$	$ATiO_2 - 4$
15	0.048075	0.048062	0.046175	0.04573	0.045408
30	0.077682	0.074356	0.068590	0.06247	0.057290
45	0.253962	0.22315	0.198313	0.183151	0.171339
60	0.344184	0.328330	0.303150	0.272650	0.23523

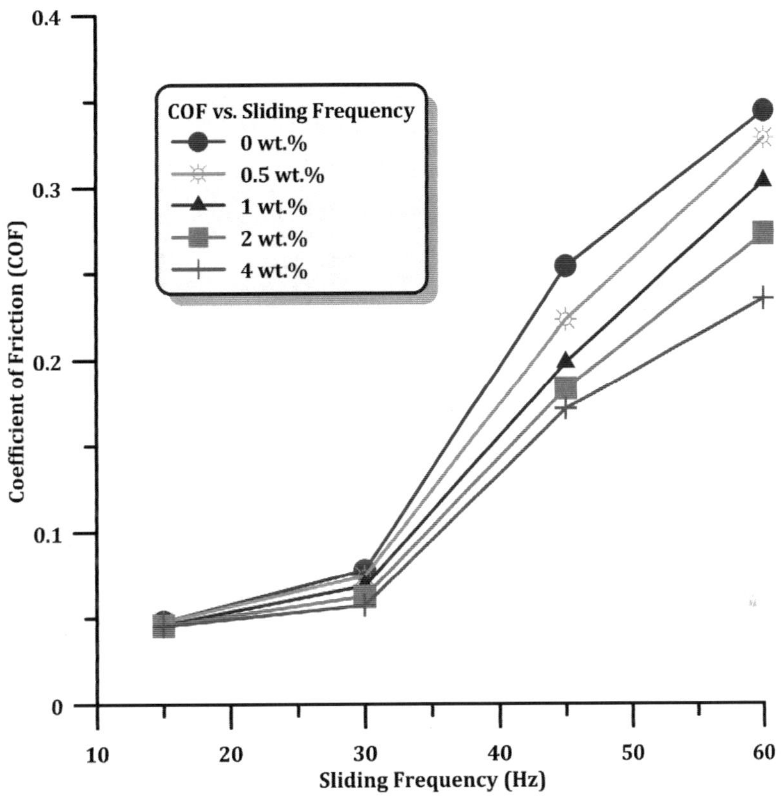

Figure 2. COF as function of reciprocating frequency

In general, the COF increases with the increase in sliding frequency for any of the samples. It is clear from the Fig.2 that the rate of increase in the value of COF as the sliding frequency increases from 30 Hz to 45 Hz is more than any other consecutive intervals. With increase in reciprocating frequency, the temperature at the contact zone rises, so cooling time per stroke of the Si_3N_4 ball is reduced. Heat generated at the contact surface results more or increased adhesion with the ball [11] leading to increase of friction coefficient. It is also observed that as the weight percentage of TiO_2 in alumina increases, the COF decreases.

3.2. Effect of normal load on friction coefficient

In this section, the variation of COF with different normal loads (0.3 kg, 0.5 kg, 0.7 kg and 1.0 kg) at 30 Hz sliding frequency has been studied and has been shown in Fig. 3(a-e) respectively. The stable COF data of Table 3 depicts that with increase in normal load, COF increases. Similar observation is reported for alumina based ceramics by Wang et al. [8]. This is due to formation of plastic deformation zone on the material surface due to larger value of normal load. Moreover, with increase in TiO_2 in alumina, the COF decreases for all the cases. COF of different samples under observation are plotted as a function of normal load in Fig. 4.

Table 3. COF of materials at different normal load at constant sliding frequency of 30 Hz

Normal load (gm)	$ATiO_2-0$	$ATiO_2-0.5$	$ATiO_2-1$	$ATiO_2-2$	$ATiO_2-4$
300	0.061038	0.059317	0.05305	0.04874	0.0427
500	0.077682	0.07436	0.06859	0.06263	0.05529
700	0.141231	0.132787	0.11551	0.092543	0.070873
1000	0.205162	0.198183	0.18132	0.166795	0.150881

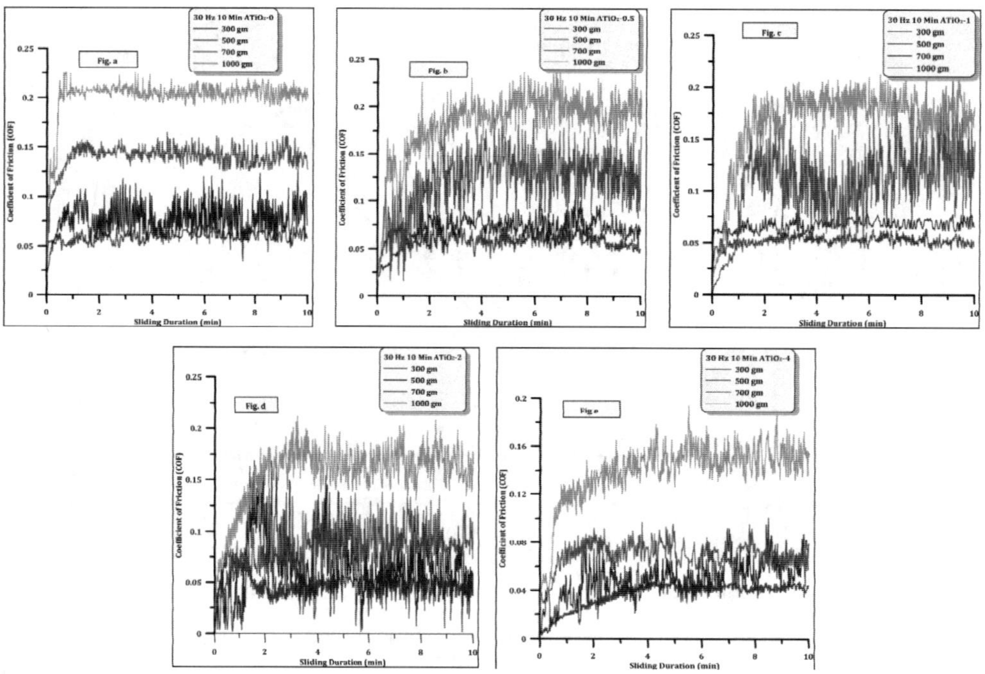

Figure 3. Variation of COF with normal load at 30 Hz sliding frequency for (a) $ATiO_2-0$ (b) $ATiO_2-0.5$ (c) $ATiO_2-1$ (d) $ATiO_2-2$ (e) $ATiO_2-4$

4. Conclusions

The normal load and sliding frequency indeed have a considerable effect on the friction behavior of nano-crystalline TiO_2 based alumina ceramics. For the samples under observation, with constant normal load of 500 gm, with increase in sliding frequency the value of COF increases, this is due to more adhesion of Si_3N_4 ball to the ceramic specimen. The rate of increment in COF from 30 Hz frequency to 45 Hz is more as compared to other consecutive frequency intervals. As the TiO_2 weight percent in alumina ceramics increases, the value of COF decreases for any

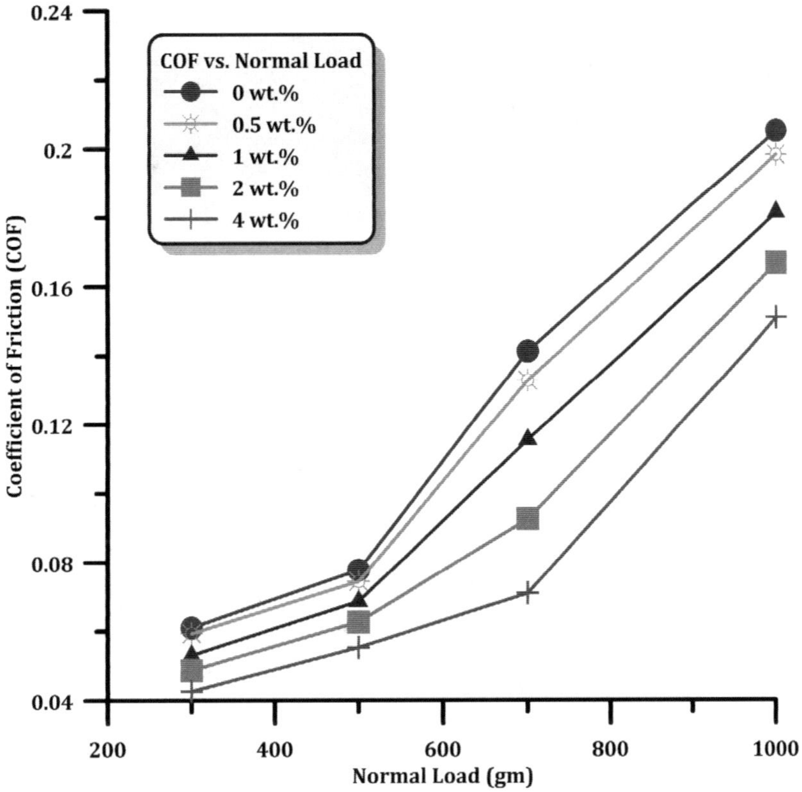

Figure 4. COF as function of normal load

constant sliding frequency level at 500 gm normal load. On the other hand, for the samples under observation, with constant sliding frequency of 30 Hz, with increase in normal load the value of COF increases. This is due to formation of plastic deformation zone on the material surface due to larger value of normal load. As the TiO_2 weight percent in alumina ceramics increases, the value of COF also decreases for any constant normal load at 30 Hz sliding frequency. This study opens a new area of research on the ceramic engine components, cutting tool and other relevant field of work.

References

[1] Ajayi O O and Ludema K C 1988 *Wear* **124** 237
[2] Kato K and Adachi K 2002 *Wear.* **253** 1097
[3] Lee S W, Morillo C, Lira-Olivares J, Kim S H, Sekino T, Niihara K and Hockey B J 2003 *Wear* **255** 1040
[4] Cho S J, Hockey B J, Lawn B R and Bennison S J 1989 *J. Am. Ceram. Soc.* **72** 1249
[5] Zum Gahr K H, Bundschuh W and Zimmerlin B 1993 *Wear* **162** 269
[6] Kerkwijk B, Winnubst A J, Verweij H, Mulder E J, Metselaar H S C and Schipper D J 1999 *Wear.* **225** 1293
[7] Sarkar P, Modak N and Sahoo P 2016 *Int. J. Surf. Eng. Interdiscip. Mater. Sci.* **4** 25
[8] Wang J, Cheng Y, Zhang Y, Yin Z, Hu X and Yuan Q 2017 *Ceram. Int.* **43** 14827
[9] Rodriguez-Suarez T, Bartolomé J F, Smirnov A, Lopez-Esteban S, Torrecillas R and Moya J S 2011 *J. Eur. Ceram. Soc.*. **31** 1389
[10] Parchovianský M, Balko J, Švančárek P, Sedláček J, Dusza J, Lofaj F and Galusek D 2017 *J. Eur. Ceram. Soc.* **37** 4297
[11] Yin Z, Yuan J, Huang C, Wang Z, Huang L and Cheng Y 2016 *Ceram. Int.* **42** 1982

A contribution concerning trends in modelling of Wire-micro EDM of Titanium alloys- A review

Rajesh Porwal[1], Ranu Maurya[2]
[1]Faculty of Mechanical Department, Shri Ramswaroop Memorial University, Lucknow, India.

[2]Student of Mechanical Department, Shri Ramswaroop Memorial University, Lucknow, India.

E-mail: porwal.me@gmail.com

Abstract. The era is shifting towards the manufacturing idea of converting raw form of material into finalized part in minimum step and lesser machining operation. Manufacturing processes like 3D printing and additive manufacturing are achieving the criteria but they requires high capital investment and are sophisticated in its operation which need next to zero possibility of error. Alloying of materials yield high derivative properties like hardness, temperature resistance and extreme strength to weight ratio like in Titanium, Nickel and their alloys and super alloys. Specialized feature at micro level, complex and intricate geometries in the part are difficult to produce via traditional machining processes than non-contact machining process at cost effectiveness. This paper reviews the modelling related studies done in the processing of Titanium alloys through Wire- micro electrical discharge machining (Wire-μEDM) process. Most of the research done in the Wire EDM domain, experimented with the wire of copper and brass with diameter range in microns, thus resulting micro-machining phenomenon. Since Ti alloys find the most of the applications in field of aerospace, automobile, nano tech due to excellent performance at elevated temperature and pressure, than other alloying materials. The summarized should provide an overview on modelling techniques used by the re-searchers for measuring and predicting the parametric values for the optimization of WEDM process.

1. Introduction

Titanium and its alloys are known for their high mechanical properties like extreme hardness, strength, resistance to creep and fatigue and temperature resistance. And their extensive applications in industries like aerospace, ballistics, automobile, etc. to manufacture intricate parts like gears. Since these alloys are very costly, these material are to be handle with care and efficient machining operations must be utilized for their fabrication. Machining process utilizing tool for material removal are ineffective due to high tool wear and energy loss in form of friction between tool and workpiece which led to wastage of material and time both. Thus, advanced machining methods like EDM, LBM, WEDM, etc. are required for machining at faster rate with desirable quality.

Due to ease of availability and machinability of WEDM of electrically conductive material, it can almost machine every Ti alloy easily without significant tool or energy loss. Controlling and monitoring of input variables in WEDM is comparatively easier than other processes. In order to achieve higher machining rate different configuration of WEDM are used along with the change in composition of tool wire material. Response factors like machining rate, surface integrity, wear ratio and dimensional accuracy are highly dependent on the input variables like pulse on/off time, spark gap voltage, current, etc. as shown in figure 1 [1-3].

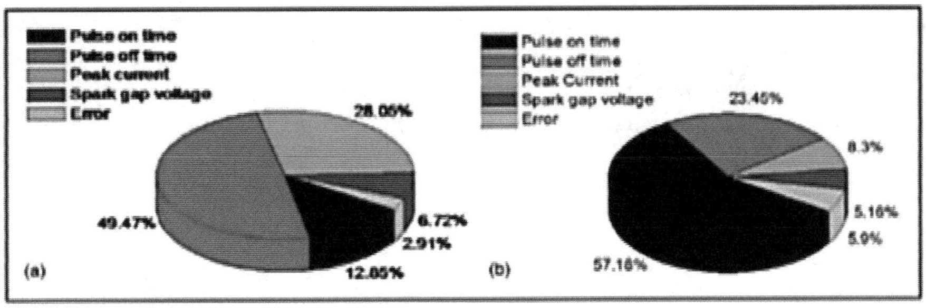

Figure 1. Effect of input variable on (a) Wire wear ratio (b) MRR, SR [3]

2. Mechanism of Wire micro EDM

The material removal mechanism of wire-micro EDM is similar as that of Wire EDM. Similarly this process utilizes tool electrode is in the form of electrical conductive wire through which spark generation take place for erosion of material from the work, refer figure 2. Wire in this process is continuously fed through the spool in tension. Wire diameter range is in the order of 0.1mm to 0.02mm as to remove material of minimal width or for internal cutting. The movement and automated feed of wire is generally programmable through CNC along with the service of guides in order to inhibit wire lag and wire breakage. Flushing system are implied for the application and pressure control of dielectric fluid required in the process for removal of debris along with heat dissipation.

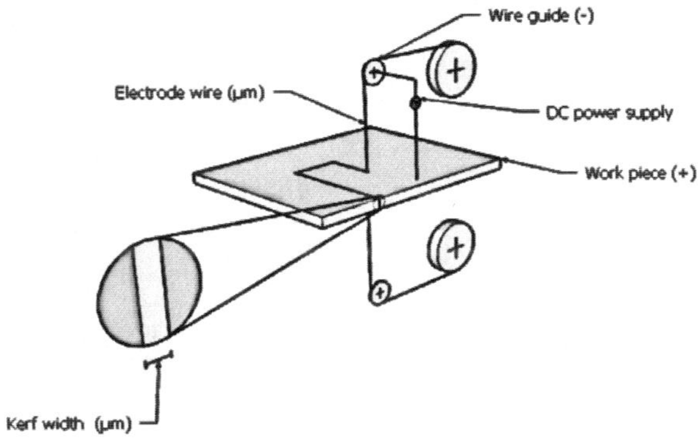

Figure 2. Schematic diagram of Wire-EDM

3. Parametric Studies of Wire-μEDM

Process parameters of Wire μ-EDM are discharge current, voltage, feed rate, capacitance, wire tension, pulse on time(T_{on}), type of circuit, etc. which were being used by the different researchers for evaluation and optimization of performance characteristics. The deviation in performance characteristics like geometrical accuracy, surface roughness and cutting rate was observed, by Bergs, with respect to discharge current, pulse interval time and flushing condition and found that accuracy is affected by increased flushing due to wire vibration and lead to high cutting rate while pulse interval time and discharge current had no significant effect. Surface roughness was lowered by decreasing the current and increasing the pulse interval time and flushing pressure than standard [4]. Tsai et al. describes the effect of vibration of tool wire with the help of auxiliary device on the machining rate and relates that increased vibration increases the removal rate of material additionally machining depth in-creases with discharge time due to unusual debris flow [5].Beside this, Han et al. used ultrasonic assisted WEDM which describes the vibration of workpiece along horizontal direction and its effect which increases MRR and surface roughness while reduces kerf width [9]. Increment in feed rate decrease the kerf width, increase the MRR and surface roughness was increased has been discussed by Alias et al. using brass wire electrode with constant current of 4A and 6A in order to find best combination of process inputs and optimize cost with quality [6-7]. Majumder et al. also discussed effect T_{on}, T_{off}, wire feed and tension on machining time, Kerf width plus surface roughness and concluded that response parameters were increase with increase in T_{on} while decreases with T_{off} and wire feed and tension had minimum effect on them [8]. Study of surface integrity with low speed WEDM using brass wire was experimentally studied by Gong et al. with the comparison of trim cuts to machined main cut for the investigation of surface roughness and found out to be fine of about 0.67μm. This was possible by decreasing the discharge current, T_{on}, wire speed and increasing wire tension [10].

4. Modelling related Studies

Chalisgaonkar et al. used coated and uncoated zinc electrodes for the trim cuts on commercial pure Titanium (CPTi) and evaluated variation of the MRR through mathematical modelling using Buckingham's π theorem and ANOVA. It predict values of MRR over significant process parameters like peak current, wire offset and wire type, shown in figure 3 [11].

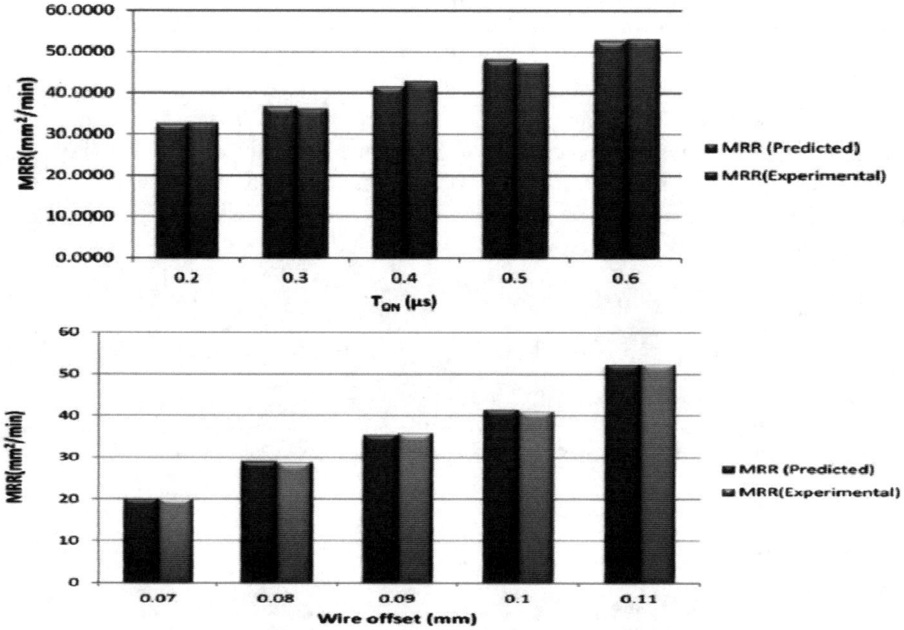

Figure 3. Comparative bar chart between predicted and experimental values of MRR for T_{on} and Wire offset [11]

Mathematical linearized equation model using logarithmic transformation was used by Raj et al. for predicting surface roughness and MRR using two configuration of electrode wire i.e. brass and molybdenum (Mo). Input parameters were pulse current, T_{on} and T_{off}. The maximum test errors for surface roughness were 12.005% and 10.78% for brass and Mo wire and for MRR, 27.65% for brass wire and 11.04% for Mo wire. Regression model depicts the deviation of actual surface roughness and MRR with anticipated values and indicate fit of the model [12]. Analysis of machining parameters of WEDM process were done by the Kumar et al. using RSM modelling in order to obtain relationship between machining rate and overcut. A micro-model has been utilized for predicting MRR of the process using dimensional analysis. Spark gap voltage, current, T_{on}, T_{off}, wire tension and feed were taken as input parameters. The model was efficient in showing the closeness in the predicted values and experimented values across T_{on} and peak current obtained using brass wire and pure titanium specimen, refer figure 4 [13].

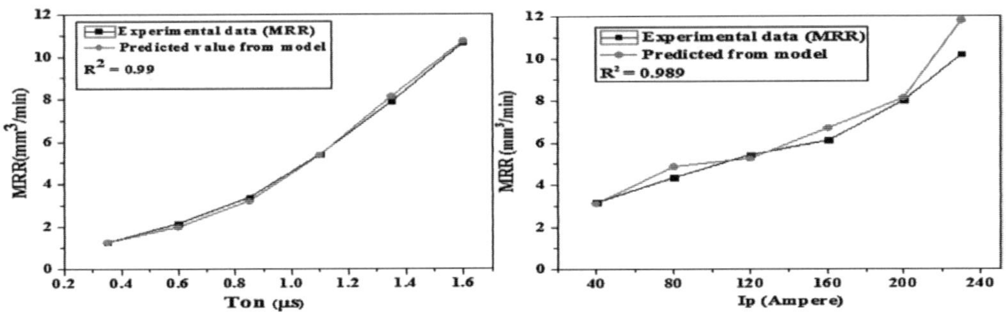

Figure 4. MRR values proposed by experimental and micro-model for T_{on} and peak current [13]

Kumar et al. discussed the RSM for obtaining MRR and wire wear ratio through WEDM of pure titanium and zinc coated brass wire. Input parameters were peak current, T_{on}, T_{off}, wire tension and feed. Regression technique was used to compare the experimental value and predicted values obtained through modelling. It also visualizes the impact of input variables over MRR signifying T_{on} and T_{off} as the peak significant factors [14]. Liao et al. adopted back propagation neural network (BPNN) for building the relationship between machining parameters like Specific discharge energy, Ton, Toff and servo voltage on the cutting speed, spark frequency, groove width and surface irregularities during the WEDM using brass wire and Ti6Al4V specimen. BPNN outperformed other modelling techniques yielding more accurate result with lesser error percent of 2.74 [15].Process parameters like discharge current, T_{on}, wire tension, wire speed and flushing pressure were adopted to obtain performance parameters like Surface roughness and micro hardness during WEDM of NiTi smart alloy was described by Majumder et al.. A general regression neural network (GRNN) was adopted for modelling and visualized through regression analysis of experimental and predicted values, as shown in figure 5 [16].

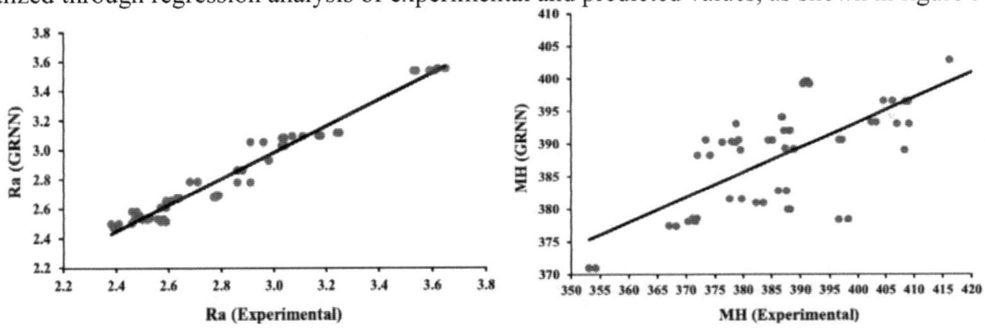

Figure 5. Relationship among Experimental and predicted value (a)Arithmetic mean roughness, (b)Micro hardness [16]

Sarkar et al. proposed second order polynomial mathematical model with RSM to weigh the effect of process constraints on machining criteria like cutting speed, sur-face integrity and dimensional shift during WEDM of ϒ-TiAl alloy with brass wire. Here log transformation technique effectively fit the model to give response value of surface roughness to 95.6% with error value of 0.07571 which was within permissible limits [17].During WEDM of TiC /Fe in situ metal matrix composites with uncoated brass wire of 250μm, Saha et al. proposed normalized radial basis function network (NRBFN) with enhanced k-means clustering technique for modelling of process using T_{on}, T_{off}, wire feed rate and average gap voltage on the performance entities like kerf width and cutting speed. The enhanced k-means clustering techniques yield superior responses based on dynamic adjustments rather than traditional k-means clustering technique which focuses on local optimality. The model signified better fit of response parameters in enhanced technique rather than with the traditional approach [18]. Saha et al. adopted BPNN model for obtaining functional relationship between process parameters and responses like kerf width and cutting speed by using non-dominated sorting algorithm (NSGA-II). Fitness evaluation of BPNN can be seen in figure 6 [19].

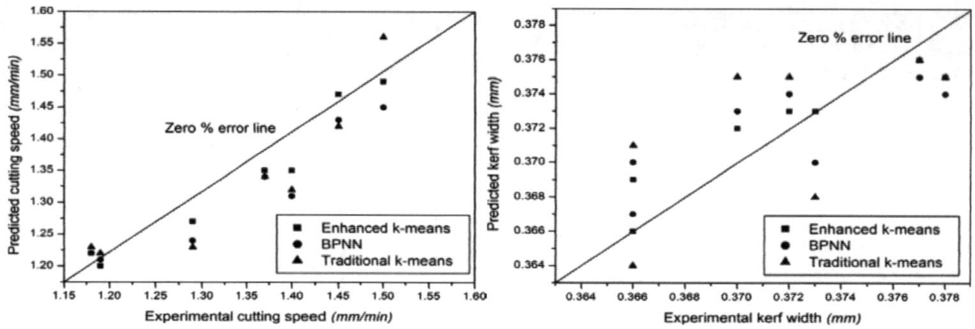

Figure 6. Model comparing experimental and predicted values of cutting speed and kerf width [19]

Artificial neural network with feed forward propagation is used to model to predict performance parameters like cutting speed, surface finish and wire offset for all potential combination of process factors by Sarkar et al.. Major observation was that surface roughness increases with increase in cutting speed and 27 parametric combinations were found to fit the responses from the total combination of 15625 [20]. Ghodsiyeh et al. proposed a second order mathematical model of RSM for determining the relation between response output and variable input parameters during machining of Ti6Al4V through zinc coated brass wire of 250μm in 5-axis WEDM. Responses machining rate and surface irregularity were predicted within permissible limit when compared to experimental value showing minute deviation [21].Machining of shape memory alloy $Ni_{49.4}Ti_{50.4}$ by WEDM for manufacturing the dynamic compression plate using zinc coated brass wire of 250μm. A finite element model was used for depicting the shape recovery near the WEDMed surface of NiTi alloy. To manufacture dynamic compression plate, process parameters were T_{on} 115 machine units, T_{off} 40 machine units, spark gap set voltage 90 V, wire feed 6 m/min and wire tension 6 machine units. The improvement in performance characteristics is done with the implication of Grey Relation Analysis and experimentation [22]. Yusoff et al. proposed orthogonal based ANN modelling along with multiple polynomial regression (MPR) for selecting and recognition of network parameters during the WEDM of Ti–48Al intermetallic alloy with brass wire of diameter 200μm. Due to the special logical characteristic of ANN which was based on the input and output training data denotes the integration of the two models. ANN–MPR predicted improved ideal machining performances and input factors, as shown in figure 7 [23].

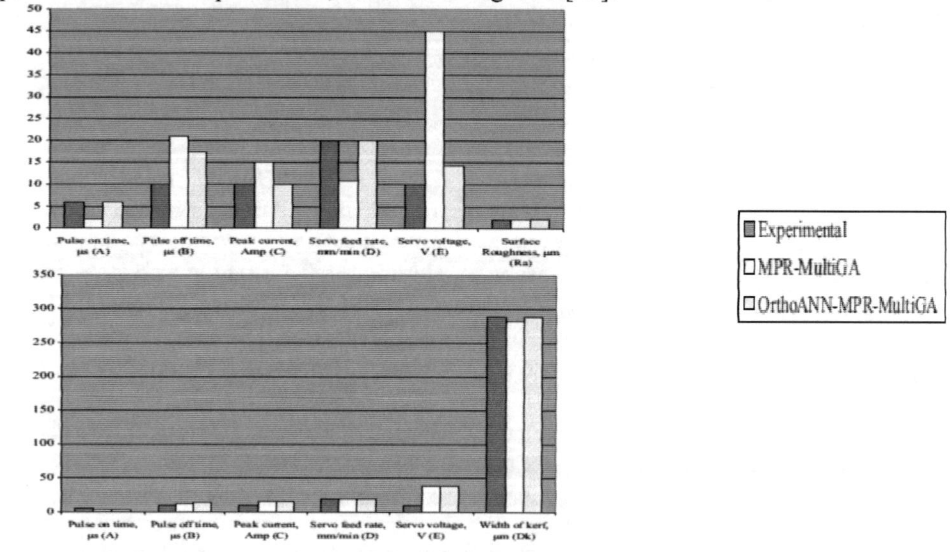

Figure 7. Comparison between experimental, MPR MultiGA and Ortho ANN-MPR-MultiGA on R_a and Kerf width [23]

5. Conclusion

From the survey of these modelling papers on WEDM process, we can conclude that WEDM process can be modelled using both analytical and mathematical model maintaining the optimal limit of performance outputs. And, following other conclusion were,

- Performance characteristic of WEDM, like MRR, SR, cutting speed and kerf width were mostly affected by the T_{on}, T_{off}, discharge current and servo voltage.
- With the increase in T_{on} and current MRR increases and surface roughness value increases with increment in T_{on} and wire feed.
- ANN modelling yields the better result than any other mathematical model for predicting the performance characteristics.
- Hybridization of ANN along with other models like GA, NSGA-II and MPR yields much better result than the ANN alone.
- With the mixing of two algorithms, the number of optimal solution increases rapidly which satisfies the limiting constraints.

References

[1] Chaubey S K, and Jain N K 2017, *Mater. Manuf. Process.* **33**(14), 1568-1577.
[2] Chen Z, Zhang G, and Yan H 2018, *Precision Eng.* **54**, 51-59.
[3] Kumar A, Kumar V, and Kumar J 2018, *J. Process. Mech. Eng.* **232**(1), 108-126.
[4] Bergs T, Tombul U, Herrig T, Klink A, and Klocke F 2018, *Procedia Manuf.* **18**, 130-137.
[5] Tsai M Y, Fang C S, and Yen M H 2018, *Int. J. Adv. Manuf. Technol.* **97**, 297-304.
[6] Aliasa A, Abdullaha B, and Abbas N M 2012, *Procedia Eng.* **41**, 1806-1811.
[7] Aliasa A, Abdullaha B, and Abbas N M 2012, *Procedia Eng.* **41**, 1812-1817.
[8] Majumder H, and Maity K 2017, *J. Adv. Manuf. Sys.* **16**(2), 81-99.
[9] Han G, Soo S L, Aspinwall D K, and Bhaduri D 2013, *Adv. Mater. Res.* **797**, 315-319.
[10] Gong Y, Sun Y, Wen X, Wang C, and Gao Q 2016, *Int. J. Adv. Manuf. Technol.* **88**, 197-207.
[11] Chalisgaonkar R, and Kumar J 2014, *Eng. Sci. Technol., An Int. J.* **18**, 152-134.
[12] Oliver S, Raj N, and Prabhu S 2016, *Australian J. Mech. Eng.* **15**(3), 198-209.
[13] Kumar A, Kumar V, and Kumar J 2014, *J. Braz. Soc. Mech. Sci. Eng.* **37**(2), 689-721.
[14] Kumar A, Kumar V, and Kumar J 2018, *J. Process. Mech. Eng.* **232**(1), 108-126.
[15] Liao Y S, Chuang T J, and Yu Y P 2014, *Int. J. Adv. Manuf. Technol.* **70**, 251-258.
[16] Majumder H, and Maity K 2018, *Applied Soft Computing* **70**, 665-679.
[17] Sarkar S, Sekh M, Mitra S, and Bhattacharyya B 2008, *J. Mater. Processing Technol.* **205**, 376-387.
[18] Saha P,Tarafdar D, Pal S K, Saha P, Srivastava A K, and Das K 2009, *Int. J. Adv. Manuf. Technol.* **43**, 107-116.
[19] Saha P,Tarafdar D, Pal S K, Saha P, Srivastava A K, and Das K 2012, *Applied Soft Computing* **13**, 2065-2074.
[20] Sarkar S, Mitra S, and Bhattacharyya B 2006, *Int. J. Adv. Manuf. Technol.* **27**, 501-508.
[21] Ghodsiyeh D, Lahiji M A, Ghanbari M, Shirdar M R, and Golshan A 2012, *Res. J. Applied Sci., Eng. & Technol.* **4**(17), 3154-3161.
[22] Takalea A M, and Chouguleb N K 2019, *Mater. Sci. Eng.* **97**, 264–274.
[23] Yusoff Y, Zain A M, Amrin A, Sharif S, Haron H, and Sallehuddin R 2017, *Artif. Intell. Rev.* **52**(1), 671-706.

IOP Conf. Series: Materials Science and Engineering **653** (2019) 012012 doi:10.1088/1757-899X/653/1/012012

Thermal Management of Battery Pack for Hybrid Vehicles Using PCM

Pant.N and Racherla.V

Department of Mechanical Engineering, Indian Institute of Technology, Kharagpur, Kharagpur, India

E-mail: pantneelaish0@gmail.com

Abstract. Advancement in thermal management system which can be adopted to absorb heat generated in Li-ion battery pack for hybrid vehicle during charge and discharge cycles, by keeping the battery pack in optimum range of 15°C-25°C. Factors such as parasitic power, additional weight of cooling unit, temperature rise and cell temperature difference play a vital role. PCM (Phase Change Material) are compounds having high thermal conductivity and latent heat storage. They go through phase change when they absorb or release heat. The basic design is to manufacture a cooling jacket using phase change material which absorbs heat during the day and rejects it at night. The result show decreases in temperature by 1.5°C, additional increase in weight of battery pack by 17.5%, no parasitic power consumption, increase in safety and compactness to applications.
Keywords: Hybrid vehicle · Lithium-ion battery · PCM(Phase Change Material) · Thermal conductivity · Latent heat · Cooling jacket.

1. Introduction

Energy conservation by adaptation of energy saving products has become a major field of research and development. Hybrid vehicles consist of large battery packs which provide energy but also release a lot of heat during charging and discharging cycles. Ye and Pesaran [1,2] study showed that the most optimum temperature range for working under safety limits for lithium-ion battery to increase its battery life is 15°C to 25°C. During manufacturing of Li-ion cells, extremely flammable and volatile electrolytes are used [3], which might pose threat due to rise in temperature of the pack. Although high density battery pack is unavoidable but safety and durability play a more important and crucial role. Extreme conditions [4] and thermal runaway [5] also pose a major threat for the pack. Thus a well-designed and effective thermal management system is required for effective cooling for battery packs. There are different methods that can be adopted which have their own advantage and disadvantages. Air cooling is the easiest approach, which works effectively in small packs but fails in case of battery pack size adopted in hybrid vehicle due to formation of heat zones leading to overheating [6] as shown by CFD of cylindrical cells [7]. Heat exchangers especially metal foam based were used by Giuliano et al. [8] to study the effects of forced air cooling. The development of thermal model based on liquid cooling was very effective in optimizing the temperature, by using fluids like oil and water [9,10,11], but its complexity, cost, and potential leakage make manufacturers hesitate to use it. The use of PCM is another method of advanced cooling. PCM or Phase Change Material are compounds that have high latent heat and undergo phase change in order to absorb or remove heat, over thousands of life cycles.

1.1. Objective

The objective of this paper is to develop an effective method for absorbing heat generated by the lithium ion cells of the pack during charging/discharging and keep them close to optimum range, thus resulting in increase in life cycles of the cells keeping in mind factors like manufacturer feasibility, parasitic power consumption, and easy replacement.

List of symbols

Nomenclature

H	Heat generated (J/s)
I	Current (Amp)
R	Resistance (ohm)
A	Area of heating pad (m^2)
T_i	Initial temperature (°C)
T_f	Final temperature (°C)
m	Mass of PCM (kg)
L	Latent heat (kJ/kg)
h	Heat transfer coefficient (W/m^2)
ϵ	Emissivity (W/m^2)
σ	Stefan boltzmann's constant (5.67×10^{-8} $W \cdot m^{-2} \cdot K^{-4}$)

2. Battery Pack

The battery pack (dimensions: 20cm × 12cm; weight: 2.525kg) consists of 27 Li-ion cells, with each cell having a nominal voltage of 3.6V, at 2.2A of current supply. The total energy dissipated by each cell is ∼7W. Since the battery pack has uneven heat generation, due to different amount of heat released by the cells at a given time and points, a uniform heating pad is manufactured with 2 terminals, which acts like a single cell which can dissipate an equal amount of energy at all given time and points. The pad is made by taking a phosphate resistance wire (resistance ∼ 1.4Ω) that is embedded inside a Silicon-Aluminum sandwich casing (dimensions: 20cm × 12cm) which creates uniform temperature over the entire surface. Fig.1 and Fig.2 shows outside and inside pictures of the heating pad. When current is made to flow through the pad, heat is generated. Fig.3 and Fig.4 show the theoretical and actual relationship between time and rise in temperature for heating the pad at different current supplies.

Figure 1: Outside of heating pad

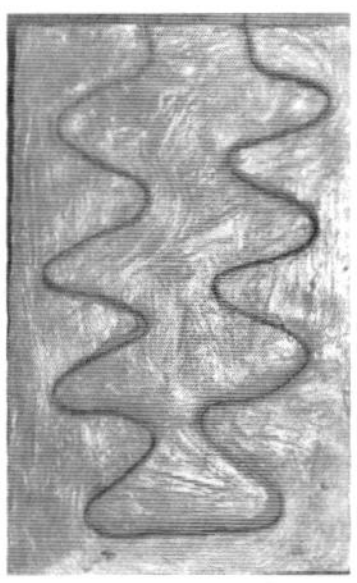

Figure 2: Inside of heating pad

IOP Conf. Series: Materials Science and Engineering **653** (2019) 012012 doi:10.1088/1757-899X/653/1/012012

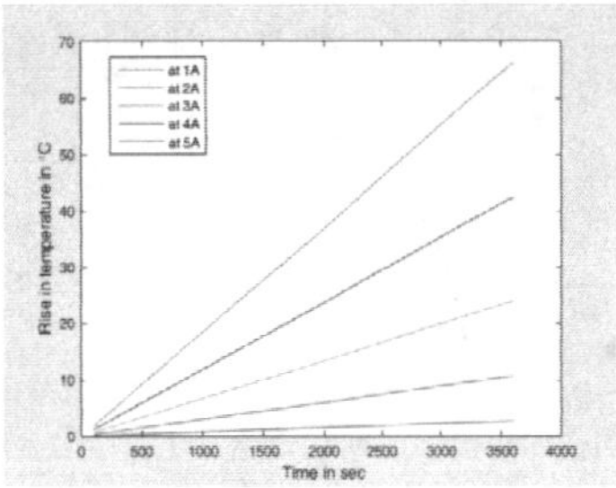

Figure 3: Theoretical time vs rise in temperature at constant current for heating pad

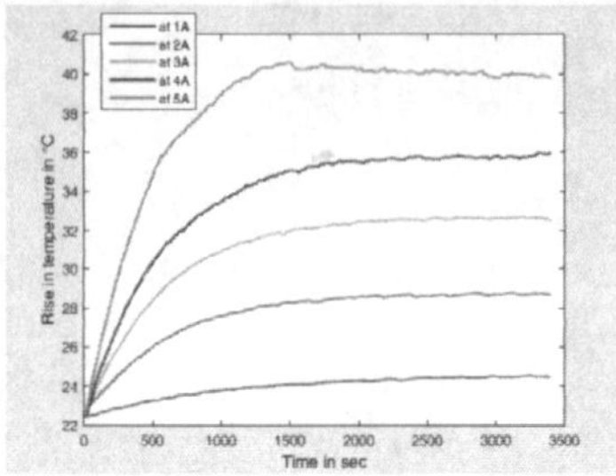

Figure 4: Actual time vs time at given current for heating pad

3. Thermal Management System

The thermal management involves manufacturing of a cooling jacket which is filled with PCM (latent heat: 190 kJ/kg) which absorbs heat during the day and rejects it at night. The pack is sealed and made leak proof, Table 1 shows specification of PCM. Fig.5 represents cooling jacket (Dimensions: 20cm × 20cm; Weight 0.700kg) manufactured by taking plastic sheets of thickness (0.2mm) that are filled with PCM, and sealed using a laminating machine. The pouch also consist of a valve to remove excess air and create a vacuum. This is done as a protective measure as hot air puts more pressure on the jacket walls during heating phase that may lead to bursting of jacket in extreme conditions.

Table 1: Properties of PCM

Property	Value	Unit
Melting Temp	29.0	°C
Freezing Temp	29.0	°C
Latent Heat	190	kJ/kg
Liquid Density	1530	kg/m
Solid Density	1830	kg/m
Specific Heat	2.3	kJ/kg·K

Figure 5: PCM jacket attached to plate

4. Calculations

During charging/discharging cycle, heat generation takes place inside the cell. Similarly, when the current is made to flow through the heating pad, heat generation takes place. The total heat generated (H) by the heating pad is calculated by

$$H = I^2R \tag{1}$$

After some time, the pad temperature reaches a constant value as shown in Fig.4, but continuously emits a constant amount of heat for the remaining time. The heat emitted by heating pad in the form of convection and radiation per second is given by

$$I^2R = 2hA\Delta T + 2\epsilon\sigma A\left(Tf^4 - Ti^4\right) \tag{2}$$

Mass of PCM (m), required to absorb the heat emitted by the heating pad in order to keep the pack in constant temperature is calculated by

$$Q = mL \tag{3}$$

Using above stated equation a linear relationship is observed between mass of PCM required and heat generated . The maximum total heat produced by the heating pad in 5 hours at 2.2A current supply is 18 KJ/kg-K, and mass of PCM required to absorb the heat is 0. 6419kg.

5. Experimental Setup

The pack is connected to a data acquisition (DAQ) system which consists of sensor (thermocouple), DAQ measurement hardware, and a computer with programmable DAQ system. Fig.6 represents the experimental setup.

Figure 6: Experiment setup

The heating pad terminals are connected to a constant current source of 2.2A for 5 hours, and rise in temperature is recorded. Now the heating pad is covered with PCM filled cooling jacket which covers the entire surface in such a way that it sticks to the surface tightly and constant current of 2.2A is made to flow through the pack for 5 hours, Fig.7 represents time and rise in temperature for heat pad covered with and without PCM jacket

6. Results

The results show that temperature of heating pad covered with PCM jacket remained ~1.5°C less than the actual temperature for a constant current supply of 2.2A for 5 hours. Increase in temperature is inversely proportional to the life cycles of a lithium ion cell. Factors such as increase in internal resistance leads to high temperature profile for the cells. The PCM jacket when attached to the pack increases the pack weight by 17.5% but lowers the temperature profile for the pack by ~1.5°C, thus bringing it closer to the optimum range as a result increasing life cycles of the pack. Also, PCM acts as a safety measure in case of short circuit/fire in the pack. Other factors such as manufacturer feasibility, no parasitic power consumption, easy replacement lay an upper hand as compared to the conventional methods. PCM jacket absorbed all the heat generated in the form of convection and radiation and underwent phase change, though minor amount of PCM crystal were still visible inside the jacket.

7. Conclusion

The use of PCM jacket lower the temperature. Fig.7 represents the difference between with and without the use of PCM jacket. A clear difference of 1.5°C can be observed. There is no parasitic power consumption, additional weight of pack was increased by 17.5%. The replacement cost is low, with high life cycles. PCM are inflammable thus enhancing safety in case of fire, in battery pack. The use of PCM jacket also provides compactness to application.

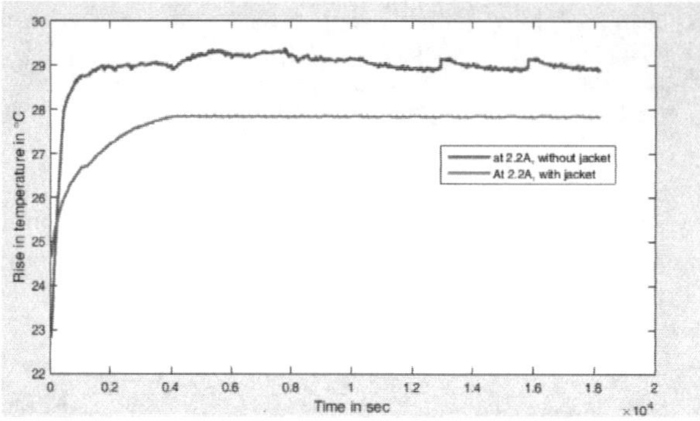

Figure 7: Rise in temperature with time at constant current supply of heating pad with and without PCM jacket

References

[1] Y. Ye, L.H. Saw, Y. Shi, A.A.O. Tay, Numerical analyses on optimizing a heat pipe thermal management system for lithium-ion batteries during fast charging, Appl. Therm. Eng. 86 (2015) 281-291.
[2] A. Pesaran, S. Santhanagopalan, G. Kim, Addressing the impact of temperature extremes on large format Li-ion batteries for vehicle applications (presentation), Tech. rep., National Renewable Energy Laboratory (NREL), Golden, CO, (2013)
[3] M.R. Giuliano, A.K. Prasad, S.G. Advani, J. Power Sources 216 (2012) 345e352.
[4] T. Tamura, T. Hachida, K. Yoshida, N. Tachikawa, J. Power Sources 195 (2010) 6095e6100
[5] Q. Wang, P. Ping, X. Zhao, G. Chu, J. Sun, C. Chen, J. Power Sources 208 (2012) 210e224.
[6] R. Kizilel, R. Sabbah, J.R. Selman, S. Al-Hallaj, J. Power Sources 194 (2009) 1105e1112.
[7] D. Ghosh, P.D.Maguire, D.X. Zhu, Design and CFD simulation of a battery module for a hybrid electric vehicle battery pack, Tech. Rep. 2009-01-1386, SAE, 2009.
[8] M.R. Giuliano, S.G. Advani, A.K. Prasad, J. Power Sources 196 (2011) 6517e6524.
[9] Y.S. Choi, D.M. Kang, Prediction of thermal behaviors of an air-cooled lithium-ion battery system for hybrid electric vehicles, J. Power Sources 270 (2014) 273-280
[10] G.H. Kim, A.A. Pesaran, 22nd International Battery, Hybrid and Fuel Cell Electric Vehicle Conference and Exhibition, Yokohama, Japan, 2006
[11] T.M. Bandhauer, S. Garimella, T.F. Fuller, A critical review of thermal issues in lithium-ion batteries, J. Electrochem. Soc. 158 (2011) R1-R25.

IOP Conf. Series: Materials Science and Engineering **653** (2019) 012013 doi:10.1088/1757-899X/653/1/012013

Tribological characteristics of Calophyllum ester based bio-lubricant – commercial engine oil blends

Kodanda Rama Rao Chebattina[1], V.Srinivas[1*], G.V.S.Pranay[1] , P. Vishwanath Krishna[1] and S.P.Jagadhi[1]

[1]Department of Mechanical Engineering, GITAM,Visakhapatnam India
[1*]vsrinivas@gitam.edu, [1]kodandaram.nitw@gmail.com

Abstract:

The main purpose of this research is to investigate the Tribological characteristics of Calophyllum Ester based Bio lubricant – commercial engine oil blends. The Bio lubricants used in this study are esters derived from Calophyllum seeds. The fatty acid content reveals the presence of saturates which form feed stock to esters. The physicochemical properties evaluated as per ASTM standards are found to be in an acceptable range. TMP (Trimethylolpropane) ester is blended with HP-RACER4 (Engine oil) in 5%, 10%, 15%, and 20% volume percentages respectively and tested for Anti-wear and Anti-Frictional properties using a four-ball tribo-tester. It was found that, although both the esters offer a very low coefficient of friction, the main drawback is their failure at higher loads. The blending of commercial lubricant with bio-lubricant resulted in lower friction and wear characteristics of the commercial lubricant. The optimum percentage of bio-lubricant in the commercial lubricant is found to be 20% which gave the desired results.

Keywords: Bio lubricant, Coefficient of friction, Tribological properties, Epoxide

1. Introduction

Several attempts have been made to identify the appropriate alternative for mineral oils in internal combustion engines such as controlled combustion technologies, usage of lightweight material and exhaust gas treatment. However, only a few researchers [1-10] have analyzed and reviewed the prospects of ester-based bio lubricants as an alternative to engine oil. The vegetable oils when blended gives competitive performance compared to the mineral oils at a relatively lower cost. They also can be used widely for industrial applications. It has been reported by researchers on how many trees are being left unexploited, which yield oil bearing materials and the studies suggested that the esters can be well utilized for new class lubricant formulations. The Non-edible Vegetable oil based Bio lubricant acts as the best alternative to Mineral oils and synthetic oils because of its Low toxicity, Low friction coefficients and High Biodegradability. The presence of hydrogen atoms on the β-carbon atom in an ester molecule makes them oxidation-prone, which restricts them to lower temperatures and leads to poor thermal stability. However, it can be overcome by using peresters of sugar instead of glycerol esters which improves the stability [2,4]. The emphasis was laid on improving the anti-wear, anti-friction and lubrication properties of bio lubricant, thereby increasing the reliability with reduced cost [5].

1.1 Present studies

Several studies were made on the tribological properties of bio-lubricants as a replacement for mineral oil based lubricants. Calophyllum inophyllum oil, which is used in these studies, is a derivative of the nuts of Tamanu tree which yields 70-75% of greenish-yellow inedible oil. The average oil yield is 11.7

kg-oil/tree which is easily extractable and is a cheaper bio-lubricant feedstock. The amicable physic-chemical characteristics and high fatty acid contents like linoleic acid (28%), oleic acid (38%) makes it much preferable.

The advantage of bio-lubricant over mineral oil is their bio-degradability and low friction coefficient due to the presence of esters. The main drawback of bio-lubricant is their failure at loads beyond 50 kgf on a tribometer. Commercial lubricants could be able to withstand loads over 120 kgf. The blending of bio-lubricant and commercial lubricant is done to attain both low friction and high load resistance. The bio-lubricant is blended to commercial lubricant in 5 %, 10%, 15 %, and 20 % and tested for tribological property enhancement.

2. Materials and Methodology

2.1 Materials used

Calophyllum seeds were sourced from plantations in GITAM, Visakhapatnam. AR grade aluminium oxide, hexane, toluene, and ethyl acetate were purchased from Finar, India. Hydrochloric acid, sodium thiosulphate and sodium sulphateare of EP grade and were purchased from Fisher Scientifics, Mumbai, India. p-TSA (Para toluene sulphonic acid) was purchased from AVRA labs, Hyderabad, India.

2.2 Methodology

2.2.1 Oil Extraction method

The Calophyllum nuts were gathered, dried and finely powered. The oil was extracted in a Soxhlet apparatus using hexane as a solvent. Later, hexane was evaporated and obtained crude oil was dried under low-pressure conditions.

2.2.2 Preparation of Calophyllum fatty acid

The TLC (Thin layer chromatography) equipment was used for the preparation of Calophyllum fatty acid. Calophyllum oil (300 g) and aqueous sodium hydroxide solution (54 g in 675 ml of water) were stirred mechanically for 4 h at 80-90°C using hexane and ethyl acetate (90:10 v/v) as a solvent. Later, the reaction mixture was cooled to room temperature and neutralized with concentrated HCl. Then it was extracted with ethyl acetate and washed twice with water. Finally, 270 g of the product was obtained by drying ethyl acetate layer over anhydrous sodium sulphate under vacuum. The fatty acid composition has been evaluated and the composition is given the table. 1. It has been found that the fatty acids are saturated and can be easily blended into esters.

Table 1. Fatty acid compositions of Calophyllum inophyllum oil

S.No	Fatty acid	Composition (%)
1	Palmitic (16.0)	15.23
2	Stearic(18.0)	15.93
3	Oleic(18:1)	37.57
4	Linoleic(18:2)	27.78
5	Others	3.4

2.2.3 Synthesis of calophyllum trimethylol propane ester (TMP ester)

Calophyllum fatty acids (270 g, 1.0112 mol), trimethylol propane (43.76 g, 0.3262 mol), p-TSA (2.7 g based on 1% weight of fatty acid) and 250 ml of were reacted using hexane and ethyl acetate as solvent. After completion of the reaction, the reaction mixture was cooled to room temperature and toluene was removed under vacuum. After drying the mixture, aqueous sodium bicarbonate solution and ethyl acetate were added to the reaction mixture and stirred for half an hour. The organic layer was separated washed twice with water, dried over anhydrous sodium sulphate and concentrated under vacuum. The crude ester was purified by basic alumina column chromatography using 95% hexane and ethyl acetate as eluent. The details of the reaction are shown in Figure 1.

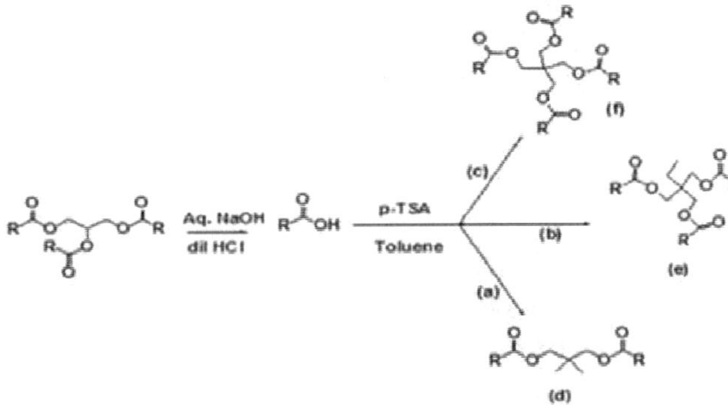

Fig. 1. Preparation of various esters of Calophyllum fatty acids

2.2.4 Physico-chemical properties of bio-lubricant

The physicochemical properties of bio-lubricants play an important role in maintaining tribological properties. Any deterioration of properties due to poor synthesis can lead to poor performance. The synthesizing method is optimized by evaluation of properties. The properties are as listed in Table 2 and within the acceptable range.

Table 2. Physico-chemical properties of Calophyllum polyol esters

S.No	Properties	Methods	Trimethylolpropane (TMP)ester
1	Colour	-	Transparent
2	Moisture (%)	-	0.1
3	Acid number(mgKOH/g)	AOCS Cd 3d -63	0.22
4	Density at 27^{0}C (g/cc)	ASTM D854	0.9954
5	Flash point (°C)	ASTM D92-01	194
6	Fire point (°C)	ASTM D92-01	204
7	Cloud point (°C)	ASTM D2500	-1.0

2.3 Experimental methods

2.3.1 Bio lubricant sample preparation

The standard oil, HP-RACER4 was used as a base lubricant and for comparison of results. The base oil is then blended with 5%, 10%,15% and 20% by volume of ester based Calophyllum inophyllum bio lubricant. An Ultrasonic bath sonicator is used to blend the composition of base oil and bio oil properly.

2.3.2 Tribological property evaluation

The tribological properties of a lubricating sample are measured using the 4 Ball tribometer a per ASTM standards. The fourballtribometer operates in a sliding or rolling motion where a single stainless steel ball (designated as top ball) rotates onto three stainless balls immersed in lubricant forming a lubricant film as shown in Figure 2. The test of wear resistance was carried out as per ASTM D 4172 – 94 standard test procedure. Thetest load (392N or 40Kgf) was applied with lubricant temperature maintained at 75±2 °C and speed of rotation at 1200rpmfor a duration of one hour.
Friction test is conducted as per ASTM D 5183in which initially "wear in" is carried out for one hour at a speed of 600 RPM and at a load of 40 kgf, maintaining oil temperature at 75°C. After wear- in, the lubricating oil used in the test is discarded and the balls are cleaned without removing the setup. The ball pot is filled with Fresh lubricant sample on the same worn test balls and the test is carried out as per above conditions with a starting load of 10 kgf. The load is successively increased by 10 kgf in 10 min interval until seizure load which is a sharp rise in the frictional Torque is observed. The seizure load is an important factor in determining the effectiveness of the lubricant.

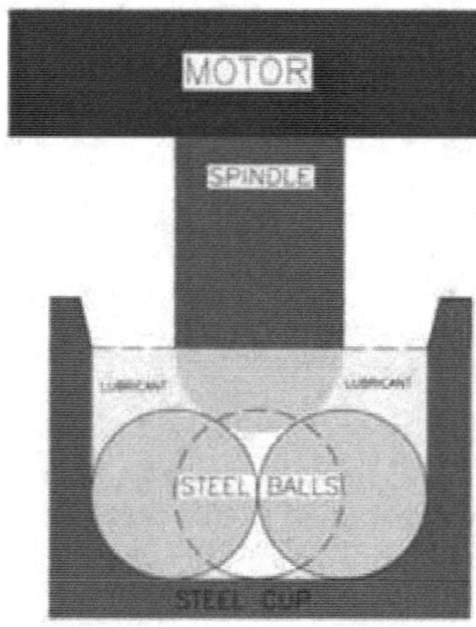

Fig. 2. Schematic of four balls in tribometer

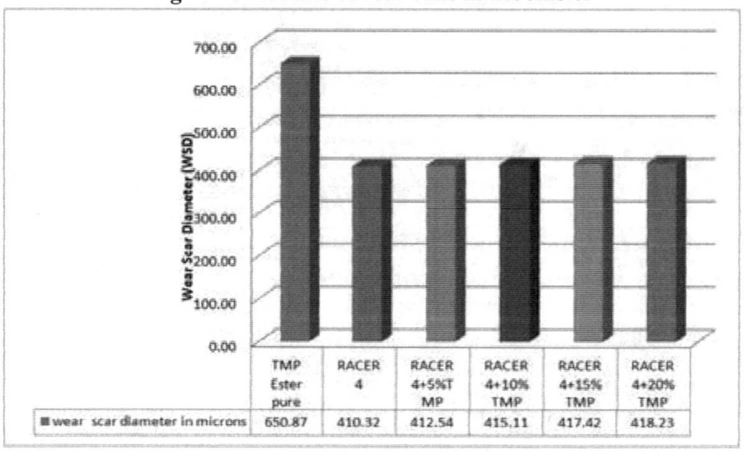

Fig. 3. Comparison of wear scar diameters of different test samples.

3. Results and Discussions

The results of the wear test conducted as per ASTM D4172 are as given below for different test samples. From Figure 3, it can be observed that the average wear scar of bio lubricant is very high. But when blended with commercial lubricant, the wear scar reduced due to the combined effect of bio lubricant and additives in the commercial lubricant.

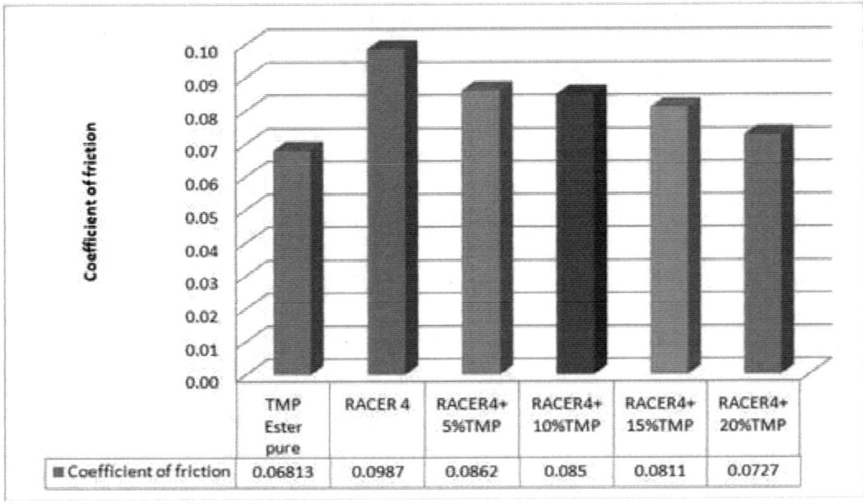

Fig. 4. Comparison of the coefficient of friction of base oils and blends

Bio lubricant blended with commercial lubricant could decrease the friction coefficient of commercial lubricant with marginal improvement in the seizure load. From Figures 4 and 5 it is observed that Racer4 has a very high coefficient of friction when compared to TMP ester. Pure TMP ester has a relatively lower coefficient of friction but has very less seizure load. Blending of base oil with pure TMP ester led to improvement in seizure load from 60 kgf (TMP ester) to 120 kgf (both 10% and 20% blends). Although wear scar diameters of 10% and 20% blends are almost equal, the coefficient of friction of 20% blend is comparatively low. Hence, Racer4+20% TMP ester blend satisfies all the requirements and is very much preferable.

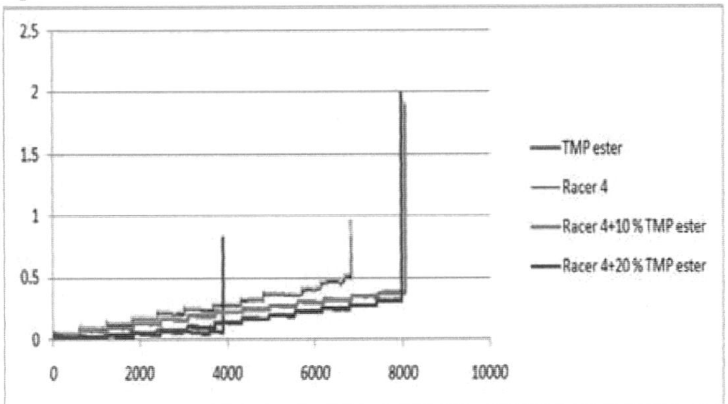

Fig. 5. Graph plotted between Frictional torque and data points

Conclusions

1. Fatty acid ester based bio-lubricant has been successfully derived from Calophyllum - inophyllum seeds which meet the requirements mentioned in SAE standards.
2. The physicochemical properties were studied as per standard ASTM methods and it was found that the physicochemical properties of ester-based bio-lubricant blends are in acceptable ranges.

3. The blends of commercial lubricant and bio-lubricant gave lower friction coefficients and performed very well in reducing wear. It is observed that Racer4+20% TMP ester blend maintains approximately similar wear scar diameter values as compared to the other test samples. While it exhibits dominant seizure load and a relatively low coefficient of friction, which makes the above-mentioned blend optimal.
4. The amicable fatty acid content helped significantly in reducing wear and friction rate. Presence of oleic acid emulsion and stearic acid played a vital role in controlling abrasion rate and maintaining a comparatively low coefficient of frictions.

Acknowledgments

The authors deeply thank HPCL for allowing the use of their facilities in testing and characterization. The authors thank the authorities of GITAM (Deemed to be University), India for their support extended for conducting research.

References

[1]. Kamalakar K, Rajak A K, Prasad R B N and Karuna M S L 2013Industrial Crops and Products **51** 249-57
[2]. Havet L, Blouet J, Valloire F R, Brasseur E and Slomka D 2001 Wear **248** 140–46.
[3]. Imran A, Masjuki H H, Kalam M A, Varman M, Hasmelidin M, Al Mahmud K A H, Shahir S A and Habibullah M 2013 ProcediaEngineering **68** 78 – 85.
[4]. Zulkifli N W M, Azman S S N, Kalam M A, Masjuki H H, Yunus R andGulzar M 2016 Tribology International B **93** 555-62.
[5]. Syahrullail S, Nuraliza N, Izhan M I, Abdul H M K and RazakaMd D2013 Procedia Engineering **68** 158 – 165.
[6]. Mobarak H M, Mohamad N E, Masjuki H H, Kalam M A, Mahmud Al KAH, Habibullah M and Ashraful A M 2014 Renewable and Sustainable Energy Reviews **33** 34–43.
[7]. Fox N J and Stachowiak G W 2007 Tribology International **40** 1035–46.
[8]. Sripada P K, Sharma R V and Dalai A K 2013 Industrial Crops and Products **50** 95– 103.
[9]. Salih N, Salimon J, and Yousif E 2011 Industrial Crops and Products **34** 1089– 96.
[10]. Heikal E K, Elmelawy M S, Salah A K and N M Elbasuny 2017 Egyptian Journal of Petroleum **26** 53–59.

IOP Conf. Series: Materials Science and Engineering **653** (2019) 012014 doi:10.1088/1757-899X/653/1/012014

Thermo-physical properties of solar thermic fluids dispersed with carbon nanotubes

Dr. V. Srinivas[1[0000-0002-6686-8780]], S.Raviteja[2], S.Jaikumar[3], V.Swathi[4] and DNV Sravya[5]

[1, 2,3,4,5] Department of Mechanical Engineering, GITAM (Deemed to be University), Visakhapatnam India

*Corresponding author. *Email: vsrinivas@gitam.edu*

Abstract. The paper examines the influence of dispersion of multi-walled carbon nanotubes on the thermo-physical properties of working fluids used in solar thermal engineering systems. Thermo-physical property evaluation is very vital to evaluate the performance of solar thermal systems. Ethylene glycol and a commercially reachable thermic fluid therminol 62 are preferred as based fluids. Nanofluids are prepared by dispersing 0.1 %, 0.25 %, 0.5 % and 0.1 % weight fraction of surface modified Multi-walled carbon nanotubes. The nanomaterials are surface modified before dispersion in the thermic fluids and extreme care is taken to prevent property deterioration during surface modification. Thermal conductivity of test fluids is evaluated in the temperature range of 100 to 300 °C while viscosity is evaluated in the range of 50 to 150 °C s. A 15 to 20 % increase in the thermal conductivity is observed with the dispersion of carbon nanotubes. The viscosity is significantly increased in the low-temperature ranges while the change is minimal above 70. The data could be predicted successfully using artificial neural networks. .

1. Introduction

The overexploitation of fossil fuel leads to environmental degradation and global warming and thus eco-freedom energy technology needs to be developed. Nowadays, solar energy is one of the key sources for sustainable development for the conversion of thermal and electrical energy using solar collectors and solar panels. However, the higher operating cost and high operating cost and lower efficiency involved in solar energy conversion are limiting its utility. The energy conversion solar thermal system is possible using a heat exchanging fluid (absorber fluid or thermic fluid). Common thermic fluids are ethylene glycol, ethylene glycol-water mixtures, silicone based fluids, and Hydrocarbon oils etc. Heat transfer fluids dispersed with nanoparticles improve the thermal and optical properties of the conventional fluids thus augmenting their performance. Several studies have been made concerning the improvement of thermal conductivity with the use of nanofluids. Among them, carbon nanotubes due to their high thermal conductivity of 2200 W/mk are the best option for thermic fluids. The usefulness of nanofluid as thermic fluid depends on the type of material, size, type of fluid, additives, shape of the particle, temperature, particle concentration, and the stability of nanofluid. The present investigations are meant to see the influence of multi-walled carbon nanotubes on the thermo-physical properties of solar thermic fluids. In the present study ethylene glycol based solar thermic fluids and commercially available thermic fluid therminol 62 are dispersed with

Multiwalled carbon nanotubes and investigated for property enhancement. Artificial neural networks are used to simulate the results in the choicest range.

2. Experiment and Methodology

2.1. Materials

In this work, the multi-walled carbon nanotubes are produced from M/s Cheap Tubes Inc., USA. The MWCNTs is 20-40 nm in diameter with a length of 25 microns and a purity of 95 %. The remaining chemicals are used in this study are EP grade and the surfactant used is AR grade which is purchased from M/s Sigma Aldrich India Pvt limited. Therminol 62 is selected as hydrocarbon-based thermic fluid and ethylene glycol as polar thermic fluid. Figure1 shows the image of MWCNTs on an HRSEM.

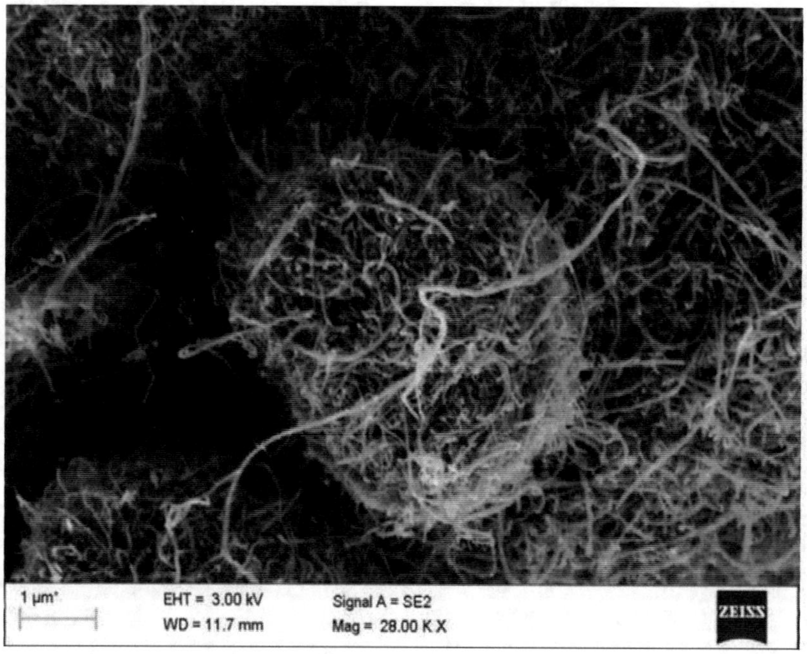

Figure1 - HRSEM image of pristine long length entangled MWCNTs.

2.2. Surface assisting of MWCNTs and preparation of nanofluids

Being denser than liquids, MWCNTs have a tendency to form substantial particle clusters thereby leading to poor dispersion stability. To improve the stability, it is required to modify the surface of MWCNTs with a surfactant to create stearic repulsions between individual nanotubes and make assist them to remain stable in the liquid. For the stabilization of nanoparticles in the liquid, a surfactant is used. The SPAN 80 (Sorbitan monooleate) surfactant accompanied by a hydrophilic-lipophilic balance of 4.6 was used. The surfactant adsorbs on the MWCNTs surface decreasing its energy on the surface and thus preventing agglomeration and settle down the nanoparticles. In the case of ethylene glycol-water mixtures, the CTAB (cetrimonium bromide) is used for surface modification. To prepare surface assisted MWCNTs, surfactant and Multiwalled carbon nanotubes are mixed in a ratio of 2:1 and ultra-sonicated in methyl alcohol for 30 minutes to create a mechanochemical reaction. This reaction covers the surface of the MWCNTs with a surfactant to create repulsion between individual tubes. The surface assisted MWCNTs in 0.125, 0.25, 0.5 and 1 weight per cent are dispersed in therminol and ethylene – glycol-water mixtures by processing it in a sonicator for 40 minutes.

3. Evaluation of physicochemical properties

3.1. Thermal conductivity
The conduction rate of heat is higher for high thermal conductivity material compared to low thermal conductivity material. The higher thermal conductivity materials are used in sink applications. The temperature of the material affects its thermal conductivity. Thermal Conductivity is measured using Hot DiskTM Thermal analyzer TPS 500. Hot Disk uses Transient Plane Source (TPS) which allows rapid and accurate measurement of Thermal Conductivity. The measurements are carried out at various temperatures ranging from 100 to 260oC.

3.2. Dynamic Viscosity
Viscosity is the resistance to the flow of fluid and higher viscosity of fluid requires more pumping power for circulating the fluid in the solar loop. Experimental results can predict the influence of dispersing MWCNTs on the viscosity of nanofluids at various mass fractions and over a range of temperatures. Wells-Brookfield Cone & Plate Viscometer is used to evaluate the absolute viscosity of test fluids. The absolute viscosity of thermic fluids dispersed with MWCNTs is measured in the temperature range of 750C and 120 0C to verify whether the fluid, follows the Newtonian law.

4. Results and discussion

4.1. Thermo—physical property evaluation of nanofluids
The trends of variation of viscosity with temperature for ethylene glycol-water mixtures are shown in Figures 2 -4. The variation of thermal conductivity for therminol 62 based fluids is shown in Fig 4. There is a noteworthy improvement in thermal conductivity with the dispersion of nanomaterials. Pure Ethylene glycol and therminol dispersed with MWCNTs showed higher improvement in thermal conductivity compared to ethylene glycol-water mixtures.

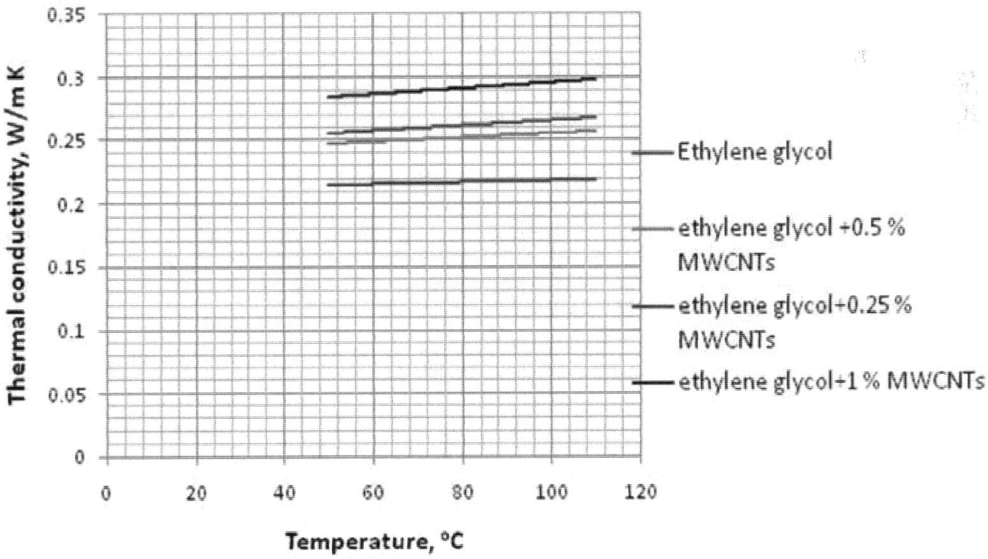

Figure 2- Variation of thermal conductivity with temperature

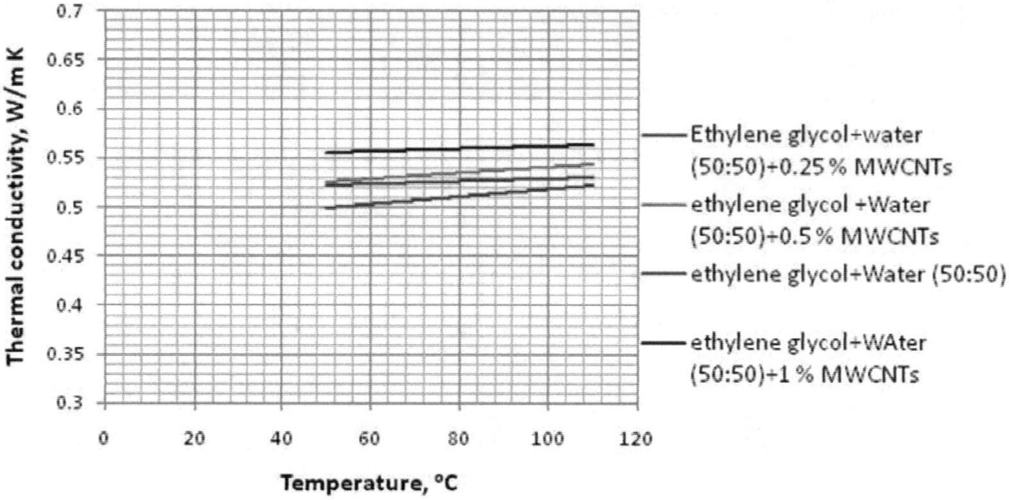

Figure 3 - Variation of thermal conductivity with temperature

The variation of viscosity of test fluids with temperature is shown in Figures 5&6. It can be seen from the Fig.5 that at lower temperatures, the significant change in viscosity was observed. Conversely, the increase in viscosity was observed marginally at higher temperatures owing to lower mass fraction of CNTs in the nano fluid preparation.

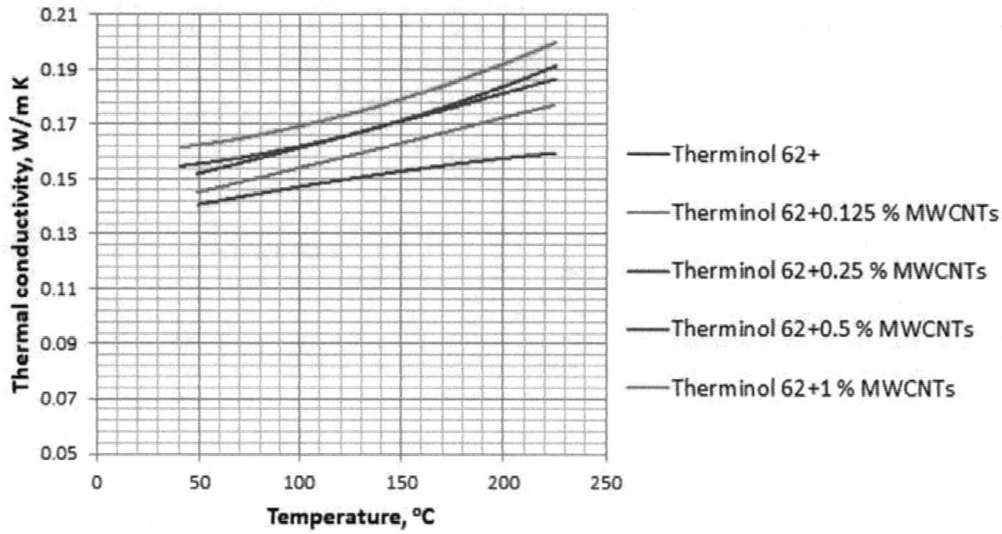

Figure 4 - Variation of thermal conductivity with temperature

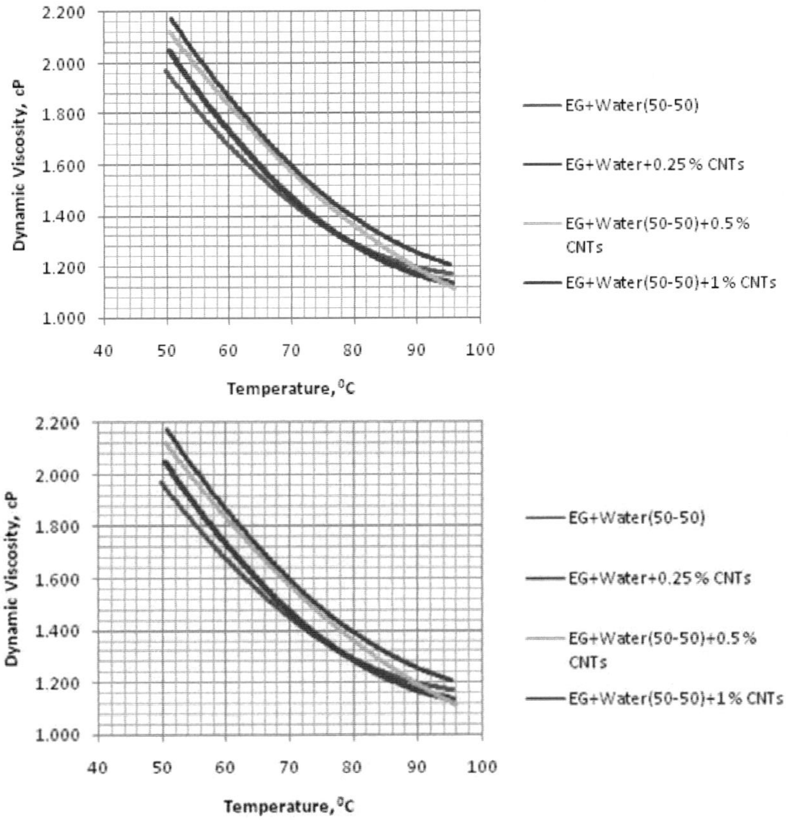

Figure 5 - Variation of dynamic viscosity with temperature for ethylene glycol water mixtures

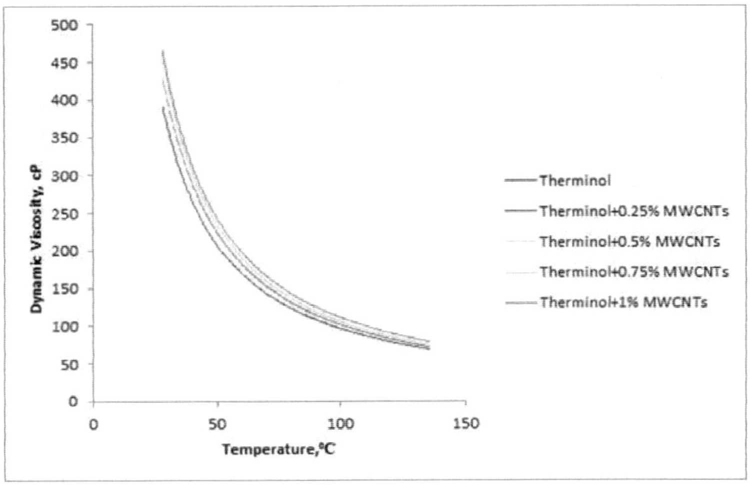

Figure 6 - Variation of dynamic viscosity with temperature for Therminol

IOP Conf. Series: Materials Science and Engineering **653** (2019) 012014 doi:10.1088/1757-899X/653/1/012014

4.2. Analysis of data using artificial neural networks

Several researcher reported thermal conductivity but vast inconsistencies were found from the reported results. Hence a good mathematical model to describe the thermal conductivity of nanofluids is essential. Artificial neural networks (ANN) contain connected units called artificial neurons which work in a similar fashion as that of axons in a human brain where each neuron can transmit a signal to another neuron. The function of the neural network is the computational output of all neurons and the neural network is developed by the set of inputs and outputs in the network. In this study, a backpropagation algorithm is used to evaluate the error involvement of each neuron during processing of data . In this study, the experimental values are trained using MATLAB with a set of input and output values by developing a network to predict the error. The developed network consists of input values as the volume percentage of ethylene glycol, the mass fraction of MWCNTs, temperature and thermal conductivity as output values. The results are shown in the following graphical Figures 7 to 9.

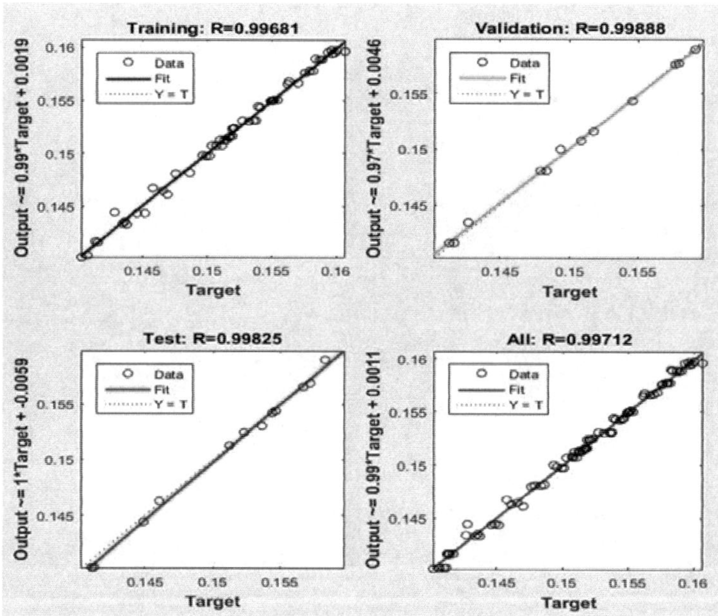

Figure 7 - Validation of data of therminol 62 at different mass fractions of MWCNTs and temperatures using artificial neural network

IOP Conf. Series: Materials Science and Engineering **653** (2019) 012014 doi:10.1088/1757-899X/653/1/012014

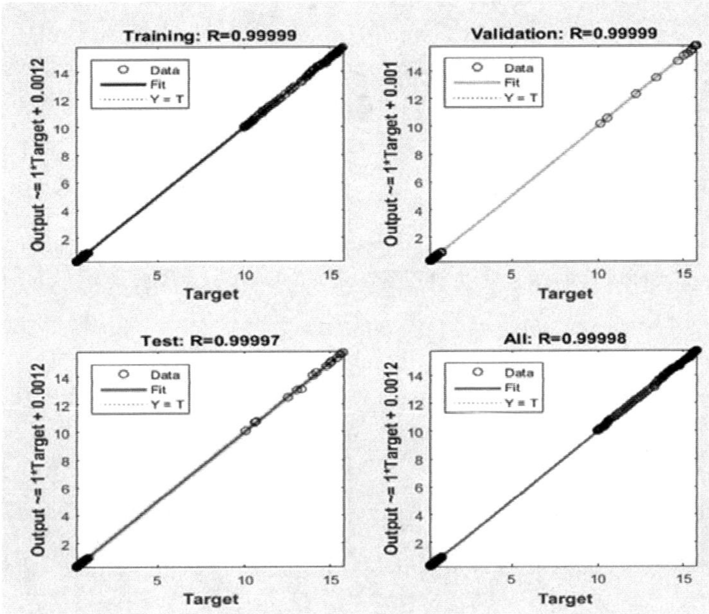

Figure 8 - Validation of data of thermal conductivity of ethylene glycol-water mixtures at different mass fractions of MWCNTs, temperatures and volume percentage of ethylene glycol using artificial neural network

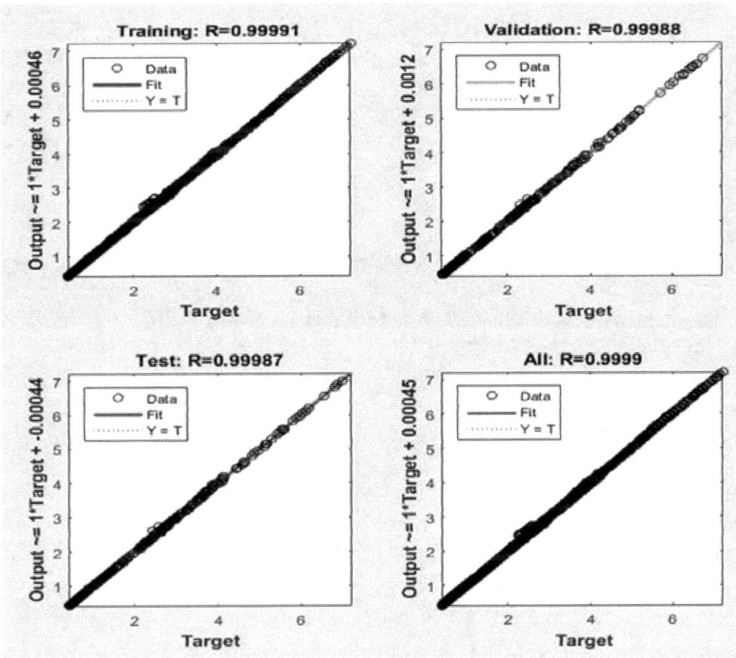

Figure 9- Validation of data of dynamic viscosity of ethylene glycol-water mixtures at different mass fractions of MWCNTs and volume percentage of ethylene glycol using artificial neural network

5. Conclusions

1. Dispersion of MWCNTs in thermic fluids improved the thermal properties without compromising viscous properties.
2. The thermal conductivity significantly increased with the increase in mass fraction and the improvement is in the range 5 % to 20 %.
3. There is no deterioration of dynamic viscosity with the dispersion of MWCNTs indicating good pumpability of nanofluids.
4. The data has been validated using artificial neural networks and the ANN is able to predict the thermal conductivity and viscosity at different temperatures, mass fractions for a wide range of data.

References

[1] Alim M A, Abdin Z, Saidur R, Hepbasli A, Khairul M A, Rahim N A 2013 *Energy Build* **66** 289–296.
[2] Assael M J, Chen C F, Metaxa I N & Wakeham W A 2004 *International Journal of Thermophysics* **25 (4)**, 971-985.
[3] Assael M J, Metaxa I N, Arvanitidis J, Christofilos D & Lioutas C 2005 *International Journal of Thermophysics* **26 (3)** 647-664
[4] He Q, Zeng S, Wang S 2015 *Applied Thermal Engineering Journal* **88** 165–71.
[5] Kameya Y, Hanamura K 2011 *Sol Energy* **85** 299–307.
[6] Meibodi M E, Sefti M V, Rashidi A, Kalal H S 2010 *Int. Commun Heat Mass Transfer* **37.3** 319–323
[7] Meibodi S S, Kianifar A, Niazmand H, Mahian O, Wongwises S 2015 *Int. Commun Heat Mass Transfer* **65** 71–75.
[8] Tian Y, Zhao CY 2013 *Applied Energy* **104** 538–553.
[9] Yousefi T, Veysi F, Shojaeizadeh E, Zinadini S 2012 *Renew Energy* **39** 293–8
[10] Yousefi T, Shojaeizadeh E, Veysi F, Zinadini S 2012 *Solar Energy* **86** 771–779.
[11] Zamzamian A, KeyanpourRad M, KianiNeyestani M, Jamal-Abad M T 2014 *Journal of Renewable Energy* **71** 658–64.
[12] Tahereh B. Gorji, A.A. Ranjbar 2017 *Journal of Renewable and Sustainable energy* **72** 10–32 .
[13] Nathan Hordy, DelphineRabilloud, Jean-Luc Meunier, Sylvain Coulombe 2014 *Journal of solar energy* **105** 82-90.
[14] Z Said, M H Sajid, R Saidur, G A Mahdiraji, and N A Rahim 2015 *Journal of Numerical Heat Transfer* **67** 1010-1027.
[15] Saidur R, Meng T C, Said Z and M Hasanuzzaman A Kamyar 2012 *International Journal of Heat And Mass Transfer* **55** 5899-5907.
[16] Gorji T B and Ranjbar A A 2017 *A review on optical properties and application of nanofluids in direct absorption solar collectors (DASCs)* 72 pp 10-32

Acknowledgment

The authors recognize the support received from Hindustan Petroleum Corporation Ltd., India for conducting the tests. The authors also acknowledge the support from IIT Madras, IIT Kharagpur ,and Osmania University, Hyderabad in characterization. The authors honestly show gratitude to the management of GITAM University, India for the support extended.

Effect of sintering atmosphere on the electrical properties of lanthanum doped barium titanate: an impedance spectroscopic study

S Praharaj[1], A Singha[1], N Mallick[2] and D Rout[1, 3]

[1]School of Applied Sciences (Physics), KIIT Deemed to be University, Bhubaneswar 751024, Odisha, India
[2]Kalinga Institute of Social Science Deemed to be University, Bhubaneswar 751024, Odisha, India
[3]droutfpy@kiit.ac.in

Abstract. The electrical properties of 0.1 mol% La-doped $BaTiO_3$ was investigated by impedance spectroscopy after sintering the sample in air and H_2 at 1350 °C. H_2-sintered sample showed more semiconducting nature than air-sintered one due to a higher level of oxygen loss accompanied by reduction of Ti^{4+} to Ti^{3+}. The overall resistance of the sample decreased with rise in temperature and change of sintering atmosphere from oxidizing to reducing. The higher value of activation energy obtained from Arrhenius activated hopping frequency f_H for H_2-sintered sample describes the delayed onset of frequency-dispersion in the conductivity curves than the air-sintered sample. The conductivity relaxation was studied in the light of frequency-dependence of the imaginary part of electric modulus and scaling studies of the real and imaginary parts of Z* elucidating the thermally activated relaxation dynamics.

Keywords: $BaTiO_3$, Conductivity, Impedance, Electric modulus.

1. Introduction

Barium titanate constitutes one of the most versatile oxide electroceramics due to its unique ferroelectric behavior, high ε, high voltage tunability, PTCR effect, piezoelectricity and pyroelectricity. Owing to its wide use in various applications including multilayer capacitors, piezo actuators, temperature sensors and tunable circuits, the requisite properties are constantly modified by various aliovalent dopants at A- and B-sites, among which substitution by trivalent rare-earths is prominent [1-5]. Rare-earths (Yb^{3+}, Gd^{3+}, La^{3+}) having an intermediate size and charge compared to Ba^{2+} and Ti^{4+}, may diffuse either to the A- site or B-site by various doping mechanisms [6]. La^{3+} is being large in size, exclusively substitutes at the A-site avoiding self-compensation and/or B-site doping. It may possibly comprise of 3 A-site doping mechanisms [6]: i) $3Ba^{2+} \rightarrow 2La^{3+}$ with the creation of A-site vacancies ($Ba_{1-a}La_{2a}TiO_3$), ii) $Ba^{2+} + \frac{1}{4}Ti^{4+} \rightarrow La^{3+}$ with the creation of Ti vacancies ($Ba_{1-b}La_bTi_{1-b/4}O_3$) and iii) donor doping mechanism via electronic compensation: $Ba^{2+} \rightarrow La^{3+} + e^-$ ($Ba_{1-c}La_cTiO_3$). As reported earlier, at lower dopant level ($x_c < 0.3$ mol%), electronic compensation infuses semiconductor behavior while at higher dopant level, a cross-over to ionic compensation (by Ti vacancies) again infuses an insulating behavior of the ceramic [5]. Thus it is inferred that the electrical properties of La-doped $BaTiO_3$ may be highly influenced by the processing conditions such as sintering temperature, oxygen partial pressure (oxidizing/inert/reducing atmosphere) during sintering and dopant concentration [5, 6]. The complex impedance spectroscopy (CIS) is a powerful tool to investigate the electrical behavior of such samples at all temperatures. However most of the reports in the literature have been conducted up to 400 °C but to the best of our knowledge, studies at elevated temperatures are limited. Hence in this work, a detailed impedance and ac conductivity analysis of 0.1 mol% La-doped $BaTiO_3$ (BT), sintered in air (oxidizing) and H_2 (reducing) atmospheres has been carried out above 400 °C. The ceramic samples were prepared by conventional solid state reaction using $BaTiO_3$ (Fuji Titanium, Kanagawa, Japan) and La_2O_3 (Aldrich) high purity powders. The green pellets were cold-isostatically pressed at 200 Mpa and sintered at 1350 °C for 24 h in air and H_2 atmosphere. The impedance measurement was carried out by HP4192A impedance analyzer (0.1 kHz to 5 MHz) at high temperature (400-700 °C).

2. Results and Discussion

2.1 Electrical Behavior

Fig. 1 (a) and (b) represent conductivity $\sigma(\omega)$ vs frequency (*f*) plot of La-doped BT [$\sigma(\omega) = \omega\epsilon_0\varepsilon''$. ω is the applied frequency, ϵ_0 is the permittivity of vacuum and ε'' is the dielectric loss]. The figure shows two frequency regions i. e. low-frequency plateau region identifying dc conductivity (σ_{dc}) and frequency-dispersive ac conductivity (σ_{ac}) at higher *f*-values. The plateau may be interpreted as the uninterrupted hopping of the charge carriers assumed to move along the infinite lattice of identical potential wells. A crossover from plateau to a frequency dispersion region occurs at hopping frequency f_H. Above f_H, the conductivity starts increasing rapidly which may be explained in terms of the ratio of successful to unsuccessful hops [7]. The onset of frequency dispersion at f_H, shifts to the higher frequency side with a rise in temperature suggesting thermally activated conductivity relaxation [8] which interestingly delayed in H_2 atmosphere compared to air. The frequency response of conductivity can be well explained by the Jonscher's power law [9]: $\sigma(\omega) = \sigma_{dc} + \sigma_{ac}$, $\sigma_{ac} = Af^S$, $\sigma(\omega) = \sigma_{dc}[1 + (\frac{f}{f_H})^S]$, $f_H = (\frac{\sigma_{dc}}{A})^{1/S}$ where A is a temperature-dependant pre-exponential factor and S is a dimensionless frequency exponent. σ_{dc} is obtained from the extrapolated fitted curves while σ_{ac} is extracted by subtracting σ_{dc} from $\sigma(\omega)$. The σ_{dc} values indicate the increase of conductivity by ~2 orders in magnitude in H_2 exhibiting n-type semiconductor behavior [10] at elevated temperatures. This is attributed to higher O_2-loss accompanied by partial reduction of Ti^{4+} to Ti^{3+} as [11]: $O^{2-} \rightarrow V_O^{\cdot\cdot} + 2e' + \frac{1}{2}O_2$ and $2Ti^{4+} + 2e' \rightarrow 2Ti^{3+}$. The activation energies calculated from the Arrhenius plots [Fig. 1(c), (d)] (i.e. E_{dc} and E_{fH}) revealed $E_{fH,\ hydrogen} > E_{fH,\ air}$ that might be the reason for the delayed onset of *f*-dispersion in the conductivity curves of H_2-sintered sample. The E_{dc} values suggest thermal excitation of charge carriers from the second ionization of O_2-vacancies [12].

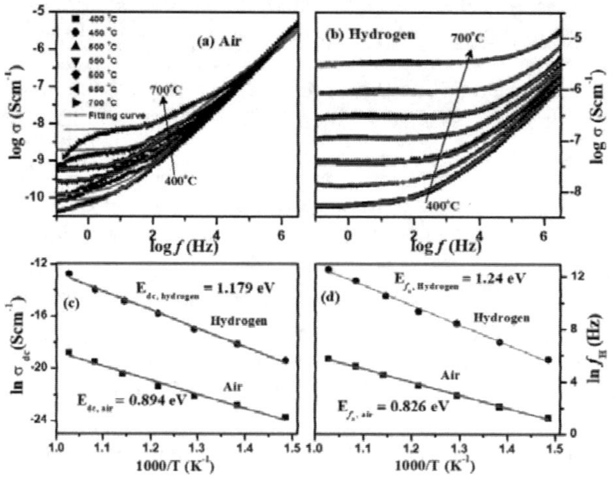

Fig. 1. Conductivity spectra of La-doped BT samples sintered in (a) air and (b) H_2; corresponding Arrhenius plots of (c) σ_{dc} and (d) f_H.

Fig. 2 (a) and (b) display asymmetric semicircles with centers lying below Z'-axis indicating the distribution of relaxation time as a departure from the Debye-type of relaxation. Asymmetry in the curves may be due to the overlapping of two depressed semicircles ascribed to grain and grain boundary contributions at high and low frequencies. The radii of the semicircles decrease with temperature indicating a fall in resistivity [11] and a dramatic fall in resistivity is seen in H_2-sintered ceramics (inset). The Z' vs log *f* plot [Fig. 3 (a) and (b)] indicates a gradual decrease of Z'-values decrease with frequency and temperature signifying an improvement in conductivity.

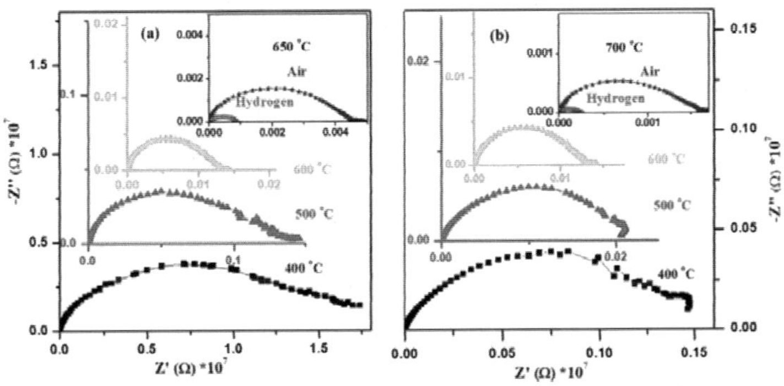

Fig. 2. Complex Impedance plots for the sample sintered in (a) air and (b) H_2; insets represent a comparative plot of both atmospheres at 650 and 700 °C

At higher *f*-values, all the curves merge which is probably due to the release of space charge [13] as a result of the reduction in the barrier potential with a rise in temperature. Further the inset to Fig. 3 (a) and (b) suggests a drop in Z' as the sintering atmosphere changes from oxidizing to reducing. This trend is further supported by Z'' vs log *f* plot in Fig. 3(c) and (d). The air-sintered sample [fig. 3(c)] is characterized by a single relaxation peak (corresponding to grain interior) while in H_2-sintered [fig. 3(d)], two peaks (corresponding to grain interior at higher *f*-values and boundaries at lower *f*-values) are seen. The relaxation phenomena can be understood by the presence of immobile species/electrons at low temperatures and mobility of defects/vacancies ($V_O^{\cdot\cdot}$) at higher temperatures. The overall resistance (active conduction through the grain boundaries) is obeying the relation [5]: $Z''_{max} = \frac{R}{2}$. In the studied temperature range, the La-doped BT shows NTCR behavior. It may also be marked that the fall of resistance with temperature is much higher in H_2 than in air [inset to Fig. 3(c) and (d)] which might be due to the formation of more O_2-vacancies in reducing atmosphere.

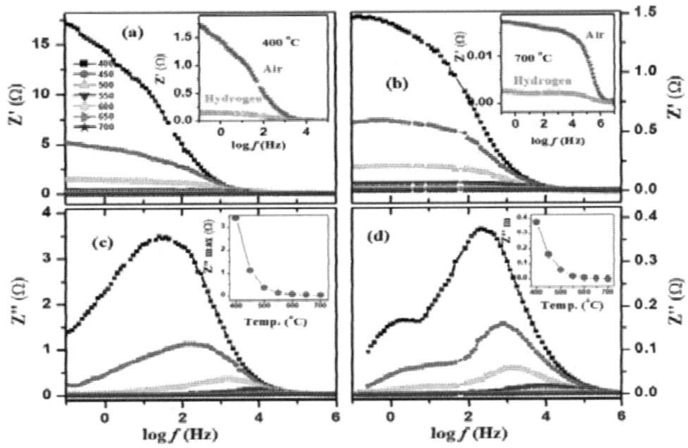

Fig. 3. Z' and Z'' vs log *f* plots for samples sintered in air and H_2 at different temperatures

Besides, the Z"-peaks shift towards the higher *f*-side with temperature indicating a decrease of the relaxation time ($\tau = {}^{1}/_{2\pi f_z}$) and the shifting may be elucidated in terms of increase of the hopping rate of the charge carriers/vacancies with temperature [14]. The activation energy for relaxation [inset: (fig. 4(c)] is obtained ~1 eV and is ascribed to conduction through O_2-vacancies. Generally, the activation energy in perovskite structure varies with O_2-vacancy concentration inversely at higher temperatures [15]. Hence, it is inferred that the O_2-loss is higher in case of H_2-sintered sample facilitating more conduction through the boundaries.

2.2 Conductivity Relaxation

Fig. 4(a) and (b) shows the frequency variation of the imaginary part of electric modulus (M") and the peaks shift to the higher *f*-side with rising temperature due to the decrease of relaxation time. The asymmetric peaks suggest a distribution of relaxation time and a thermally activated relaxation mechanism. The low *f*-range below the peak frequency f_M corresponds to the long-range mobility of the charge carriers while the *f*-range above f_M to the short-range mobility [16].

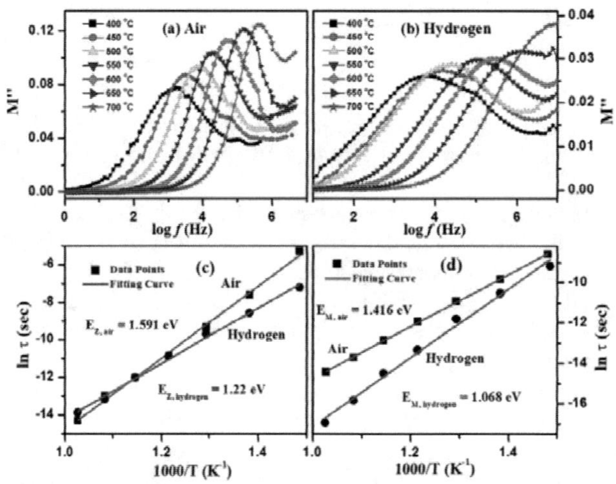

Fig. 4. M" vs log *f* plots of La-doped $BaTiO_3$ sintered in (a) air and (b) H_2 at 400 – 700 °C; Arrhenius plots of τ extracted from frequency variation of (c) Z" and (d) M"

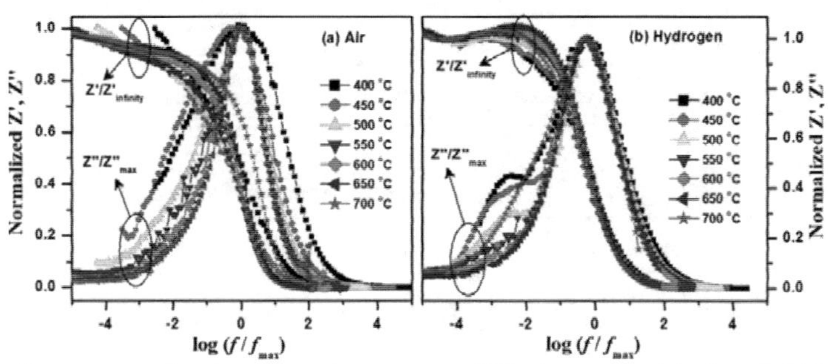

Fig. 5. Scaling curves of real and imaginary parts of Z* sintered in (a) air and (b) hydrogen.

The activation energy [Fig. 4 (d)] is found to be less in H_2-sintered sample than air-sintered. The scaling of Z' and Z" with frequency (Fig. 5) further inferred thermally activated relaxation dynamics in air-sintered ceramic as the curves are dispersed instead of getting completely collapsed into a single master curve. However, in H_2-sintered sample, a significant dispersion is observed on the lower *f*-side. This advocates the thermally activated dynamics which affect the movement of the charge carriers across the grain boundary while on the higher frequency side the relaxation phenomena is independent of temperature.

3. Conclusion

The effect of sintering atmospheres on electrical properties of La-doped BT was investigated by CIS technique. H_2-sintered sample exhibits n-type semiconducting behavior and both samples show NTCR behavior in the studied temperature range. The onset of frequency dispersion in the conductivity curves is delayed in reducing atmosphere compared to oxidizing evidenced by a higher value of activation energy. The conductivity relaxation is analyzed using electric modulus formalism.

4. References

[1] Tian Z, Wang X, Shu L, Wang T, Song T H, Gui Z and Li L 2009 Preparation of nano $BaTiO_3$-based ceramics for multi layer ceramic capacitor application by chemical coating method *J. Am. Ceram. Soc.* **92** 830-833

[2] Sauer H A and Fisher J R 1960 Processing of positive temperature coefficient thermistors *J. Am. Ceram. Soc.* **43** 297-301

[3] Heywang W 1971 Semiconducting barium titanate *J. Mater. Sci.* **6** 1214-1224

[4] Jonker G H 1964 Some aspects of semiconducting barium titanate. Solid State Electron **7** 895-903

[5] Morrison F D, Sinclair D C and West A R 2001 Characterization of La doped barium titanate ceramics using impedance spectroscopy *J. Am. Ceram. Soc.* **84** 531-538

[6] Freeman C L, Dawson J A, Chen H R, Ben L, Harding J H, Morrison F D, Sinclair D C and West A R 2013 Energetics of donor-doping, metal vacancies, and oxygen loss in A-site rare earth doped $BaTiO_3$ *Adv. Funct. Mater.* **23** 3925-3928

[7] Funke K 1993 Jump relaxation in solid electrolytes *Prog. Solid State Chem.* **22** 111-195

[8] Singh I, Kim I W, Sin, B C, Ullah A, Woo S K and Lee Y 2015 Study of dielectric, ac-impedance, modulus properties of $0.5Bi_{0.5}Na_{0.5}TiO_3$-$0.5CaCu_3Ti_4O_{12}$ nanocomposite synthesized by a modified solid state method *Mater. Sci. Semicond. Process.* **31** 386-396

[9] Jonscher A K 1997 The 'universal' dielectric response *Nature* **267** 673

[10] Smyth D M 2000 The defect chemistry of metal oxides. Oxford University Press, New York, NY

[11] Zang J, Li M, Sinclair D C, Jo W and Rodel J 2014 Impedance spectroscopy of $(Bi_{1/2}Na_{1/2})TiO_3$-$BaTiO_3$ ceramics modified with $(K_{0.5}Na_{0.5})NbO_3$ *J. Am. Ceram. Soc.* **97** 1523-1529

[12] Long S A and Bluementhal R N 1971 Ti-rich nonstoichiometric $BaTiO_3$: II, analysis of defect structure *J. Am. Ceram. Soc.* **54** 577-583

[13] Jha A K 2013 Electrical characterization of zirconium substituted barium titanate using complex impedance spectroscopy *Bull. Mater. Sci.* **36** 135-141

[14] Kumara R, Ahlawat N, Agarwal A, Sanghi S and Sindhu M 2016 Phase transformation and impedance spectroscopic study of Ba substituted $Na_{0.5}Bi_{0.5}TiO_3$ ceramics *J. Alloys. Compd*. **676** 452-460

[15] Steinsvik S, Bugge R, Gjonnes J, Tafto J and Norby T 1997 The defect structufe of $SrTi_{1-x}Fe_xO_{3-y}$ ($x = 0$–0.8) investigated by electrical conductivity measurements and electron energy loss spectroscopy (EELS) *J. Phys. Chem. Solids* **58** 969

[16] Barik B K, Choudhary R N P and Pradhan D K 2013 Dielectric and impedance spectroscopy of zirconium modified $(Na_{0.5}Bi_{0.5})TiO_3$ ceramics *Ceram. Int.* **39** 5695-5704

IOP Conf. Series: Materials Science and Engineering **653** (2019) 012016 doi:10.1088/1757-899X/653/1/012016

Characterization of raw and alkali treated prosopis juliflora fibers for potential polymer composite reinforcement

P. Madhu[1], S. Pradeep[1], M. R. Sanjay[2,3], Suchart Siengchin[3]

[1]Department of Mechanical Engineering, Malnad College of Engineering, Hassan, Visveswaraya Technological University, Belagavi, Karnataka, India.
madhu.p.gowda15@gmail.com
[2]Department of Mechanical Engineering, Ramaiah Institute of Technology, Bengaluru, Visveswaraya Technological University, Belagavi, Karnataka, India.
[3]Department of Mechanical and Process Engineering, The Sirindhorn International Thai-German Graduate School of Engineering (TGGS), King Mongkut's University of Technology North Bangkok, Bangkok, Thailand.

[1]madhu.p.gowda15@gmail.com

Abstract. The present world scenario owes for the need of materials which posses more ecological and biodegradable characteristics as well as capable of substituting man-made materials which resulted in the rapid increasing use of natural fibers. In the present investigation, the physico-chemical, thermal, tensile and morphological properties of Prosopis juliflora fibers (PJFs) are presented. Alkali treatment of PJFs was carried out at 2%, 4%, 6% and 8% (w/v) concentrations for a period of 45 min. 6% (w/v) treated PJFs results showed optimal results. These optimally treated fibers showed higher cellulose (74.42 wt. %), lower hemicellulose (2.21 wt. %), lignin (9.11 wt. %), higher crystallinity index (56.64%), crystallite size (15.76 nm), tensile strength (650 ± 6.43 MPa) and better thermal stability.

1. Introduction

Global awareness towards using more energy conservative and energy reduction processes have made researchers to look for more eco-friendly materials. Also, new environmental legislation and consumers demands have made many industries including construction, automotive, and packaging industries to look towards green resources that can replace traditional reinforcing materials like glass and carbon fibers [1-2]. With this concern, the use of natural fiber reinforced polymer composites have been found as a significant solution for the researchers to substitute for conventional man-made composite materials [3]. Also, they are preferred due to some of their some unique characteristics such as less skin irritation, cheap processing, low cost, non-corrosive in nature, biodegradability and rich availability along with their excellent mechanical properties like high specific strength, lightweight, specific modulus and flexibility. Over the years several natural fibers such as Acacia leucophloea, hemp, coir, ramie, kenaf, artichoke, kudzu, nettle, jute, chakshir, tamarind etc. have been utilized as reinforcements in the polymer matrix composites [4-6]. These composites are mainly composed of cellulose content and effect in high content of hydroxyl groups (OH) which results in their hydrophilic characteristics [7]. However the presence of pectin and wax substances in these fibers hinders this OH groups to react with polymer matrices and leads to voids and debonding between the fibers and matrices in the composites [8-9]. To overcome this problem, various chemical treatments have been used which also help in improving the interfacial bonding between the fibers and matrices. The most common type of chemical treatment being used is alkaline treatment which is also called as mercerization treatment that enhances the interlocking properties between the fibers and matrices and thus resulting in obtaining better mechanical properties from the composites. Also, alkali treatment is found as a cost effective method to modify the fiber surface [10-11]. The various physical and chemical treatments helps in reducing moisture content in the fibers as well as modifies the mechanical properties, thermal stability and surface morphology of these fibers [12]. The main aim of the present study is to identify prospective property of PJFs as a potential green composite reinforcement and understand effect of alkaline treatment on PJFs.

2. Materials and methodology

2.1. Materials and extraction of fibers

Bark fibers from Prosopis juliflora (PJ) plant are accumulated from Virudhunagar, Madurai Road, Tamil Nadu State, South India. The fibers were extracted manually by hand plucking method and then barks of the plant was dipped in tap water to facilitate microbial degradation for a time of 14 days for easy separation of fibers. Fibers were then removed from the bark by utilizing conventional combing process. [7].

2.2. Alkaline treatment

Extracted PJFs were treated under different concentrations viz., 2%, 4%, 6% and 8 % (w/v) using NaOH solution maintained at a room temperature for about 45 min.

2.3. Chemical analysis

Standard TAPPI methods were used to determine various chemical constituents such as cellulose, hemicellulose and lignin contents [3]. Sartorious electronic moisture analyzer (model MA45) was employed to measure the moisture content [2]. The wax content was evaluated by standard Conrad method [13-14]. Density was determined using melterbertoledoxsz05 balances method [15].

2.4. Fourier transform-infrared (FTIR) analysis

The FTIR spectrums of raw and treated PJFs were recorded using Shimadzu FTIR-8400S spectrometer. Two mg of PJFs were first powdered and then mixed with potassium bromide (KBr) pellets analysis. Further they were coupled with 32 scans having a resolution of 2 cm^{-1} and the range of frequency was measured in terms of wave numbers typically ranging between 4000 and 400 cm^{-1} [3,15].

2.5. X-ray diffraction (XRD) analysis

The XRD analysis of PJFs in powdered form was carried out by X'Pert Pro diffractometer system having diffracted intensity of CuKα radiation maintained at 0.154 nm wavelength, while its operating range maintained between 10^0 to 80^0 with a step size of 2θ (0.050). The crystalline index (CI) of cellulose materials were estimated by using Segal empirical expression. [14]:

$$CI = \frac{I_{200} - I_{AM}}{I_{200}} \tag{1}$$

Where,
I_{200} = Maximum intensity (200 lattice plane) maintained at 2θ angle between 22^0 and 23^0
I_{AM} = Minimum intensity of amorphous materials at an angle of 2θ maintained at 18^0.

The crystallite size (CS) was calculated by applying Scherrer's expression [16]:

$$CS_{200} = \frac{k\lambda}{\beta_{200}\cos\theta} \quad (2)$$

Where, K = Scherrer's constant (0.89), λ = Wavelength of the radiation (0.1541nm), β = Peak's full-width at half-maximum (radians), θ = Bragg angle.

2.6. Thermo-gravimetric analysis (TGA)

Thermal degradation behaviour and transformation thermograms of raw and alkali treated PJFs were assessed through Jupiter simultaneous analyzer (model STA449 F3TGA). The optimally treated fibers were exposed to atmosphere of high-purity nitrogen gas maintained at a flow rate of 20 mL/min. The measurements were performed in an alumina container maintained at a temperature ranging 30^0 C to 1000^0 C with a heat rate of 10^0 C/min [4, 17].

2.7. Tensile testing

Nikon Eclipse LV 150 polarized light microscope was used to evaluate the diameter of raw and alkali treated PJFs [19]. Tensile strength was evaluated using standard test methods and tensile testing was carried out by a standard ASTM D3379-75 method using an INSTRON UTM equipped with load cell capacity of 500 N [3, 18].

3. Results and discussions

3.1. Chemical analysis

Table 1 represents chemical composition of raw and different alkaline treated PJFs. The cellulose content of optimally treated PJFs has increased from 60.46 wt. % (Raw) to 74.42 wt. % due to the swelling in the fibers influenced by alkaline treatment. The alkaline treatment has also diminished the hemicellulose and lignin contents in treated PJFs as compared to raw PJFs. Similarly, low wax and moisture contents were found in various alkali treated PJFs enabling these fibers to be effectively used as reinforcement in hydrophobic polymer matrices by improving their interfacial bonding properties. The higher density values in various alkali treated PJFs could be due to the removal of contamination and lower wax contents.

Table 1. The chemical composition of raw and different alkaline treated PJFs.

Fibers	Cellulose (wt. %)	Hemicellulose (wt. %)	Lignin (wt. %)	Wax (wt. %)	Moisture content (%)	Density (Kg/m^3)
Raw	60.46	15.24	15.16	0.59	8.58	589
2% NaOH	62.56	11.41	13.11	0.43	7.91	610
4% NaOH	68.42	5.21	11.11	0.38	5.48	642
6% NaOH	74.42	2.21	9.11	0.24	3.48	670
8% NaOH	68.44	1.21	7.11	0.22	2.48	596

3.2. FTIR analysis

FTIR spectra peaks of raw and different alkali treated PJFs are showed in Fig. 1. The peaks were examined between 4000 and 500 cm^{-1}. From the peaks it is evident that the broad absorption bands from 3442 – 3100 cm^{-1} represented the hydrophilicity behaviour of raw and different alkali treated PJFs and is related to O-H stretching of α-cellulose. C-H stretching vibration of cellulose and hemicellulose contents are envied by two band peaks 2926 cm^{-1} and 2847 cm^{-1} respectively [3, 15]. The presence of wax or wax like gist was noticed from a visible peak 2342 cm^{-1} that demonstrates the C=C stretching. The two peaks 1750 cm^{-1} corresponded to the CH3COO bending of hemicellulose, while the peak 1608 cm^{-1} indicated the C-O stretching of lignin. Spectrum peak 1020 cm^{-1} attributes to the condensed hemicellulose, lignin and wax contents owing to alkali treatment of PJFs [19-20].

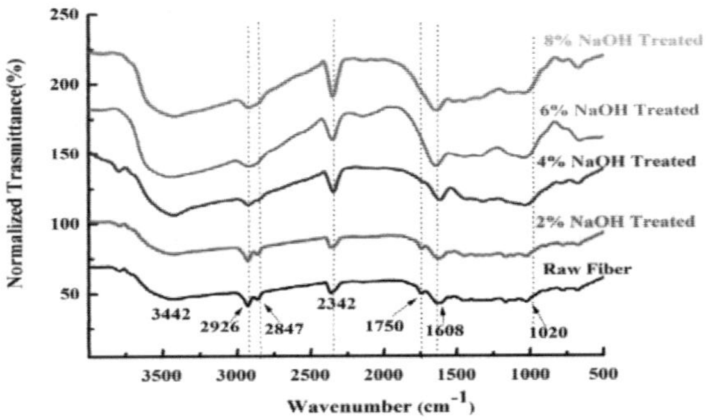

Fig. 1. FTIR analysis of raw and different alkali treated PJFs.

3.3. XRD analysis

Fig. 2 reveals the XRD patterns of raw and different alkali treated PJFs. The main crystalline peak for different alkali treated PJFs was observed at $2\theta = 22.67^0$ while the peaks at $2\theta=15.07^0$ and $2\theta=16.90^0$ symbolized the typical diffraction pattern of cellulose-I. The slightest intensity peaks were seen at $2\theta = 18.12^0$. These peaks specified the existence of larger amount of amorphous contents in PJFs. The CI values of raw and alkali treated PJFs were established using Eqn. (1) and are of the order: 6% NaOH treated PJFs (56.64%) > 8% NaOH treated PJFs (54.42%) > 4% NaOH treated PJFs (52.52%) > 2% NaOH treated PJFs (48.97%) > raw PJFs (47.3%). This improvement of CI values in treated fibers were mainly due to (i) chemical treatment which diminished the amorphous contents as well as the non-crystalline structure in the fibers; (ii) rearrangement of crystalline structure of the fibers. The CS of raw and treated PJFs was evaluated using Eqn. (2) is in the order: 6% NaOH treated PJFs (15.76 nm) > 8% NaOH treated PJFs (14.43 nm) > 4% NaOH treated PJFs (13.43 nm) > 2% NaOH treated PJFs (12.15 nm) > raw PJFs (9.38 nm). From this order, it is evident that optimally treated PJFs show higher CS values compared to raw PJFs which makes the optimally treated PJFs to have good moisture absorption and chemical reactivity characteristics [21].

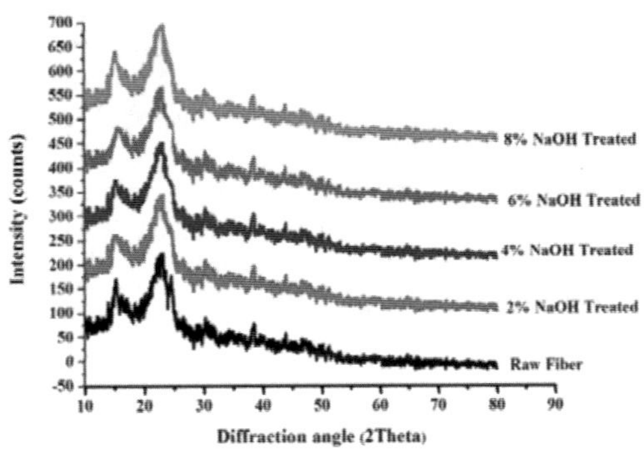

Fig. 2. XRD diffractograms of PJFs.

3.4. TG analysis

The primary thermograms of the raw and different alkali treated PJFs are depicted in Fig. 3. The thermal degradation process arises in three different stages and it was noticed that all the thermograms showed an initial weight loss at a temperature of 100^0 C attributed to the loss in absorbed moisture of fibers. A weight loss of around 10% was observed at this temperature range. To start, first stage of degradation happened at temperature ranges 50^0 C to 125^0 C and 75^0 C to 125^0 C for raw and alkali treated PJFs and corresponds to the elimination of moisture content and other volatiles. The loss of weight in these temperature ranges was about 30%. The second stage occurred at the temperatures 220^0 C to 300^0 C and 220^0 C to 330^0 C for raw and treated PJFs and corresponded to the medium temperature degradation stage indicating the removal of hemicelluloses and some amount of lignin. Third and final stage of temperature ranging in between 300^0 C to 400^0 C represented the degradation of lignin and α-cellulose. Around this temperature, it was seen that approximately all the lignin and cellulose contents were pyrolyzed. The decomposition of lignin extended through the entire temperature ranging up to 700^0 C [6, 15].

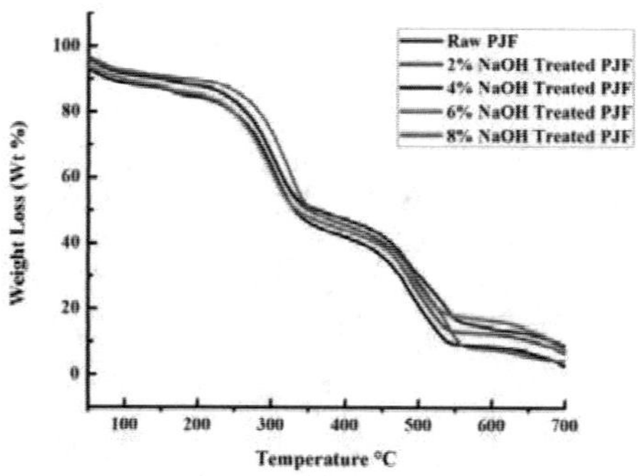

Fig. 3. TG curves of raw and different alkali treated PJFs.

3.5. Tensile testing

The chemical composition of any natural fiber will have a significant effect on the tensile strength and is depends on percentage of cellulose content present on it. Table 2 gives the data of various tensile properties of raw and different alkali treated PJFs. From the results it was clear that the tensile strength and percentage of elongation at break of the treated PJFs were superior to raw PJFs. This was mainly because of the rupturing of alkali sensitive bonds occurring between the cellulose, amorphous hemicellulose and also due to stress distribution between interfibrillar regions [22]. The diameter of alkali treated PJFs has lessened as compared to raw PJFs because of the inward swelling in the cell walls of alkali treated PJFs and in turn resulting in better tensile strength [23]. All above results prove that the alkali-treated PJFs could be regarded as better reinforcement material in green composites.

Table 2. Tensile properties of raw and different alkali treated PJFs.

Fiber	Diameter (µm)	Tensile strength (MPa)	Elongation at break (%)
Raw	20	570 ± 8.4	1.55 ± 0.12
2% NaOH	18.4	590 ± 6.4	1.78 ± 0.08
4% NaOH	17.42	632 ± 7.6	2.28 ± 0.06
6% NaOH	16.54	650 ± 6.43	3.54 ± 0.06
8% NaOH	14.65	640.52 ± 6.4	3.22 ± 0.07

4. Conclusions

In this study, raw and different alkaline treated PJFs were exemplified through the chemical analysis, FT-IR, XRD, thermogravimetric, and mechanical testing methods. The PJFs treated with 6% (w/v) NaOH treatment was discovered to have optimal properties which was justified from their high cellulose content, lower amorphous contents (hemicellulose, lignin and wax) and lower moisture content which improves interfacial bonding between fiber and matrix making them better reinforcements in polymer matrices. FT-IR spectroscopy revealed the lessening of amorphous contents in treated PJFs. XRD analysis revealed the improved crystallinity index and crystallite size in optimally treated PJFs thereby improving the crystalline structure and moisture absorption in these fibers. Alkaline treatment improved cellulose degradation temperature of PJFs which was conferred by thermal stability study and also density increased due to alkaline treatment. Finally, from this analysis it can be concluded that PJFs have potential characteristics to be effectively used as reinforcement materials in manufacturing of composite applications.

References

[1] Sathishkumar TP, Navaneethakrishnan P, Shankar 0. Tensile and flexural properties of snake grass natural fiber reinforced isophthallic polyester composites. Composites Science and Technology. 2012 Jun 8;72(10):1183-90.
[2] Seki Y, Sarikanat M, Sever K, Durmuşkahya C. Extraction and properties of Ferula communis (chakshir) fibers as novel reinforcement for composites materials. Composites Part B: Engineering. 2013 Jan 1;44(1):517-23.
[3] Hyness NR, Vignesh NJ, Senthamaraikannan P, Saravanakumar SS, Sanjay MR. Characterization of new natural cellulosic fiber from heteropogon contortus plant. Journal of natural fibers. 2018 Jan 2;15(1):146-53.
[4] Jawaid MH, Khalil HA. Cellulosic/synthetic fibre reinforced polymer hybrid composites: A review. Carbohydrate polymers. 2011 Aug 1;86(1):1-8.
[5] Sanjay MR, Madhu P, Jawaid M, Senthamaraikannan P, Senthil S, Pradeep S. Characterization and properties of natural fiber polymer composites: A comprehensive review. Journal of Cleaner Production. 2018 Jan 20;172:566-81.
[6] Madhu P, Sanjay MR, Senthamaraikannan P, Pradeep S, Saravanakumar SS, Yogesha B. A review on synthesis and characterization of commercially available natural fibers: Part-I. Journal of Natural Fibers. 2018 Apr 6:1-3.
[7] Saravanakumar SS, Kumaravel A, Nagarajan T, Moorthy IG. Effect of chemical treatments on physicochemical properties of Prosopis juliflora fibers. International Journal of Polymer Analysis and Characterization. 2014 Jul 4;19(5):383-90.
[8] Ray D, Sarkar BK. Characterization of alkali-treated jute fibers for physical and mechanical properties. Journal of Applied Polymer Science. 2001 May 16;80(7):1013-20.
[9] Mwaikambo LA, Ansell MP. Mechanical properties of alkali treated plant fibres and their potential as reinforcement materials. I. Hemp fibres. Journal of materials science. 2006 Apr 1;41(8):2483-96.
[10] Valadez-Gonzalez A, Cervantes-Uc JM, Olayo RJ, Herrera-Franco PJ. Effect of fiber surface treatment on the fiber–matrix bond strength of natural fiber reinforced composites. Composites Part B: Engineering. 1999 Apr 1;30(3):309-20.
[11] Das M, Chakraborty D. Evaluation of improvement of physical and mechanical properties of bamboo fibers due to alkali treatment. Journal of applied polymer science. 2008 Jan 5;107(1):522-7.
[12] Reddy KO, Ashok B, Reddy KR, Feng YE, Zhang J, Rajulu AV. Extraction and characterization of novel lignocellulosic fibers from Thespesia lampas plant. International Journal of Polymer Analysis and Characterization. 2014 Jan 31;19(1):48-61.
[13] Madhu P, Sanjay MR, Senthamaraikannan P, Pradeep S, Siengchin S, Jawaid M, Kathiresan M. Effect of Various Chemical Treatments of Prosopis juliflora Fibers as Composite

Reinforcement: Physicochemical, Thermal, Mechanical, and Morphological Properties. Journal of Natural Fibers. 2018 Oct 20:1-2.

[14] Segal LG, Creely JJ, Martin Jr AE, Conrad CM. An empirical method for estimating the degree of crystallinity of native cellulose using the X-ray diffractometer. Textile Research Journal. 1959 Oct;29(10):786-94.

[15] Saravanakumar SS, Kumaravel A, Nagarajan T, Moorthy IG. Investigation of physico-chemical properties of alkali-treated Prosopis juliflora fibers. International Journal of Polymer Analysis and Characterization. 2014 May 19;19(4):309-17.

[16] Roy A, Chakraborty S, Kundu SP, Basak RK, Majumder SB, Adhikari B. Improvement in mechanical properties of jute fibres through mild alkali treatment as demonstrated by utilisation of the Weibull distribution model. Bioresource technology. 2012 Mar 1;107:222-8.

[17] Rwawiire S, Tomkova B. Morphological, thermal, and mechanical characterization of Sansevieria trifasciata fibers. Journal of Natural Fibers. 2015 May 4;12(3):201-10.

[18] Ramanaiah K, Ratna Prasad AV, Chandra Reddy KH. Thermal and mechanical properties of sansevieria green fiber reinforcement. International Journal of Polymer Analysis and Characterization. 2011 Nov 1;16(8):602-8.

[19] Fiore V, Valenza A, Di Bella G. Artichoke (Cynaracardunculus L.) fibres as potential reinforcement of composite structures. Composites Science and Technology. 2011 May 31;71(8):1138-44.

[20] Sathishkumar TP, Navaneethakrishnan P, Shankar S, Rajasekar R. Characterization of new cellulose sansevieria ehrenbergii fibers for polymer composites. Composite Interfaces. 2013 Nov 1;20(8):575-93.

[21] Pan MZ, Zhou DG, Deng J, Zhang SY. Preparation and properties of wheat straw fiber-polypropylene composites. I. Investigation of surface treatments on the wheat straw fiber. Journal of applied polymer science. 2009 Dec 1;114(5):3049-56.

[22] Mukherjee A, Ganguly PK, Sur D. Structural mechanics of jute: the effects of hemicellulose or lignin removal. Journal of the Textile Institute. 1993 Jan 1;84(3):348-53.

[23] Van de Weyenberg I, Ivens J, De Coster A, Kino B, Baetens E, Verpoest I. Influence of processing and chemical treatment of flax fibres on their composites. Composites Science and Technology. 2003 Jul 1;63(9):1241-6.

IOP Conf. Series: Materials Science and Engineering **653** (2019) 012017 doi:10.1088/1757-899X/653/1/012017

Investigation of the Machining Performance of Neem Oil as a Dielectric Medium of EDM: A Sustainable Approach

Shirsendu Das[1*] *Swarup Paul*[1] *Biswanath Doloi*[2]
[1,]Production Engineering Department NIT Agartala, India-799046.
[2]Production Engineering Department Jadavpur University Kolkata, India-700032

Abstract: The Electric Discharge Machining is one of the popular non-traditional approaches which has huge utility in industrial sectors, mostly in die-making industries and as well as in bio-medical fields. But more power consumption, evolvement of unhygienic gases, development of non- biodegradable wastes, etc. are some curtail issues, which leads the poor sustainability of the EDM operation. In this approach, transesterified neem is introduced as a new and green dielectric medium of EDM, and its impacts on EDM responses are examined. Here, machining is carried out between 4 A to 10 A ranges of input current, 100 μ-sec to 200 μ-sec of pulse ON time and 10 μ-sec to 30 μ-sec of pulse OFF time using both kerosene and transesterified neem as a dielectric medium. The responses like MRR, surface roughness, and microscopic views of machined surfaces are analyzed to check the viability of transesterified neem as a dielectric medium of EDM. It is experimentally observed that transesterified neem as a dielectric medium ensured 6.2 % to 15.6 % more removal rate and 12.25 % to 15.45 % less surface roughness than kerosene.

1. Introduction

Electric or electro Discharge Machining (EDM) is one of the most widely using non-traditional machining approach, where thermo-electric fusion is responsible for metal erosion. Because of its easy accessibility and low cost Kerosene is commercially used as a dielectric medium for last few decades. But considering the future insufficiencies of hydro-carbon based fossil oils and the issue of evolvement of unhygienic monoxides during EDMing, diverge a new trend to search a better substitute without compromising with the machining performance. According ISO-14000, sustainability is one of the premier issue so far machining or manufacturing is concerned which balanced both the commercial and environmental aspects. Valaki et al. used jatropha bio-dielectrics to machine steel with copper tool and observed that MRR varies proportionally with the increment of pulse current like conventional dielectric medium and maximizes up to 50 mm^3/ min where as in case of conventional it is limited up to 30 mm^3/ min. The variations of MRR with gap voltage and pulse off time are almost same in case of jatropha and kerosene. Basically in case of both dielectric medium MRR is decreasing. But a significant variation is observed in case of MRR variation with pulse ON time. In case of kerosene it first increases up to peak value of 15 mm^3/ min and then it reduces. In case of jatropha it gradually increases and maximize up to 20 mm^3/min. Regarding the issue of surface properties, it is observed that if the pulse current and pulse ON time increase, indentations on the machined surface also increase i.e poor surface finishes have been achieved for both jatropha and kerosene. But for jatropha the maximum value of surface roughness (12 microns) is less than kerosene (16-17 microns). Both the kerosene and jatropha show almost similar types of surface properties when the gap voltage is increased. But in case of the variation of pulse OFF time jatropha shows much better surface integrity with less surface indentations (8 microns) compared to kerosene (12 microns) [1]. Apart from this high electrostatic field (i.e Voltage) and liberated heat make the channel so reactive that it continuously emits toxic monoxides and harmful micro-waste during the machining. So along with industrial aspects machining should be environment friendly so that it can maintain the sustainability to. Recently some multi-criteria based simulation efforts have been found in this context which may balance both the commercial and environmental issues [2]. Dielectric flushing is one of the prime activity of fluidize medium which wash out micro-chips along with heat absorption, otherwise these micro-particle clotted inside the spark gap and collapse the plasma channel. Vegetable oils have higher magnitude of density than kerosene oil so, while vegetable oils move with high flushing velocity can wash out microchips more efficiently than kerosene because of their high momentum and large impact on the particles. It is observed that most of the vegetable oils have remarkably high break down strength, which implies, they required high external voltage to channelize charged particle within the spark gap. Valaki et al. used both traditional dielectric and vegetable based dielectric and observed vegetable based oil had shown significant contribution in terms of impressive MRR and surface properties with comparatively less tool wire. During the experimentations of waste vegetable oils with copper tool and steel work piece it is observed that the nature and trend of the graph of MRR with respect to pulse current is almost similar to jatropha but the only difference is it maximizes up to 90 mm^3/min. Similar trend of jatropha had also been observed while gap voltage and pulse OFF gap are varied and MRR significantly falls up to its minimum value. Slightly different trend has been found while pulse ON gap had varied and it is observed MRR becomes constant beyond the limit of 50 μSec but before that it gradually increased with pulse ON timing. On the other hand comparative analysis of MRR for kerosene and canola oil with high/ low energy setting had shown that experimentally canola had shown better response than kerosene. Both canola and waste vegetable oil had been introduced as new dielectric medium in this aspect [4, 5].Sunflower oil, which has slightly low flash point and high dielectric constant also used as dielectric medium of EDM with different range of pulse current and pulse time gap and found that, in every aspect it had shown less tool consumption than kerosene oil as well as canola impressive MRR and surface integrity. Sunflower oil also shown almost similar characteristics and it is slightly lower than canola but much better than kerosene in experimental stage. Canola also shows impressive surface integrity with lesser roughness than kerosene so far EDMing is concerned [6]. Impressive responses are observed in terms of electrode consumption or tool wire for these experimented vegetable oils with compare to kerosene. Here, in this work neem vegetable oil is introduced as a new dielectric fluid of EDM operation and the effects of input variables are observed on response parameters like MRR and SR. Apart from these SEM analysis also performed to check the micro-structures of the EDMed surface with both the dielectric mediums.

2. Experimental Methodology

2.1. Preparation of dielectrics

Neem is a vegetable based oil derived from Azadirachta indica tree which is enriched with very high quantity of Free Fatty Acids (FFA). During transesterification process of this FFA-oil first acidic and then alkaline types of catalyst are added and finally nano-ester and glycol are formed [7]. To prepare suitable blend of Neem oil for the operation of EDM, first it is heated up to 75-80^0 C and then methanol solution is added with NaOH as catalyst. After that H_2So_4 catalyst is added in the solution. After 20-24 hour precipitation glycol and methyl-esters are formed. Then the ester is collected in a separate container and used as dielectrics after few chemical processing. The physio-chemical properties of the prepared dielectric are presented in table.1.

Table. 1: Properties of kerosene and neem dielectrics.

Sl. No.	*Properties*	*Kerosene*	*Neem Blend*
1.	Density (gm/cm^3)	0.802	0.868

2.	Viscosity (cm^2/Sec)	1.219	4.3-4.4
3.	Thermal Conductivity (W/mK)	0.128	0.15
4.	Specific Heat (kJ/kgK)	2.01	1.962
5.	Break Down Voltage (kV)	48	27
6.	Dielectric Constant	2.113	3.18

2.2. Experimental procedure

Here titanium alloy 'Ti6Al4V' and electrolytic copper bar of diameter 10 mm have been used as work piece and tool materials. Sparkonix EDM machine assisted with Servomax pulse generator is used for experimental purposes. A small chamber of dimension 30cm×20cm×15cm and a separate pump arrangement are preferred for experimentations. During the experiment the 4 A to 10 A current, 100 μ-sec to 200 μ-sec T_{ON}, 10 μ-sec to 30 μ-sec T_{OFF}, 15 kg/ cm^2 flushing pressure and positive tool polarity were preferred. The obtained results are listed in table. 2.

Table.2: Experimental results

Sl.No	Pulse current (A)	T_{ON} (μ-sec)	T_{OFF} (μ-sec)	MRR-Kerosene (mm^3/ min)	SR-Kerosene (μ-m)	MRR-Neem (mm^3/ min)	SR-Neem (μ-m)
1.	4	100	10	2.837	3.046	2.602	2.523
2.	4	150	20	3.021	3.958	2.897	3.407
3.	4	200	30	3.647	4.572	3.526	4.095
4.	7	100	20	4.214	5.153	4.475	4.608
5.	7	150	30	4.579	5.424	4.807	4.897
6.	7	200	10	5.627	6.538	5.753	5.736
7.	10	100	30	5.082	5.901	5.682	5.049
8.	10	150	10	5.496	6.479	6.179	5.625
9.	10	200	20	6.138	7.385	7.095	6.723

3. Results and Discussions

3.1. Impact of process parameters on the responses

Fig. 1 has represented the variation patterns of removal rate with respect to process parameters for both kerosene and transesterified neem. Fig 1(a), represents the trends of MRR with respect to current for both the dielectrics when the other process parameters like gap voltage, pulse ON and OFF times are 100 V, 200 μ-sec and 20 μ-sec.

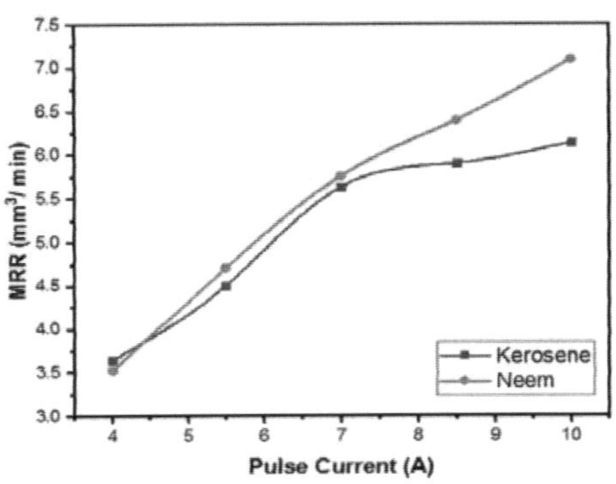

(a)

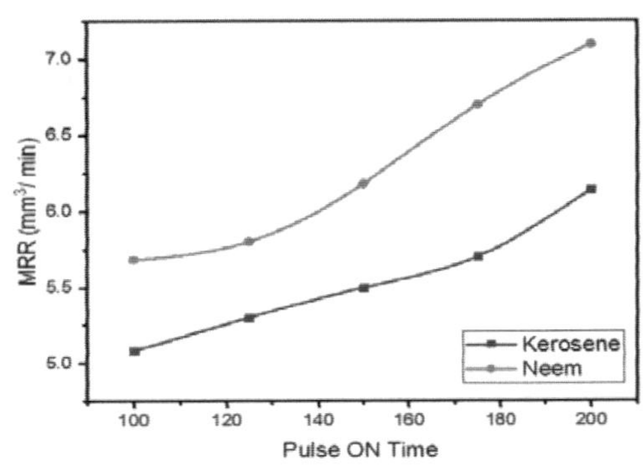

(b)

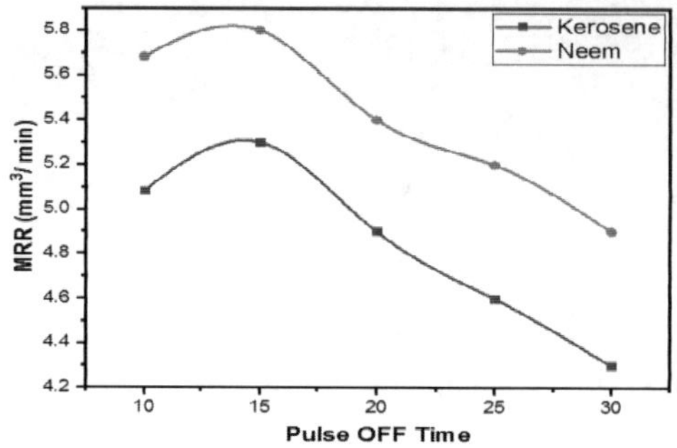

(c)

Fig 1: Variations of material removal rate with pulse current (a), with pulse ON time (b) and with pulse OFF time when gap voltage is 100 V.

On the other hand, variation of MRR with pulse ON time is shown in fig 1(b), when other process parameters are 100 V (Gap voltage), 10 A (pulse current) and 20 μ-sec (pulse OFF time). The increment in MRR may be due to the availability of huge discharge energy with the increased value of current and pulse ON time. The obtained results claim that neem dielectrics exhibits better MRR than kerosene. This is may be due to higher oxygen content, temperature dependent conductivity and comparatively higher density of this oil than kerosene which helps in quick completion of melting and vapourizing [1,3,5]. The variations of roughness with process parameters have shown in fig 2. In each case, transesterified neem possesses less surface roughness than kerosene when experiments are performed under same conditions. Basically, higher viscosity of neem oil ensures the adhesive wetness of the surface which causes comparatively better surface finish. In first case (fig. 2a), variation of SR is plotted with respect to pulse current when the other process parameters like pulse ON time, pulse OFF time and gap voltage are 200 μ-sec, 20 μ-sec and 100 V.

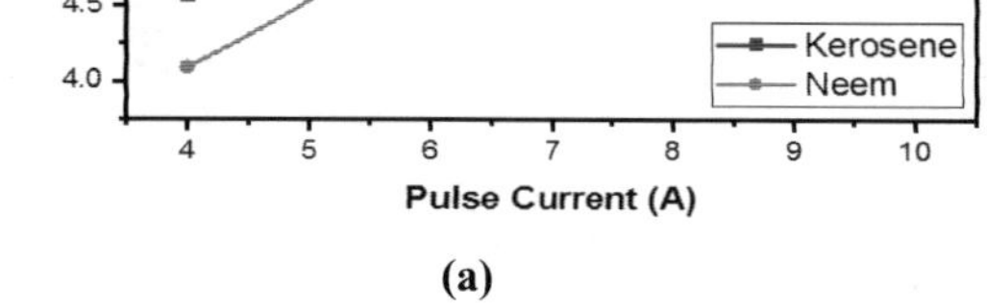

(a)

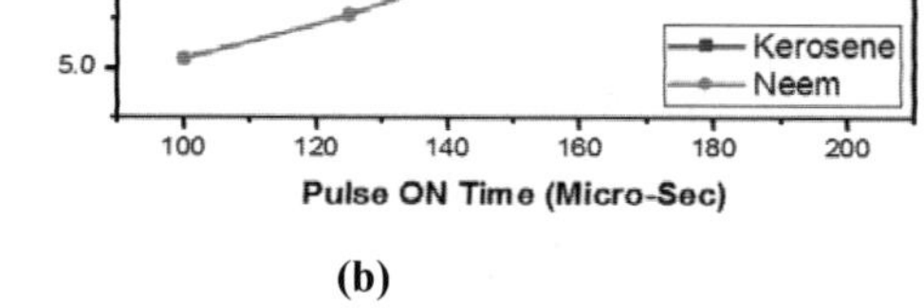

(b)

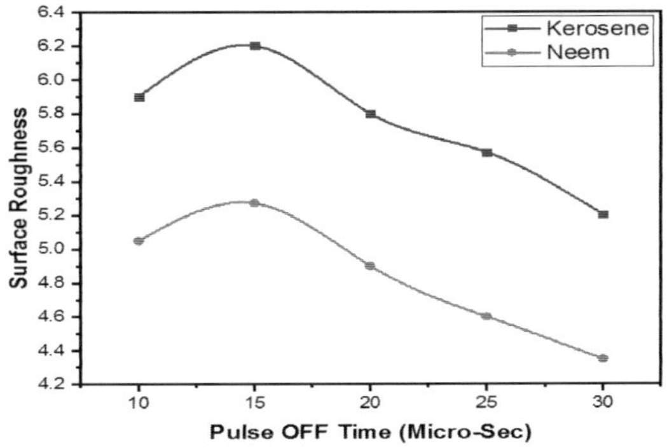

(C)

Fig 2: Variations of surface roughness with pulse current (a), with pulse ON time (b) and with pulse OFF time when gap voltage is 100 V.

Here the SR is proportionally increases with the pulse current because the available spark energy increases with pulse ON time. On the other hand fig. 2(b), represent the variation of SR with pulse ON time when pulse current, pulse OFF time and gap voltages are 10 A, 20 μ-sec and 100 V. In this case of SR increases proportionally with pulse ON time because of rapid increase in spark energy. But for fig. 2(c), significant reductions are observed with pulse OFF time, because the available discharge energy reduces with the gradual increments in pulse OFF time. But in all three cases the surface roughness developed with neem dielectric are comparatively lesser than kerosene. This is may be due to higher viscosity and decent dielectric constant of the used medium which ensures better energy confinement, wettability and impressive susceptibility [1,3].

3.2. Microstructure and elemental compositions

Fig. 3 shows the SEM images of the surfaces machined with both the dielectrics at 100 V of gap voltage, 7 A of pulse current, 200 μ-sec of pulse ON time and 20 μ-sec of pulse OFF time. The obtained images apparently reveal the presence of irregularities like indentations, scattered tiny holes, indiscipline overlapping of craters. Comparative analysis of these two surfaces clearly disclosed that the surface processed with neem has comparatively better surface integrities than that of the kerosene.

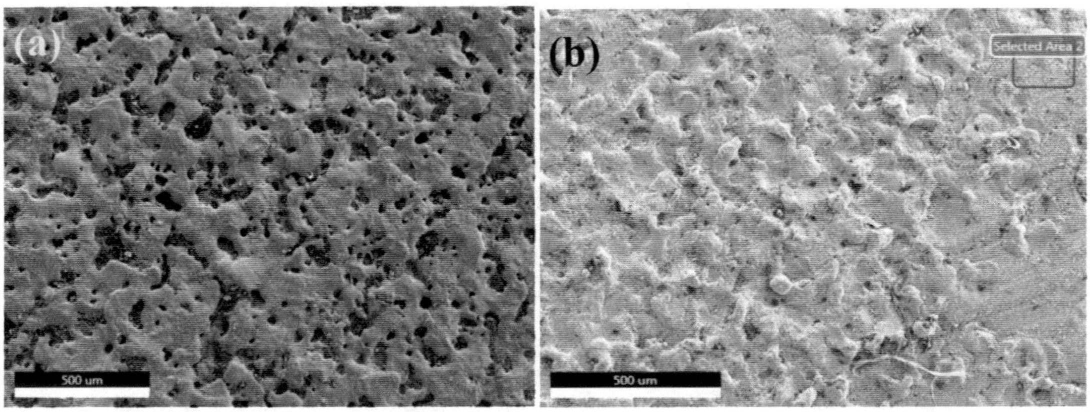

Fig 3: SEM images of the EDMed surfaces processed with kerosene (a) and neem (b).

IOP Conf. Series: Materials Science and Engineering **653** (2019) 012017 doi:10.1088/1757-899X/653/1/012017

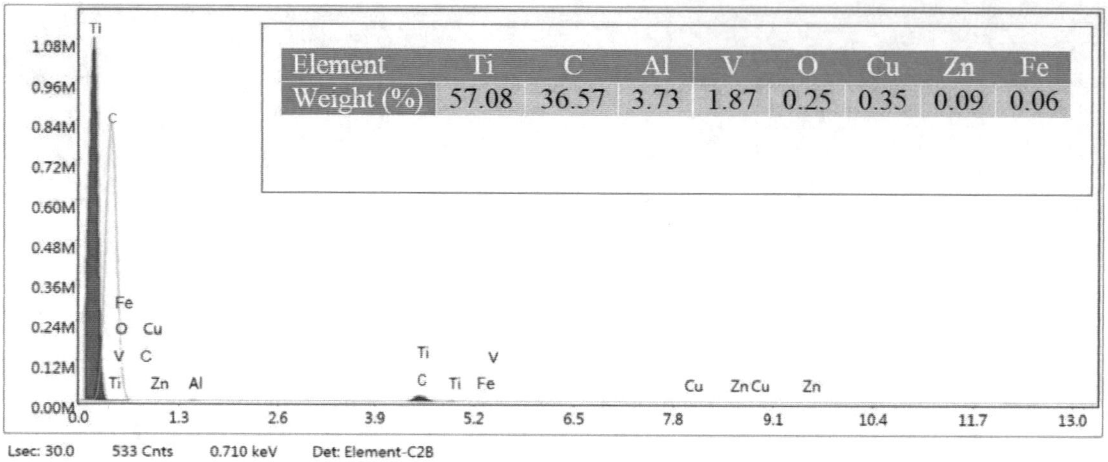

Element	Ti	C	Al	V	O	Cu	Zn	Fe
Weight (%)	57.08	36.57	3.73	1.87	0.25	0.35	0.09	0.06

Fig. 4: EDX analysis of the machined surface processed with kerosene.

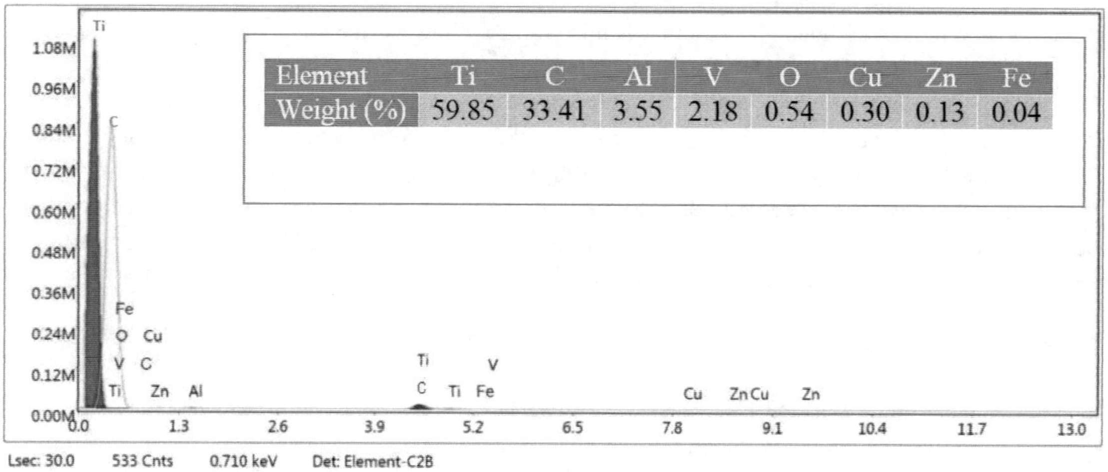

Element	Ti	C	Al	V	O	Cu	Zn	Fe
Weight (%)	59.85	33.41	3.55	2.18	0.54	0.30	0.13	0.04

Fig. 5: EDX analysis of the machined surface processed with neem.

Apart from these EDX analysis is perform to check the percentile amount of migrated particles from tool and dielectrics. Fig. 4 and 5 shows the EDX patterns of EDMed surface using kerosene and neem as dielectric medium. It is noticed that 36.57 % of C, 0.35 % of Cu, 0.09 % of Zn and 0.06 % of Fe have been migrated from dielectrics and tool materials if kerosene is used as dielectrics. For neem the trends of material migrations are almost similar with kerosene just slightly deviate in percentages. Therefore the migration of 33.41 % of C, 0.30 % of Cu, 0.13 % of Zn and 0.04 % of Fe have been observed for the sample which were processed with neem dielectrics.

4. Conclusions

This article addressed different issues like non-biodegradability of developed wastes, the emergence of gases due to the partial burning and oxide formation at high machining temperature. Apart from these, the machining performances of transesterified neem has been evaluated here. So based on the observations following conclusions can be drawn.

(1) Because of the high flash point and fire point, the chances of partial burning are less in the case of neem oil than kerosene. So neem dielectrics can minimize the probable chances of gas emission during machining and helps to transform an oxide-free hygienic breathing environment.
(2) Vegetable oils are biodegradable, so the semi-solid waste which will develop after machining can be degraded by natural processes without the addition of extra chemical reagents. So the neem dielectrics take care of the environment and human health, which is very much expected so far sustainability of EDM operation is concerned.
(3) Comparative analysis of the machining performances of these two dielectrics gives a clear view of removal rate and surface properties under the same machining environment. It is observed that within 4.5 A to 10 A ranges of pulse current transesterified neem possesses 6.2 % to 15.6 % more removal rate and 12.25 % to 15.45 % less surface roughness than kerosene.
(4) The elemental composition of the EDMed specimen claims the immigration of carbon, copper, zinc, and iron from the dielectrics and tool materials. It is noticed that the migration of C and Cu is less in the case of the sample processed with neem.

Therefore, based on the outcomes, transesterified neem can be recommended as a potent substitute for conventional kerosene considering the sustainability issues of EDM.

References

[1] Valaki J B, Rathod P P, Sankhavara C D 2016 Journal of Manufacturing Processes p.151–160.
[2] Garg A, Lam J S 2016 Journal of cleaner production p. 1588-1601.
[3] Valaki J B, Rathod P P, 2015 Int. J. Adv. Manuf. Technol. p. 1509-1518.
[4] Kibria G, Sarkar B R, PradhanB B, Bhattacharyya B 2010 Int J Adv Manuf Technol p. 557–570.
[5] Ng P S, Kong S A, S. Yeo S H 2016 Int. J. Adv. Manuf. Technol. p. 2549- 2556.
[6] Djibril D, Mamadou M, Gérard V 2015 Research Journal of Chemical Sciences p. 52-58.
[7] Nair J N, Kaviti A K, Daram A K 2016 Egyptian Journal of Petroleum.

IOP Conf. Series: Materials Science and Engineering **653** (2019) 012018 doi:10.1088/1757-899X/653/1/012018

Effect of Mn doping on the Optical and Electrical Properties of TiO_2 thin film prepared by Unconventional Sol-gel route

[1]Maya Devi and [2]M.R.Panigrahi
[1] School of Applied Sciences, KIIT Deemed to be University, Bhubaneswar, India
[2] Dept. of Physics,VSSUT, Burla, India.
mdevifpy@kiit.ac.in mrpanigrahi@vssut.ac.in

Abstract. Semi-transparent nanostructured Mn doped titanium dioxide thin films have been prepared by modified sol-gel route starting from the powder reagents. The films deposited on plane glass slides is annealed at 350 ^{0}C. The characterizations of the films are done with X-ray diffraction, Scanning electron microscopy, UV-Vis spectroscopy and Hall measurement. Although annealing transformed the structure from tetragonal body centered to tetragonal primitive, the preferential growth have been reported along the (300) orientation of the diffracting planes. The optical transmittance of the film increases with annealing temperature. The optical parameters like refractive index and extinction coefficients are calculated from the transmission and reflection spectra, and the dispersion in the refractive index has been discussed in terms of single oscillator Wemple and Di Domenico model.

Key words: Mn doped titanium dioxide, Optical transmittance, Optical parameters, Dispersion in refractive index

1. Introduction:

There has been a demand of high performance, low cost semiconductors; which can transform maximum part of solar light to electric energy. TiO_2 is an environmentally benign semiconductor, which has drawn maximum attention due to its chemical stability and wide applications starting from photocatalytic to sensing nature [1]. But, the main demerit is its high band gap which limits its efficiency to harness only 5% of the solar radiation approximately. So, attempts are being made to dope several foreign metals and non metals to maximise its absorption into visible region [2, 3]. Studies show that the properties of the doped titania mainly depend on the concentration and structure of the nature of incorporated metal ion. Out of all the transition metals Mn doping is of much interest because of its high theoretical capacitance value due to reversible oxidation state that changes during charging and discharging process [4,5] and the capability to change the magnetic properties by suitable doping procedure[6]. Theoretical calculation also shows that Mn doping induces curvy intermediate bands which are similar as regular arrangement of quantum wells or dots resulting in significant reduction in the band gaps [7, 8]. So in this work it is tried to prepare the Mn doped TiO_2 thin films with low concentration of doping (0.01%) and to study their microstructure, optical and electrical properties. Here in this paper different optical parameters of the studied material are presented since refractive index and energy band gap are the important parameters for the determination of optical and electronic behaviour of semiconductors used for optoelectronic applications. Sol-gel process is usually a wet chemical process in which the liquid precursors are taken and added with suitable solvents and by aggregation process the solid phase formation is mainly controlled by polymerization and polycondensation process, whereas in modified sol-gel method the precursor is added with a chelating agent like acetic acid/ citric acid containing -COOH group which act as binding ligands and form polynuclide complexes with metals usually by etserification process. The advantage of this process over the traditional sol-gel process is ease of fabrication, cost effectiveness and higher degree of homogeneity Thus it is chosen over the traditional sol-gel process

2. Experimental:

TiO_2 and Mn doped TiO_2 thin films were deposited on clean microscopic glass slide by unconventional sol-gel route starting from the powder precursors as reported in our previous paper [9]. High purity powder precursors of TiO_2 , MnO_2 (Merck), Acetic acid and Nitric acid were used as the starting materials. Lump free, uniform paste was made by mixing the stochiometrically calculated proportion of precursor materials with acetic acid. The clear sol was obtained by hydrolysis of the paste, with slow addition of DI water with continuous stirring. Vigorous stirring of the sol was done at 500 rpm with a magnetic stirrer, followed by refluxing with 0.1M nitric acid. Continuous stirring was done at a temperature of 180°C for almost 8-9 hours until the gel was formed. Thin films were deposited on ultrasonically cleaned glass plate, using the prepared gel by doctor blade method and were annealed at 350°C. The structural characterizations of the films were done by X-ray diffractometer (Schimandzu-6100), with a step size of 0.02° in the angular range 20-80°.

Using the UV-Vis photo spectrometer (Schimadzu-2540), the transmittance and reflectance data were recorded in 300-900nm range.The thickness of the films were measured by Filmetrics-F-10. The electrical properties of the films were measured by Ecopia HMS 3000 Hall measurement setup.

3. Result and Analysis:

The X-ray diffraction studies of the pure TiO_2 and Mn doped TiO_2 films 'figure-1' reveal that no extra characteristic peak of MnO_2 is observed which indicates the elemental scattering of Mn in the TiO_2 matrix. Due to incorporation of Mn ion the crystal structure changed from tetragonal body centered space group *141acd* to tetragonal primitive space group *p4122*. The crystallite size corresponding to plane of highest intensity is calculated by Scherrer formula [10]. The calculated structural parameters are presented in table-1. A slight increase in crystallite size is observed in the doped sample which may be due to the strain. It is observed that the intensity of the peaks decreased along with a slight shift in the peak position with Mn doping. This supports the production of defects in the lattice by successful doping of Mn ion into TiO_2 crystal lattice and thereby destroying the symmetry of TiO_2 crystal structure [11, 12].

Table-1. Structural parameters

Sample	2θ(Degree)	Lattice planes			d-spacing(Å)	Crystallite size (nm)	Lattice strain	Lattice vectors		
		h	k	l				a(Å)	b(Å)	c(Å)
Pure TiO_2	25.14	1	0	1	3.54	11.30	.807	3.79	3.79	9.52
Mn doped TiO_2	25.48	0	0	3	3.49	14.14	.639	6.17	6.17	8.56

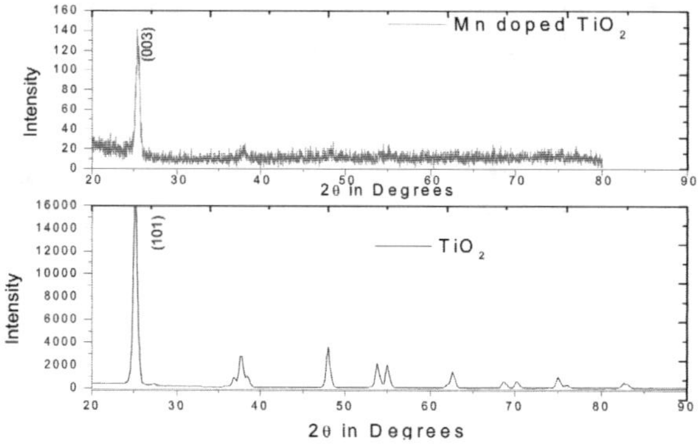

Figure.1 X-ray diffractogram of Pure and Mn doped TiO_2 samples

The optical properties of the films are studied by taking the absorbance, transmittance and reflectance data at normal incidence within the wavelength range 300-900 nm. Figure -2 shows that the optical absorption for both the pure and doped films are high below 600 nm. The red shift is observed in case of the doped film which may be due to the differences in the surface states or the effect of doping on the textural parameters. The doping of Mn ion into the matrix of TiO_2 replaces some Ti ion and thereby charge compensating vacancy get created in the different lattice points, which is also observed for transition metal doped TiO_2 [13, 14]. The cause of red shift may also be due to new electronic state in middle of TiO_2 band gap and charge transfer between the conduction band of TiO_2 and dopant [15].This property is very useful in increasing the photocatalytic activity of the sample.

The optical band gaps of the samples were calculated from the absorption spectra using Tauc plot method [16]. The optical band gaps of the films are calculated by considering the direct transition. By extrapolating the linear part of Tauc plot 'figure3' and from the intercept with the energy axis the band gaps are calculated and presented in table-2. From this it is observed that the band gap of the doped sample is less than that of the pure sample. The TiO_2 valence band shape is related to the chemical state of oxygen and it is mainly occupied by the 2p states of O_2. Mn doping creates oxygen vacancy which may in turn change the peaks of valence band (VB) to platue and a doner level may be appearing near the conduction band (CB) which may be a reason of reduction in band gap.

The refractive indices (n) and the extinction coefficients (k) of the films corresponding to different wavelengths are calculated from the reflectance data using the formulae given in equation-1 and -2 respectively, where R represents the reflectance at normal incidence and α is the absorption coefficient [17]. Figures 4 and 5 show the variation of them with incident wavelength.

$$n = \frac{1+R^{0.5}}{1-R^{0.5}} \qquad (1)$$

$$k = \frac{\alpha\lambda}{4\pi} \qquad (2)$$

It is observed that the refractive index of Mn doped film is lower than the pure TiO_2 film thereby increasing its transparency. In Mn doped sample it is seen that n value is dependent on photon energy which may be due to the virtual electronic transition in the film. The extinction coefficient of the doped film is more than the pure film supporting the existence of electric dipoles and hence enhanced light absorption.

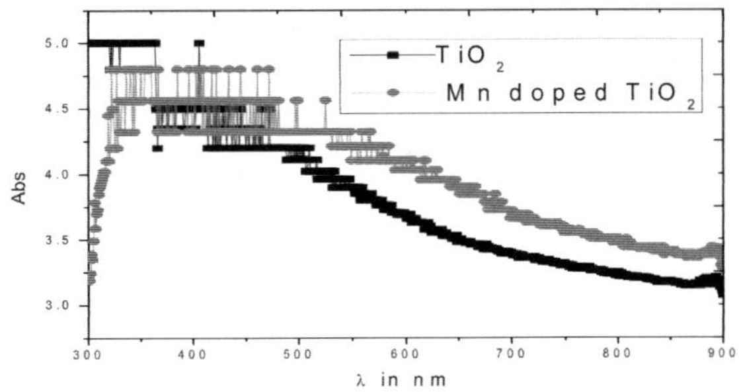

Figure.2 Absorption spectra

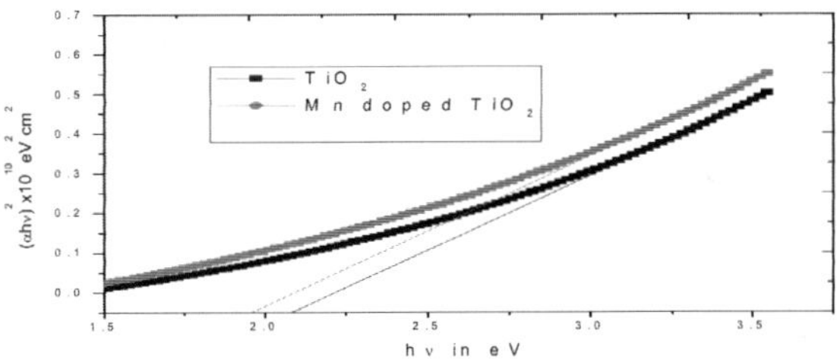

Figure.3. Tauc plot for band gap determination

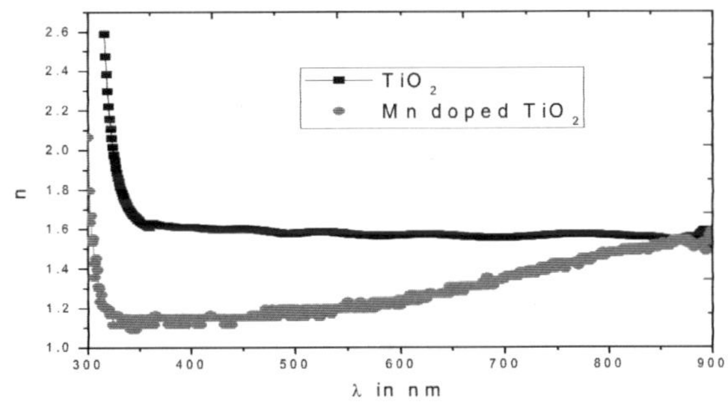

. **Figure.4** Variation of refractive indices with wavelength of incident photon

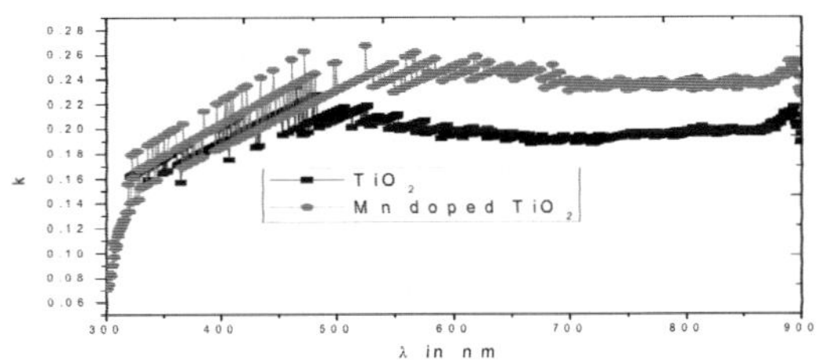

Figure.5. Variation of extinction coefficient with wavelength of incident photon

The refractive index dispersion is studied by fitting Wemple Di Domenico model which is presented in 'figure 6'.

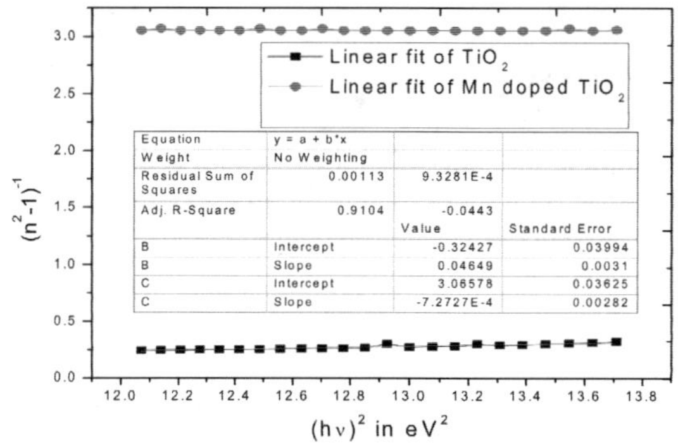

Figure.6 Wemple Di Domenico model fitting

The oscillation (E_0) and dispersion energy (E_d) is calculated plotting the linear fit between $1/(n^2-1)$ and $(hv)^2$and using the equation-3 presented in Table-2 .

$$(n^2-1) = \frac{E_d E_0}{E_0^2 - h\vartheta^2} \qquad (3)$$

It is observed that E_d which is a measure of strength of interband optical transition is higher in Mn doped sample in comparison to Pure TiO_2 sample. The oscillator energy which is a measure of average band gap is also calculated from the model fitting. The E_{opt} values obtained from model fitting is in agreement with the values obtained from Tauc plot.

Table-2. Optical band gaps

Sample	E_g in eV	E_o in eV	E_d in eV	E_o/E_d
Pure TiO_2	2.07	2.65	8.19	1.28
Mn doped TiO_2	1.98	2.09	21.58	1.06

Electrical property study of the films show that the conductivity of the doped film is more than that of the pure film as shown in table-3 which may be due to enhanced grain size and in Mn doped TiO_2 the Mn^{+2} dopant atoms may act as the electron donating centres. The mobility of the doped film may be reduced due to enhanced phonon scattering.

Table-3. Electrical Properties

Sample	Resistivity in Ohm-cm	Mobility in cm^2/Vs	Carrier concentration in cm^{-3}
Pure TiO_2	2.4×10^4	240	1.07×10^{12}
Mn doped TiO_2	3.8×10^3	94.6	1.74×10^{12}

4. Conclusion:

The structural, optical and electrical studies of the pure and Mn doped TiO_2 thin films are done. It is observed that there is a change in crystal structure with slight increase in crystallite size of the doped film. The band gap is found to be reduced in case doped film due to metal co-ordination. The band gap calculated from Tauc plot method is in good agreement with that calculated from Wemple Di Domenico method and which is approximately 5% more than the undoped one. The higher optical transition owe to the direct transition. The optical parameters such as refractive index and extinction coefficient are calculated which shows more dispersion in case of Mn doped film due to enhanced strength of interband optical transition. The electrical conductivity of the films also increased almost 85% time due to Mn doping.

References

[1] Ni M, Leung M K H, Leung D Y C and Sumathy K *Renewable and Sustainable Energy Reviews* 2007 **11(3)** 401
[2] Asahi R, Ohwaki K, Morikawa T, Aoki K and Taga Y *Science 2001 293(5528) 269*
[3] Weng H, Yang X, Dong J, Miuseki H, Kawasaki M and Kawazoe Y *Phys. Rev.B: Condens. Matter.* 2004 **69** 125219
[4] Toupin M, Brousse T and Belanger D *Chem. Mater.*2004 **16** 3184
[5] Lu X *et al.* Energy Environ. Sci 2011 **4** 2915
[6] Zhang L, Zhu L, Hu L, Li Y, Song H and Ye Z *RSC Adv* 2016 6 57403
[7] Shao G *J Phys Chem C* 2008 **112** 18677
[8] Shao G *J Phys Chem C.*2009**113** 6800
[9] Devi M, Panigrahi M Rand Singh U P *J Mat Sci: Mater Electron* 2015 **26(2)** 1186
[10] Langford J I , Louer D, Sonneveld E J. and Visser J W 1986 *Powder Diffraction* 1 211
[11] Momeni M M, Hakimian M, Kazem Pour *A. Surf.Eng.* 2016 **32** 514
[12] Shu Ye, Hyun-il Kim, Won-Chun Ohic *J. Korean Ceram, Soc.*2016 **53(3)** 343
[13] Santara B, Pal B, Giri P K *J.Appl.Physics* 2011**110.**114322
[14] Colis S, Bouaine A, Moubah R., Schimerber G, Ulhaq Bouillet C, Dinia A. et.al *J Appl. Physics* 2010 **108** 053910
[15] Wu J C S and Chen C H Journal of Photochemistry and Photobiology A 2004 **163(3)** 509
[16] Hemissi M, Adnani H A, PlenetJ C *Curr. Appl. Phys* 2009 **9** 717
[17] Hussein H F, Ghufran M S and Hashim S Sh *J.Mater. Environ. Sci* 2011 **2 (4)** 423

Probabilistic aspects of fatigue crack growth parameters under single overload in 6061-T6 aluminium alloys

Prakash Chandra Gope and Ankit Kabdwal

Mechanical Engineering Department, College of Technology,
G. B. Pant University of Agriculture & Technology, Pantnagar-263145, Uttarakhand, India

Email : pcgope@rediffmail.com
Tel: +91 9411159916; Fax: +91 5944 233338

Abstract: In the present investigation, several crack growth tests were performed on compact tension specimen under constant amplitude loading and single overload at an overload ratio of 1.8. The experimental data obtained were used to construct a-N, and da/dN vs ΔK curves and various fatigue crack growth parameters were determined and the variability associated with these parameters are addressed. Normal distribution, Weibull and extreme value distribution etc. were used to study the probabilistic nature of different crack growth parameters and best probabilistic distribution for each parameter are identified on the basis of statistical analysis. The probabilistic models of growth parameters are presented for the reliability assessment of a component under fatigue overload condition.

Keywords: Overload, Probabilistic model, Crack growth

1.0 Introduction

Fatigue crack growth under single or multiple over loading condition have been studied in past by many researchers in a deterministic manner. However, there is ample evidence which indicates variability in crack growth behaviour and different growth parameters of specimen of the same shape, size manufactured in an identical fashion from a same lot of material. Thus, fatigue crack growth is of statistical nature and it should be treated statistically. Many crack growth laws have been proposed so far to study the growth behaviour under single or multiple overloading conditions. Different crack growth models such as Willenborg, Wheeler crack growth model, Gallagher modified Willenborg crack growth model have been presented to account the crack growth retardation due to tensile overloads [1-8]. Many experimental results in support of these theories are found [9-19].

There are two main objectives of this study. One of the objectives is to study the variability in the crack growth behaviour due to application of single overload condition for 6061 –T6 aluminium alloys. The second objective is to present a probabilistic distribution of the various parameters used in modeling crack growth under such overload.

2.0 Material and Method

The commercially available 6061-T6 aluminium alloy is used in the present investigation due to its wide application in the field of aerospace, automobile, etc.. The chemical composition and mechanical properties of 6061-T6 aluminium alloy are presented in Table 1 and 2 respectively. The chemical properties are determined by EDS and the tensile properties from the tensile test.

Table 1. Chemical composition of Al 6061-T6.

Material	Al	Cr	Cu	Mg	Fe	Si	Zn
wt%	95.8	0.35	0.40	0.8	≤0.7	0.4	≤0.25

Table 2. Mechanical properties of Al 6061-T6.

Material	Ultimate tensile strength, (σ_{ut}), MPa	Yield strength (σ_{yt}), MPa	Young's modulus, (E), GPa	Poisson's ratio (ν)
Al 6061-T6	310.0	276.0	69.0	0.33

The mechanical properties of 6061-T6 aluminium alloy are determined by servo controlled universal testing machine (ADMET, USA) at 0.5 mm/min cross head speed under displacement control. The tensile specimens are made as per ASTM E8-99 standard [20] and shown in Fig. 1. Crack growth tests are conducted using compact tension specimens. Compact tension (CT) specimens were prepared from a 12.5 mm thick 6061-T6 aluminum-alloy plate. All CT and tensile specimens are prepared according to ASTM-E647 [21] and ASTM E8-99 using the EDM wire cut machine and polished to mirror finish by polishing machine. The dimensions of the CT specimen are shown in Fig. 2.

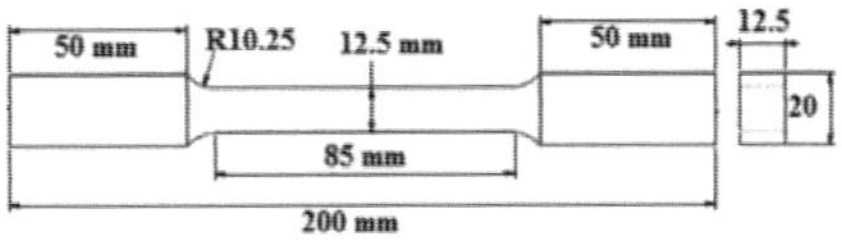

Fig. 1. Tensile specimen geometry

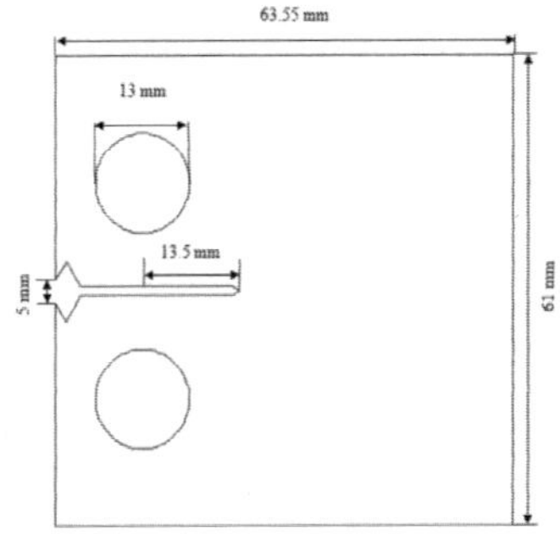

Fig. 2. CT specimen with all dimensions in mm

The fatigue crack growth tests are conducted in a ± 25 kN servo controlled hydraulic fatigue testing machine (HEICO, India) under load control mode. Pre-cracking was done with P_{max}=2 kN and P_{min}=0.2 kN and frequency of 10 Hz. An extension of crack length from 13.5 mm to 14.8 mm was made in the pre-cracking. Constant amplitude loading and single over load crack growth tests are conducted under sinusoidal signals with maximum load P_{max} = 5.0 kN and minimum load P_{min} =0.5 kN at 10 Hz. Crack growth was monitored by COD gauge. Crack growth test with single overload are conducted with an overload ratio (OLR) of 1.8 applied at 25001 cycles. Total of 30 specimens are tested under the similar test condition and $a-N$ and $da/dN-\Delta K$ curves are obtained. Detailed of the crack growth testing conditions are given in Table 3.

Table 3. Details of the crack growth testing parameters

Loading condition	Frequency (Hz)	Load (kN)		Stress ratio R	Overload (kN)	Overload ratio	Total specimen
		P_{max}	P_{min}				
Pre-crack test	10	2.0	0.2	0.1	-	-	33
CA loading	10	5	0.5	0.1	-	-	03
Tensile overload	10	5	0.5	0.1	9	1.8	30

The modelling of the crack growth data is made by frequently used Paris law [22] which states as:

$$\frac{da}{dN} = C(\Delta K)^m \tag{1}$$

a and N are the crack length and number of load cycles respectively. C and m are crack growth parameters determined from the test. The stress intensity factor range is denoted as ΔK and given by Eq. 2.

$$\Delta K = \frac{\Delta P}{BW} f(\alpha) \qquad \text{where ,} \tag{2}$$

$$f(\alpha) = [(2+\alpha)/(1-\alpha)^{3/2}]\,(0.886 + 4.64\alpha - 13.32\alpha^2 + 14.72\alpha^3 - 5.6\alpha^4)$$

and $\alpha = a/W$

Paris law is used here because of easy conversion to stochastic differential model just by considering at the model parameter as random variable or by adding a random factor into the Paris law. Yang's power law

IOP Conf. Series: Materials Science and Engineering **653** (2019) 012019 doi:10.1088/1757-899X/653/1/012019

[23], Yang and Manning model [24] are some of the probabilistic crack growth models derived from the Paris model and used by many researchers for stochastic studies of the crack growth data under constant amplitude loading. There are other models such as Wheeler model, Wallenberg model used by many researchers to address the issue of crack growth retardation or acceleration due to application of overload or under-load. To convert these deterministic models, the probability distribution of random crack size, random cycles to reach a given crack size, opening stress intensity factor, plastic zone size, etc. are required. The statistical aspect of most of the parameters used to characterise retardation effect due to overload is presented here.

3.0 Results and Discussion

a-N curves under constant amplitude (CA) loading for 3 samples are shown in Fig. 3. The Paris fitting coefficients *C* and *m* for CA and the statistical parameters mean, standard deviation and coefficient of variations are presented in Table 4. The $da/dN - \Delta K$ curves in log-log scale for three specimens are shown in Fig. 4. $a - N$ behaviour of 30 replica tests under single overload condition are shown in Fig. 5. The initial crack length, pre-cracking condition etc. are kept identical for all 30 CT specimens. It is seen that there is no significant change in *a-N* behaviour as compared to constant amplitude loading. It is just during the application of overload that the sudden increase in crack length takes place. A crack increment of 0.6 to 0.8 mm is observed due to the application of single overload of magnitude 1.8. This increment is about 1.85% to 2.47% of the average failure crack length under constant amplitude loading. The results also reveal that the slope of the a-N curves changed due to application of the overload and the increment in the crack length is seen after several load cycles. Just after the application of the overload, acceleration effect in the growth rate *da/dN* is observed but it continued for a very short period of about 10 to 20 cycles only. This period may be due to material response capability or the sensitivity of the instrument. After overcoming this period a drastic drop in the growth rate is observed which is referred as crack growth retardation. The delayed retardation is the number of cycles up to which it attains the same crack growth rate that of the constant amplitude loading. Different parameters such as *m* and C, growth rate just before and after the application of overload, retardation factor, plastic zone size and crack opening stress intensity factors are determined from all 30 tests. Retardation coefficient η is calculated from the ratio of minimum crack growth rate after the application of overload $(da/dN)_{min}$ to the crack growth rate just before the application of overload $(da/dN)_{befOL}$. Figs. 6 and 7 shows the probability distribution of few crack growth parameters.

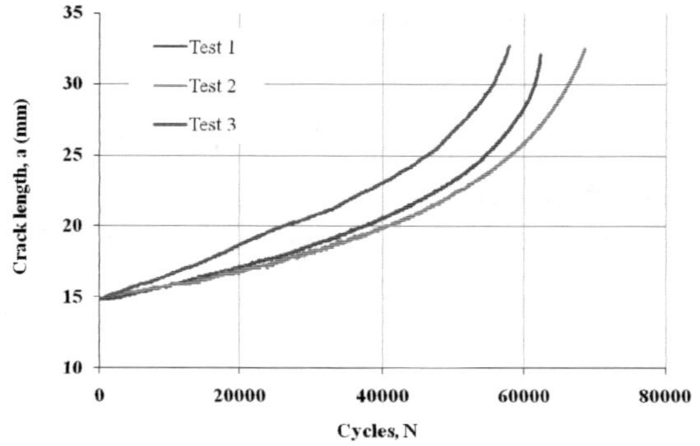

Fig. 3. *a*-N curves under constant amplitude loading at R=0.1 and Pmax = 5.0 kN for CT specimens

Based on the results of regression coefficient the best probability model for fatigue life $f(N_f)$, delayed cycles $f(N_D)$, crack opening stress intensity factor $f(\Delta K_{op})$, Paris constant $f(C)$ and exponent $f(m)$ and retardation factor $f(\eta)$ are described below. The proposed probability distributions are :

$$f_{norm}(N_f) = 1.289 \times 10^{-04} e^{-5.224\times10^{-08}(N_f - 92218)^2} \quad (3)$$

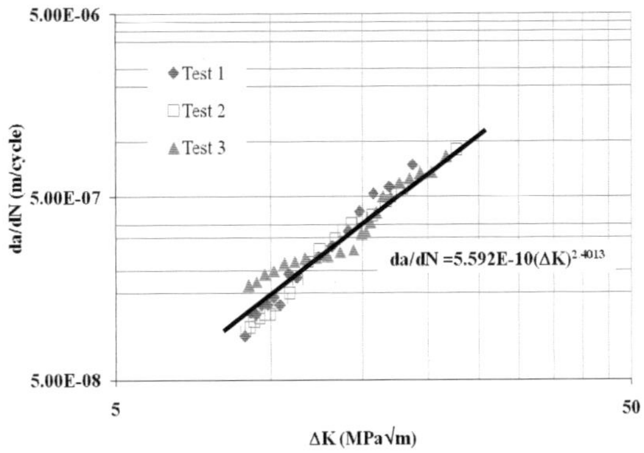

Fig. 4. *da/dN* vs. ΔK curves under constant amplitude loading (R=0.1, Pmax=5 kN)

Table 4. Crack growth parameters under constant amplitude loading and their statistics.

Parameters	Fatigue life (Cycles)	Material constant, C (m/(cycleMPa√m))	Exponent, m
Mean, μ	62887	1.027E-09	2.389
Standard deviation, σ	5361.9	1.267E-09	0.478
Coefficient of variation (%)	8.526	123.4	20.01

$$f_{norm}(C) = 5.770\times10^{09}\, e^{-5.131\times10^{29}\left(C-9.270\times10^{-11}\right)^2} \tag{4}$$

$$f_{norm}(r_p) = 28.496\exp\left[-\left(\frac{(r_p-0.899)^2}{3.92\times10^{-04}}\right)\right] \tag{5}$$

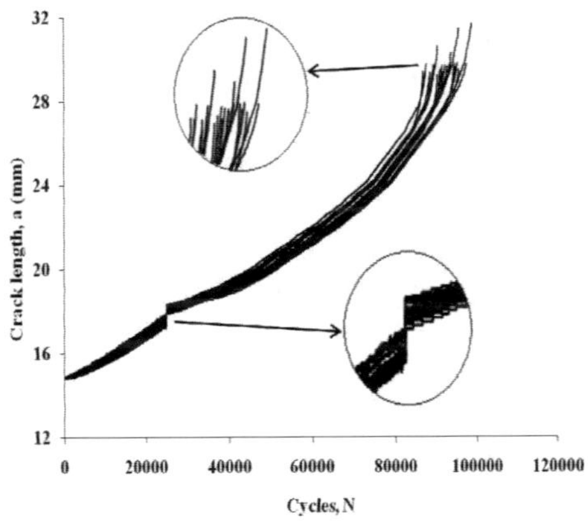

Fig. 5 *a-N* curves under single overload

$$f_{wbl}(\Delta K_{op}) = 0.015\left(\frac{\Delta K_{op}}{171.21}\right)^{1.612} e^{-\left(\frac{\Delta K_{op}}{171.21}\right)^{2.612}} \tag{6}$$

$$f_{ev}(m) = 4.983\exp\left(\frac{m-3.105}{0.2007}\right)\exp\left[-\exp\left(\frac{m-3.105}{0.2007}\right)\right] \tag{7}$$

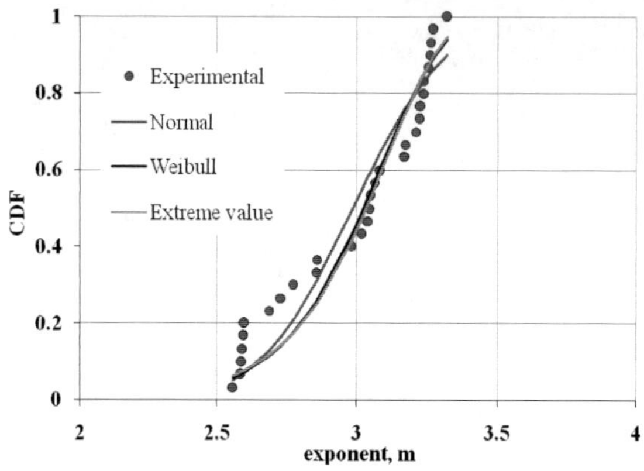

Fig. 6. Cumulative Probability Distribution (CDF) of exponent, m

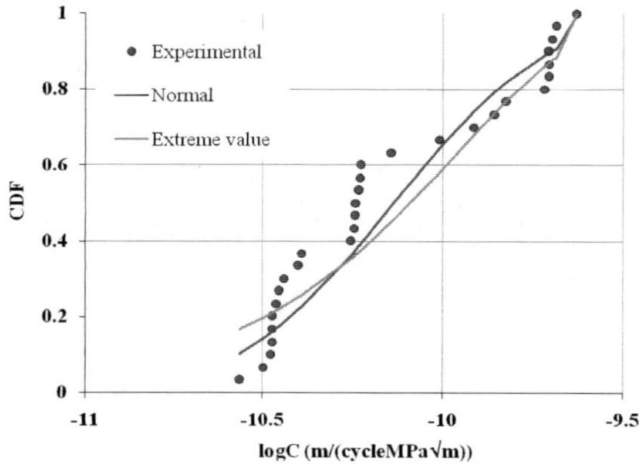

Fig. 7. Cumulative Probability Distribution (CDF) of Material Constant, C

$$f_{ev}(N_D) = 9.285 \times 10^{-04} \exp\left(\frac{N_D - 23877}{1077}\right) \exp\left[-\exp\left(\frac{N_D - 23877}{1077}\right)\right] \tag{8}$$

$$f_{ev}(\eta) = 28.571 \exp\left(\frac{\eta - 0.255}{0.035}\right) \exp\left[-\exp\left(\frac{\eta - 0.255}{0.035}\right)\right] \tag{9}$$

These proposed probability distribution functions (PDF), shown in Eq. 3-9, can be adapted in the fracture mechanics based fatigue reliability analysis under the effect of overload.

4.0 Concluding Remarks

From the present observation, following conclusions can be drawn.

- Due to application of single overload of magnitude 1.8 times the constant amplitude load, the mean fatigue life is increased by 46 %, the growth exponent by 24.88% and the Paris constant C is decreased by 9.02% as compared to the constant amplitude loading.
- The application of overload decreases the growth rate with a mean retardation factor of 0.235 and a mean delayed cycles of 23250.
- There occurs a statically significant variation in the variances of crack length and fatigue life after the overload.
- Normal probability distribution provides the best fit curve for plastic zone size, fatigue life, Paris material constant, C whereas the extreme value distribution provides the best fit curve for Paris

exponent m, delay cycles and retardation factor. Weibull can be approximated well for crack opening stress intensity factor, K_{op}.

- By knowing the distribution parameters due to application of overload, stochastic crack growth model under overloading condition can be derived from any of the deterministic crack growth model and shall be helpful to study the fatigue crack growth under overload condition under required reliability condition.

Acknowledgment

The authors wish to acknowledge the financial support from the TEQIP-III.

References

[1] Wheeler OE. Spectrum Loading and Crack Growth. Trans ASME, J Basic Engng 1972;181-86.
[2] Willenborg J, Engle RM, Wood H A. A crack growth retardation model using an effective stress concept AFFDL-TM-FBRgl-7 (Air Force Dynamics Laboratory, Dayton, OH, USA), 1971.
[3] Sadananda K, Vasudevan AK. Analysis of overloads effects and related phenomena. Int. J. Fatigue 1999; 21:S233–246.
[4] Taheri F, Trask D, Pegg N. Experimental and analytical investigation of fatigue characteristics of 350WT steel under constant and variable amplitude loadings. J Marine Struct 2003;16:69–91.
[5] Gallagher JP. A Generalized Development of Yield-Zone Models AFFDL-TM-74-28, Air Force Flight Dynamics Laboratory, Wright-Patterson Air Force Base, 1974.
[6] Ward-Close CM, Blom AF, Ritchie RO. Mechanisms associated with transient fatigue crack growth under variable-amplitude loading: an experimental and numerical study. Engng Fract Mech 1989; 32:613–38.
[7] Christensen RH. Metal fatigue. New York: McGraw-Hill, 1959.
[8] Knott JF, Pickard AC. Effects of overloads on fatigue-crack propagation: aluminium alloys. Metal Sci 1995; 11:399–04.
[9] Corbley DM, Packman PF. On the influence of single and multiple peak overloads on fatigue crack propagation in 7075 T6511 Al Engng Fract Mech 1973; 5:479-97.
[10] Matsuoka S, Tanaka K. Delayed retardation phenomenon of fatigue crack growth resulting from a single application of overload . Engng Fract Mech 1978; 10: 515-525.
[11] Bathias C, Vancon M. Mechanisms of overload effect on fatigue crack propagation in aluminium. Engng Fract Mech 1978; 10:409-24.
[12] Gope PC, Kumar H, Purohit H. Effect of tensile or compressive overload on the fatigue crack growth of friction stir welded 19501 Aluminium alloy. J Testing Eval, (Accepted), 2017
[13] Borrego LP, Costa JD, Ferreira, JAM. Crack growth behavior of AA6082 and AA6061 aluminum alloys. Procedia Engng 2014; 74: 175-78.
[14] Zhou X, Gaenser HP, Pippan R. The effect of single overloads in tension and compression on the fatigue crack propagation behaviour of short cracks. Int J Fat 2016;89: 77-6.
[15] Vardar O. Effect of single overload in FCP. Engng Fract Mech 1988; 30:329-35
[16] Carlson RL, Kardomateas GA, Bates PR. The effects of overloads in fatigue crack growth. Int. J. Fat. 1991;13(6): 453-60.
[17] Wu WF, Ni CC. A study of stochastic fatigue crack growth modeling through experimental data. Probab Engng Mech 2003;18:107–18
[18] Wu WF, Ni CC. Probabilistic models of fatigue crack propagation and their experimental verification. Probab Engng Mech 2004; 19: 247–57.
[19] Gope PC. 2016. Probabilistic model of fatigue crack propagation and estimation of probability-confidence bounded a–N curves. Int J Comput Meth Engng Sci Mech 2016; 17(4): 298-14.
[20] ASTM E8 - 99 Standard Test Methods for Tension Testing of Metallic Materials.ASTM International, West Conshohocken, PA, 1999, www.astm.org
[21] ASTM standard E647-93. Standard test method for measurement of fatigue crack growth rates. Am Soc Test Mater 1994;03.01:679–06.
[22] Paris P, Erdogan F. A critical analysis of crack propagation laws. J. Basic Eng. Trans. ASME 1963;528–34.
[23] Yang JN. Application of reliability methods to fatigue, quality assurance and maintenance. In: Schueller GI, Shinozuka M, Yao JTP, editors. Proceeding of the Sixth International Conference on Structural Safety and Reliability. Netherlands: Balkema; 1994.
[24] Yang JN, Manning SD. A simple second order approximation for stochastic crack growth analysis. Engng Fract Mech 1996;53:677–86.

Novel Use of Waste Chicken Feather Fiber for making Prototype Non-woven Mats and their Evaluation

R.B.Choudary[1]*, Madhuri Burri[1], R.Srinath[2], and Rosamma Alex[2]

1 Sasi Institute of Technology & Engineering, Tadepalligudem-534 101, AP, India
2 Rubber Research Institute of India, Kottayam-686011, Kerala, India

*rbchoudary64@yahoo.co.in

Abstract. Chicken feather is a waste product that remains after processing the chicken to obtain meat. A series of non-woven mats were made by adding several levels of chicken feather fiber (CFF) to determine the effect of CFF on mat properties. The fractions of CFF used ranged between 10 and 60% and pre-vulcanized latex was used as a binder. The mats were tested for tensile strength, resistance to tear and aging behavior. It was found that the tensile and tear strengths were similar to the latex adhesive strength. Water absorbency and thermal conductivity were also determined. These mats can be used as soil erosion mats, nursery bags, shoe insoles, thermal and acoustic insulators, etc.
Keywords: Chicken Feather Fiber, non-woven mats, latex.

1. Introduction

More than 100 million kilograms of chicken feathers are generated by the poultry industry in India every year [1]. This feather, in general, is going as a waste. The indiscriminate disposal of feather waste poses pollution problems. Scientists are exploring new methods to convert waste chicken plume into products ranging from auto parts to a variety of papers [2-4]. Research continues to find new uses for chicken feathers, most of which would substitute for petroleum-derived materials. The chicken feather can be a viable source of polymers to make fiber, films and a variety of products currently manufactured by the synthetic polymer industry. The advantages are that the feathers are derived from a sustainable source and can be processed in a very similar way to conventional synthetic polymers without being dangerous.

The feathers are composed of many parallel barbs, thin and well arranged, forming a vane on each side of a conical hollow shaft called a quill or rachis. Barb and its components known as CFF, have nodules that occur naturally at approximately 50 μm apart and are similar to fiberglass.

Ye and Broughton [5] made non-woven mats by mixing chicken feather fiber with binder fiber. They reported effective insulating properties compared to other battings made of goose down and polyester fiber.

Fuller developed air-laid and thermal-bonded non-woven fabrics with down feathers to alleviate the confinement problem of individual down in isolated products [6]. He evaluated their thermal and structural characteristics and demonstrated that they have a leading heat-to-weight ratio in the industry. Gassner et al. [7] obtained a patent for their method of manufacturing nonwovens and other products from feather fiber.

Shi et al. [8] demonstrated the feasibility of using waste feather fibers to make sheets of paper by hand. They compared their properties with the properties of handmade sheets with standard bleached eucalyptus tropical wood fibers. Wrześniewska-Tosik et al. [9] evaluated a method for preparing hydrophobic composite non-woven fabrics containing chicken feathers and synthetic fibers (PES, PP). They made non-woven fabrics in aqueous medium using a papermaking method. They presented adsorption and mechanical properties and a morphological evaluation.

George et al. [10] created latex bonded fabrics containing turkey feather fibers for use as erosion control fabrics. A comparison was made with commercially available erosion control fabrics. Ifelebuegu and Johnson [11] prepared biopolymer sorbents based on low-cost unconventional cellulose and keratin and evaluated their capacities, properties and mechanisms of oil sorption.

The technique of latex binding is widely used in the preparation of non-woven coir and animal hair mats [12] for various applications, such as upholstery and packaging. A similar technique is used in this study to prepare non-woven mats with CFF.

The car seat industry, which is one of the largest consumers of latex bonded mats [13], could become a potential user of latex-bound CFF mats. The use of CFF mats bonded with latex allows the design of very comfortable and lightweight seats that, due to the high air permeability, provide a cool seat even in hot climates. They can also be used in the manufacture of shoe insoles, door mats and delicate instrument packaging material, due to their excellent cushioning characteristics. The present study aims to find ways and means to use chicken feathers in the manufacture of chicken feather mats made with latex.

2. Materials and Methods

The dry CFF was treated with pre-vulcanized latex. Latex is natural and less dangerous to work with. It could be a better alternative to other commercial binders. Ordinary latex requires drying and curing at high temperature for complete vulcanization, while for pre-vulcanized rubber, drying is only adequate.

2.1 Preparation of Chicken Feather Fiber Mat

CFF was cut with scissors. It was disinfected by immersing in boiling water for 30 min. Later it was dried in an electric oven at 50°C. A piece of Borosilicon glass plate of 150 mm x 225 mm was used as a mold with 3 mm thick glass strips glued on all four sides. The mold was cleaned and dried. Silicone polymer was applied as release agent. Dry CFF was placed in the mold and a predefined amount of pre-vulcanized latex was poured. The mold containing CFF and latex was dried in an air oven maintained at 100°C for 2 hours. Figure 2 shows a sample of latex bonded CFF mat. For each weight fraction, three samples were taken to minimize the experimental error. The control sheets were made with total latex. To make objects such as egg tray, etc., the moderately dried mat was cut into shape and hot pressed in a mold. Figure 3 shows a cut portion of an egg tray made of a CFF mat bonded with latex.

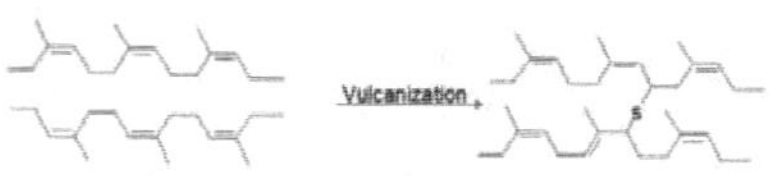

Fig. 1. Schematic representation of vulcanization

Fig. 2. Latex bonded CFF mat

Tensile, tear and moisture absorbency tests were carried out on latex bonded CFF mats.

2.2 Tensile Test

The tensile test was carried out according to the ASTM 412 standard. The tests were carried out using the universal testing machine Zwick Royal model 1474. The crosshead speed used was 460 mm/minute.

2.3 Tear Test

The tear of a rubberized product is a mechanical rupture process initiated and propagated in a site of high concentration of stress caused by a cut, defect or localized deformation. Tear samples (Fig. 4) were prepared according to ASTM D624 Type C. Tear strength was also determined using the Universal Test Machine.

Fig. 3. A portion cut out of an egg tray

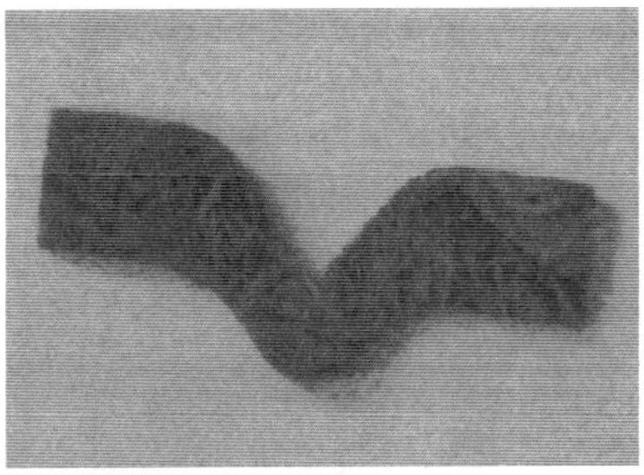

Fig. 4. Tear test sample made of latex bonded CFF mat

2.4 Water Absorbency

Water absorbency test was performed on CFF mats bonded with latex according to ASTM D579. Rectangular samples were cut from each mat with dimensions of 25.4mmx76.2 mm. The CFF mats were first dried by heating at 50°C oven for approximately 24 hours, cooled in a desiccator and immediately weighed to the nearest 0.005 g to measure the water absorption of the mats.

All samples were soaked in water for 24 hours at room temperature. The excess water on the surface of the samples was removed using a blotting paper before weighing the samples again. The percentage of increase in weight after soaking was calculated to the nearest 0.01% as follows:

$$\text{Water absorbency} = \frac{\text{Wet weight - Reconditioned weight}}{\text{Reconditioned weight}} \text{x}100 \quad \text{.....................}(1)$$

2.5 Ageing Test

The ageing of the samples was carried out at 105°C for 72 hours. The effect of ageing on the tensile strength of CFF mats was investigated. The results of several investigations have indicated that accelerated dry heating at 105°C for 72 hours corresponds approximately to normal ageing of 25 years [14].

2.6 Thermal Conductivity

The coefficient of thermal conductivity of the CFF mat was measured using the Lee method in Siddhartha Mahila College, Vijayawada, AP. 100 mm diameter disks of CFF mats were cut and stacked to form a sample. The sample was placed between the brass disk and the steam chamber. The sample was heated until a state of equilibrium was reached throughout the sample. After reaching the equilibrium temperature, the disk was heated to an additional temperature of 5°C and the temperature was recorded at half minute intervals.

3. Results and Discussion

The results of the mechanical tests performed on feather mats bonded with latex are shown in Table 1. The resistance to bursting of the craft paper is given in the footnote for comparison.

3.1 Tensile Test

Pure latex shows a high tensile strength. The tensile strength decreases as the CFF percentage increases. In general, the modulus and the strength of the rubber increase when fillers are used in particles with a particle size of less than 0.1 microns. When larger fillers or additives are used in the range of a few microns, then there is a reduction in tensile strength. Here, a comparatively low tensile strength (Fig. 5) was obtained for the CFF mats. The variation of tensile strength with fiber loading is shown in Fig. 6. The variation in elongation is shown in Fig. 7. In applications such as thermal insulation, a low strength will be sufficient since the control parameter is thermal conductivity. The pre-vulcanized latex is only act as a binder of acceptable strength.

3.2 Tear Test

It was found that the tear strength of the CFF mat was very low compared to that of the latex (Fig. 8). This could probably be due to the inadequate distribution of latex in the open molding technique used to prepare CFF mats. The variation in tear strength with fiber loading is shown in Fig. 9. Needle punching, wet laying, airless spraying, electrospinning [15] techniques can be used to make non-woven mats with greater strength.

Table 1. Mechanical properties of latex bonded CFF mats.

S No.	Fiber (%w/w)	UTS (MPa)	Elongation (%)	Tear strength (N/mm)	Water Absorbency (%)	UTS after ageing (MPa)
1	0	14.04	1612	36.03	2.04	9.70
2	10	5.23	487	24.92	17.41	2.86

3	20	2.15	208	20.28	36.15	1.60
4	30	1.97	47	18.54	48.88	1.53
5	40	1.77	31	14.79	64.77	1.49
6	50	1.72	28	9.85	87.88	1.44
7	60	0.33	19	1.97	92.08	1.25

#5-6 kg/cm^2 for 3 ply 150 gsm craft paper
#10-12 kg/cm^2 for 5 ply 150 gsm craft paper

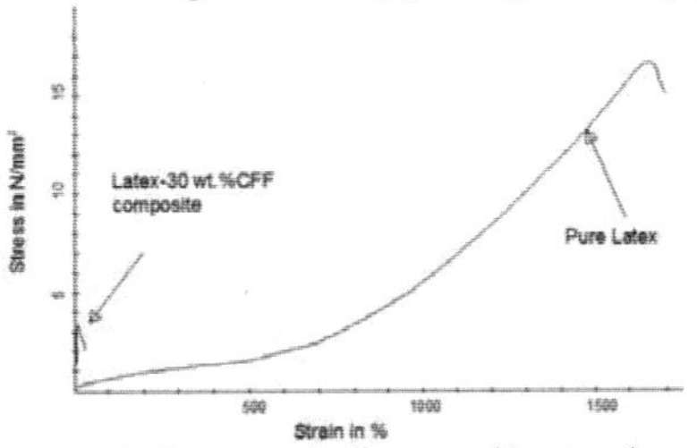

Fig. 5. Stress-strain diagram of latex and latex-30 wt.% CFF mat

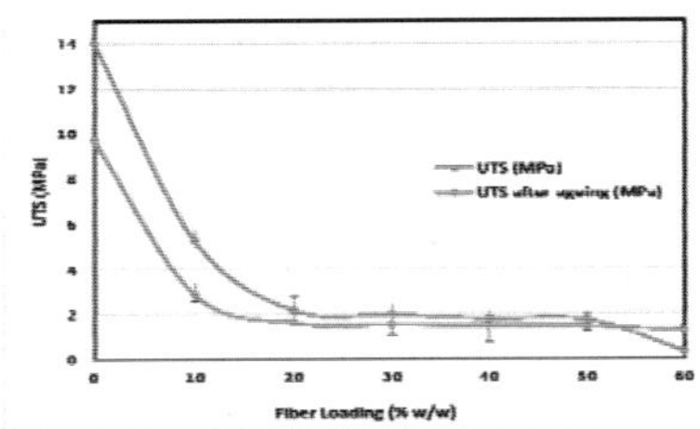

Fig. 6. Variation of UTS with fiber loading

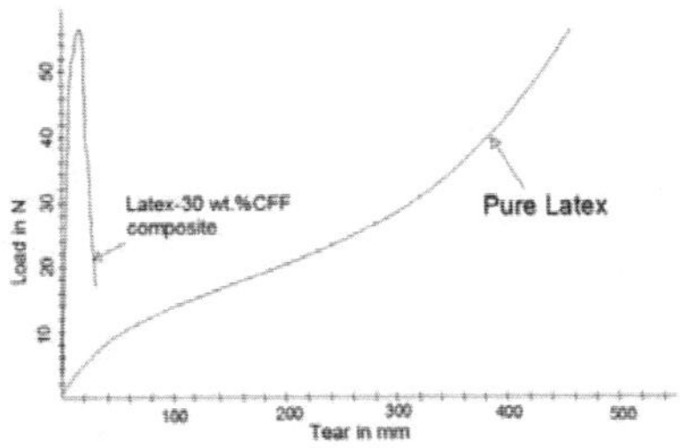

Fig. 7. Variation of Elongation with fiber loading

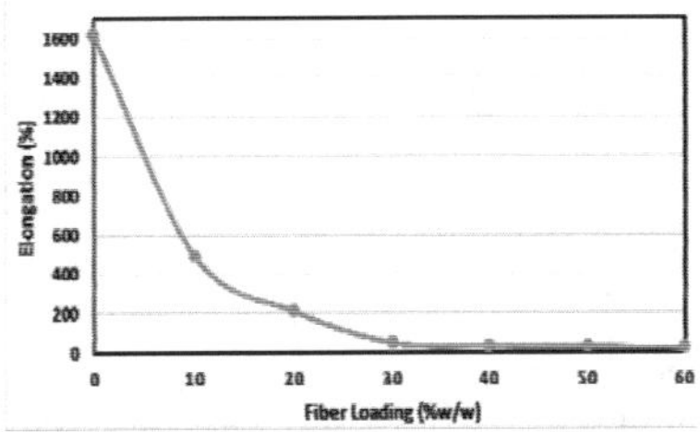

Fig. 8. Tear strength of pure latex and latex bonded 30 wt% CFF mat.

3.3 Water Absorbency

The water absorbency results (Fig. 10) indicate that the water absorbency of the CFF mats bonded with latex is low. Since the constituents are hydrophobic, these mats should have exhibited much less absorbency. The high water absorbency exhibited could be due to water trapped in the porous structure of the mat. Shi et al. [8] reported a decrease in water absorbency beyond 20% feather fiber.

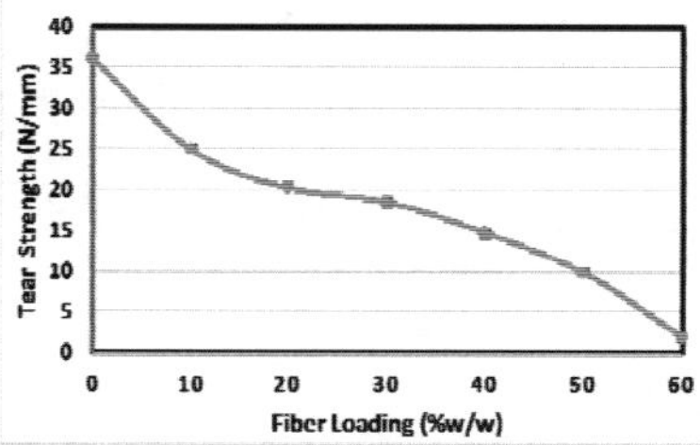

Fig. 9. Variation of tear strength with fiber loading

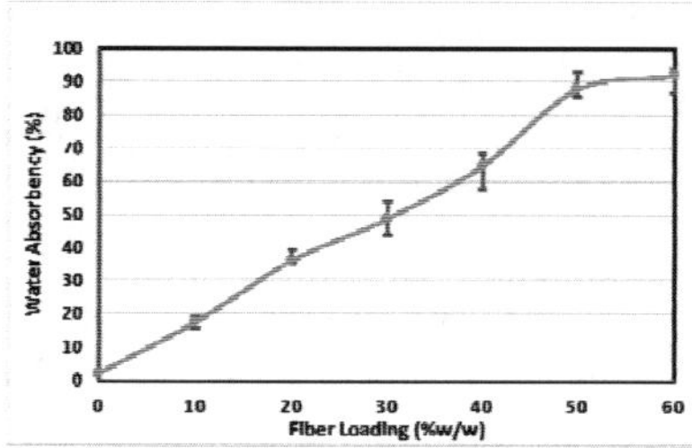

Fig. 10. Variation of water absorbency with fiber loading.

3.4 Ageing Test

The results of the accelerated ageing experiment (Fig. 6) show that the mechanical properties deteriorated at elevated temperatures. However, the results indicate that the CFF mats will retain almost 75% of the original properties even after 25 years.

3.5 Thermal Conductivity

The results of the thermal conductivity measurement are shown in Table 3. Although they are not superior to wool, rubberized CFF mats exhibit better insulation compared to pure rubber. CFF mats bonded with latex can be used as thermal insulation mats.

4. Conclusions

The careless disposal of 100 million kg of chicken feather waste generated in the poultry industry in India poses serious pollution problems. The present study reveals the strength of the CFF mat attached to the latex is equal to the adhesive strength of the latex. CFF products adhered to latex can be used in applications such as packaging, automotive upholstery, soil erosion control mats, acoustic and thermal insulation mats. The CFF mat coated with latex can be used to make nursery bags.

Acknowledgements

The authors thank the authorities of RRII, Kottayam, Kerala, for permitting them to carry out this work in their organization.

References

[1] Chicken: population, selected indicators of food and agriculture development in Asia-Pacific region, 1992-2002, Food and Agriculture Organization of The United Nations, Regional Office For Asia and The Pacific, Bangkok, October 2003

[2] Wool R 2002 *Affordable composites from renewable sources, The future is composites, A year of growth*, Poster Book, University of Delware.

[3] Schmidt WF and Barone JR 2004 *New uses for chicken feather keratin fiber* in Proceedings of National Poultry Waste Management Symposium, October 24-26, Memphis, Tenn, 99-101.

[4] Barone JR and Schmidt WF 2004 *Poultry feathers: New end uses* in Proceedings of the 38th National Meeting on Poultry Health and Processing, Delmarva Poultry Industry, Inc., Ocean City, MD, Oct. 22-24, 47-50.

[5] Ye W and Broughton RM 1999 *Chicken Feather As A Fiber Source For Nonwoven Insulation.* International Nonwovens Journal, 8, 53-59.

[6] Fuller ME 2015 *The structure and properties of down feathers and their use in the outdoor industry.* Ph.D. Thesis, The University of Leeds.

[7] George Gassner III, Walter Schmidt, Michael J. Line, Clayton Thomas and Rolland M. Walters 1998 US Patent # 5,705,030.

[8] Shi B, Shannon TG and Pelky E 2010 *Novel use of waste keratin and cotton linter fibers for prototype tissue papers and their evaluation.* BioResources, 5, 1425-1435.

[9] Tosik W, Szadkowski M, Marcinkowska M and Pałczyńska M 2012 *Feather-containing Composite Non-wovens with Barrier Properties.* Fibres & Textiles in Eastern Europe, 20, 96-100.

[10] George BR, Bockarie A, Mc.Bride H, Hoppy D and Scutti A 2003 *Utilization of Turkey Feather Fibers in Nonwoven Erosion Control Fabrics.* International Nonwovens Journal, 12, 45-52.

[11] Ifelebuegu AO and Johnson A 2017 *Nonconventional low-cost cellulose- and keratin-based biopolymeric sorbents for oil/water separation and spill cleanup: A review.* Critical Reviews in Environmental Science and Technology, 47, 964-1001.

[12] Calvert KO 1982 *Polymer lattices and their applications*, Applied Science Publishers Ltd., London.

[13] Blackley DC 1997 *Polymer Lattices*, Volume 3: Applications of lattices, II Ed. Chapman & Hall.

[14] Wilson WK and Hebert RL 1969 *Evaluation of the stability of record papers.* TAPPI, 52, 1523-1529.

[15] Chen C, Chuanbao C, Xilan M, Yin T and Hesun Z 2006 *Preparation of non-woven mats from all-aqueous silk fibroin solution with electrospinning method.* Polymer, 47, 6322-6327.

Experimental analysis of mechanical properties of stir casted aluminium-graphene nanocomposites

Arnav Shil[1], Sandipan Roy[1], P. S. Balaji[2], Jitendra Kumar Katiyar[1,*], Sumit Pramanik[1] and Anuj Kumar Sharma[3]

[1]Department of Mechanical Engineering, SRM Institute of Science and Technology, Kattankulathur, Chennai-603203, Tamil Nadu, India

[2]Department of Mechanical Engineering, NIT Rourkela, Odisha, India

[3]Centre for Advanced Studies, Dr. APJ Abdul Kalam Technical University, Lucknow-226031, India

*jitendrakumar.v@ktr.srmuniv.ac.in

Abstract. Aluminium (Al) based materials are widely used in aerospace and automobile industries due to its higher strength to weight ratio. Under fatigue loading the components are prone to failure, which is one of the key factors in designing the components. In the present study, Graphene (Gr) is added to Al at different weight percentages (1%, 2% and 3%) using stir casting. Tensile strength and fracture toughness of the composites were determined using tensile testing of Dogbone specimens and plane strain fracture toughness testing of C (T) speci-mens in mode-I, respectively. The result showed that for Al-Gr nanocomposites, with the increase in Graphene percentages, tensile strength was found to be in-creased however the fracture toughness was decreased. It is also found that the manufacturing method followed in the present work to be cost effective method for Al-Gr nano composite. The analysis of different tests suggests that the addition of 2wt% Graphene found to be the saturation point in the nanocomposites and any further addition of Graphene can lead to the degradation in the quality of composite.

Keywords: Stir casting; Tensile strength; Fracture toughness; Wettability; Ag-glomeration; FEA The first section in your paper

1. Introduction

Aluminium (Al) based alloys of 6010 grade have several important applications in various industrial sectors such as automobile, manufacturing, and aircraft or aerospace [1-2]. Several attempts have been made over the past few decades on Al composites to improve the specific strength of several components. In this context, Graphene was proved to be one of the best second phase materials. Graphene was mixed in Al at different temperatures, i.e., at 700 °C and 800 °C followed by annealing to improve the Al mechanical properties. Formation of Aluminium carbide was avoided in the composite at 700 °C, which led to higher corrosion resistance and mechanical strength with increase in Graphene percentages [3]. Manufacturing of such Al-Graphene composites by high pressure torsion (HPT) induced extremely high shear deformation in the composite, and its effectiveness could be increased by ball milling. Thus, the obtained composite was homogeneous. However, the composite was prone to have carbide formation with Al. Photoelectron spectroscopy showed the formation of Aluminium-carbon (Al-C) bonds rather than carbon-carbon (C-C) bonds [4]. To resolve this problem, modified powder metallurgy technique was used to fabricate bulk Al-Graphene composite, which induced the trans-formation of thermally expanded graphite to Graphene using pulverizing ball-mill. The powder was then compacted by cold pressing and annealed at 400 °C, 450 °C and 500 °C temperatures. Microstructure when analyzed, the result showed decreased micro-cracks with increase in annealing temperature [5]. Added Graphene into metal matrix led to the enhancement in mechanical properties such as tensile strength of the material. However, the dispersion optimization of Graphene in metal matrix was necessary to get favorable results [6]. In a study [7], tensile strength of Al-Graphene nanoflake (Al-GNF) composite made by powder metallurgy increased by 50% for 0.5wt% of GNF in the composite, due to exceptional tensile property of GNF. Furthermore, 0.5%Al-Graphene composite showed minimum impact strength, whereas pure Al2024 sample showed maximum impact strength in Charpy and Izod impact tests.

However, in another study [8], 0.75% Al-Graphene sample showed higher tensile strength than that of 0.5% Graphene content in Al matrix. Graphene was impregnated on the surface of Al by applying pressure at the locality with varying current and time to determine the extent of impregnation. Formation of Aluminium oxide at the surface of Al-Graphene assisted increase in surface hardness of the composite [9]. Hardness of the Al-Graphene composite sample with 0.5wt% Graphene showed the maximum resistance, whereas the sample with 2wt% Graphene showed the minimum resistance to indentation as Graphene couldn't replace parental atoms, reducing the resistance to indentation [10]. Al-Graphene composites manufactured by stir casting showed maximum compressive strength for 0.5wt% Graphene which was determined to be the saturation point of strength experimentally [11]. When Al and Graphene composites were hot pressed, the composites showed structural anisotropy, where Graphene was perpendicularly aligned to the applied axial pressure. Graphene particles deposited at the grain boundaries of Aluminium, as well as multilayer formation of Graphene led to increased defects in the composites [12]. Graphene nanoplatelets (GNPs) were uniformly mixed in the composite. The GNPs were found to be aligned in the direction of rolling of the composite sample. Due to high surface area of GNPs, wrinkled morphology was introduced, which led to stacking fault in the composite. It increased tensile strength of the composite, but increased brittleness due to the presence of micro cracks in the composite [13]. There was a significant improvement in adsorption of reduced Graphene oxide (rGO) attributed to formation of cationic CTA+ membrane on Al powder during sintering at 350 °C, CTAB was removed and a significant rise in tensile and hardness properties were observed at 0.3 wt% of rGO. Further addition of rGO led to brittle nature of the composite due to formation of Al4C3 [14].

An another study using Monte Carlo simulation [15] showed the fractal surface's dependence upon the inter-particle interaction energy, which helped to determine the agglomerate properties of dry nanoparticles and showed dominant Van-der-Waals force acting amongst the particle interfaces. The lower energy level of interaction led to larger fractal dimensions.

Various methods were used for manufacturing composites. It was determined that the mechanical properties of Al composite depended on the method used since the clustering of reinforcement can lead to the increase in porosity in the composite and hence can degrade the mechanical properties of the composite [16]. Similar tests were done where Cu was mixed with Graphene nano-sheets (GNS) using ball-milling and which was further sintered to fabricate composite. Since GNS got deposited at the copper grain boundaries, the tensile strength initially increased and then subsequently decreased with increase in GNS content [17].

Testing methods suggest standard practices for measurement of fracture tough-ness in plane strain condition in metallic solids; they also provide guidelines for machining the specimens according to the dimensional proportions with tolerances. Ac-cording to ASTM E399-90 (Reapproved 1997), a stringent methodology for estimation of fracture toughness is suggested, whereas ASTM B645-10 (Reapproved 2015) gives some relaxation for the same. ASTM E1820-17 suggests guidelines for deter-mining fracture toughness using R-curve, crack tip opening displacement (CTOD), fracture toughness (K), and J-integral methods [18-21]. Fracture toughness test on austenitic steel for modes - I, II and III indicated that for load applied in mode III, crack propagated at an angle of 70o [22]. FGMs were manufactured by centrifugal casting. Deposition of high density material, such as silicon carbide (SiC), at the base of centrifuge led to the variation in mechanical properties through the thickness of specimen and the crack propagated at a slower rate in the Al-rich side as compared to the SiC rich side [23].

In another study [24], the fracture toughness of Ni/ZrO_2 non-functionally graded materials (non-FGMs) increased from 6.6 to 15.7 MPa/m with increase in nickel percentages (up to 50%). It was found that

there were some local perturbations in the crack propagation of (non-FGM) due to presence of Ni, which led to heterogeneity of the composites. The fracture toughness of functionally graded material (FGM) was determined to be 31.48 MPa/m. The fracture toughness of layered Ti/TiB_2 FGM was found to be increased from 7 MPa/m for (38%Ti and 62%TiB2) to 31 MPa/m for pure Ti in the FGM as increased in TiB2 and induced its brittle nature into the compo-site [25].
The main objective of this study is to fabricate Al-Graphene nanocomposites using stir-casting technique and to study the effect on the nanocomposite on the tensile and Fracture toughness. Experimental study was conducted to determine the tensile strength and fracture toughness properties of the Al-Graphene composite using ASTM E8/E8M-16a and ASTM E399-90 (Reapproved 1997) respectively. This work provides the effect of graphene in the tensile and fracture toughness properties of the Al-Gr nano composite and further, enable the design of the material for various engineering applications.

2. Materials and Method

2.1. *Fabrication of Al/Graphene nanocomposite*

Al-Graphene composite was manufactured by stir casting method in which Al was first melted in an electric furnace at 700°C, and then Graphene nanoparticles were mixed in molten Al at weight percentages of 1%, 2% and 3%. Exochloroethane was used as a degasing agent and Coverall was used to increase wettability and to avoid oxidation of the molten metal. The molten metal was poured into rod shaped dies (length 250mm and diameter 22mm) and also in the form of square plates having (length 65mm and thickness of 25mm) and cooled at a slow rate to avoid cracks and induce ductility in the nanocomposites. The specimens were machined into Dog-bone and compact tension C (T) specimens according to ASTM E8/E8M-16a and E399-90 (Reapproved 1997), respectively.

2.2. *Tensile and Fracture toughness testing*

Dog-bone specimens were uniaxially loaded at a rate of 1mm/min on TUE-CN-400 as shown in as per ASTM E8.True stress vs true strain was obtained using WINUTM software.
Square blocks of the casted composites were machined initially on lathe and then on computer numerical control (Jyothi CNC PX10) using Siemens-Sinumerik 828D controller. Final machining was performed on the specimens using electrical discharge machining on (Ezeecut plus) to generate crack in the C (T) specimens. The specimens were then uniaxially loaded on Denison Mayes Material testing system at a loading rate of 1mm/min, as ASTM E8-399 suggests the stress intensity at the crack tip to increase at a rate of (0.55 to 2.75 MPa/m). Also stress at the vicinity of the crack tip increases up to √3 times of yield stress. The specimens were tested under plane strain loading condition in mode-I, in which the plastic zone size was neglected due to its smaller size compared to specimen size. Fracture toughness was calculated for the given condition using the following equation.

$$K_I = \frac{P}{BW^{1/2}} f\,(a/W) \qquad (1)$$

Where, P = peak load applied during testing, B = thickness of the specimen, W = width of the specimen, f (a/W) = function of (a/W) ratios, α = a/W, a = initial crack length, and

$$f(\alpha) = \frac{(2+\alpha)(0.886+4.64\alpha-13.32\alpha^2+14.72\alpha^3-5.6\alpha^4)}{(1-\alpha)^{3/2}} \qquad (2)$$

2.3. *Micrography*

Field emission scanning electron microscopy (FESEM) at high vacuum was performed on the fractured surfaces using FEI Quantra FEG 200 to analyse the grain structure and deposition of Graphene in the composite.

2.4. X-ray diffraction (XRD)

In order to identify the crystalline phases of Aluminium and Graphene in the nanocomposite GI- XRD was performed in the range of diffraction angle, 2θ= 10^0 – 90° of Cu K radiation using x-ray diffractometer (model: XPert Pro, make: PANalytical, country: UK) at normal scanning.

3. Results and Discussion

3.1. Surface Topography of Fractured Surface

The fractured surfaces of (1 wt %) Al-Graphene and (3wt %) C (T) specimens using scanning electron micrographs are shown in **Figures (1a) and (1b)**, respectively. It is observed that the slow cooling of composite led to a larger grain size and also agglomerated Graphene in the form of sheets at the grain boundaries, it can be observed in Figure (1a). The surface of 3wt% Al-Graphene shows the cleavage fracture. From this result, it can be understood that with the increase in Graphene percentages the composite becomes more brittle.

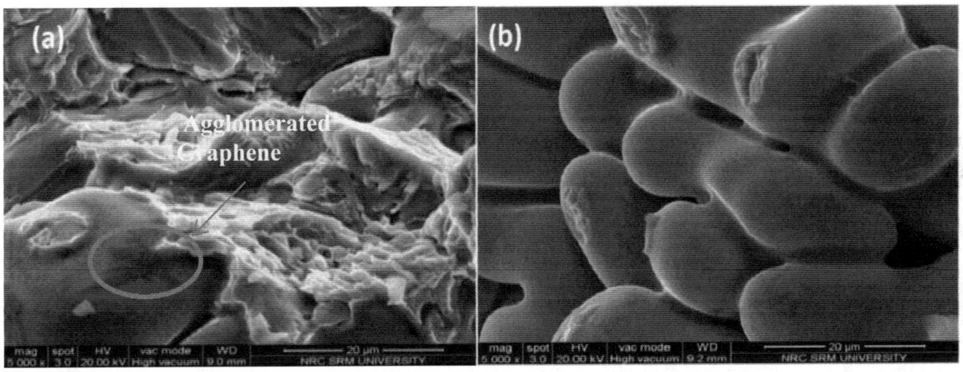

Figure 1: FESEM images of Al-Graphene nanocomposites with (a) 1wt% Graphene (showing ductile fracture) and (b) 3wt% Graphene (showing brittle fracture)

3.2. X-ray diffraction (XRD)

XRD results reveal the presence of both Aluminium (see sharp (111), (200), (220), (311), and (222) peaks in **Figure 2**) and Graphene (see sharp (100)) peak near 2θ=46° in (Figure2) peaks in the Al/Graphene (2 wt %) nanocomposite. Here, (100) peak of Graphene merged with (200) peak of Al near 2θ=46°. This indicates the formation of good bonding between Al and Graphene in the (2 wt %) Al-Graphene nanocomposite.

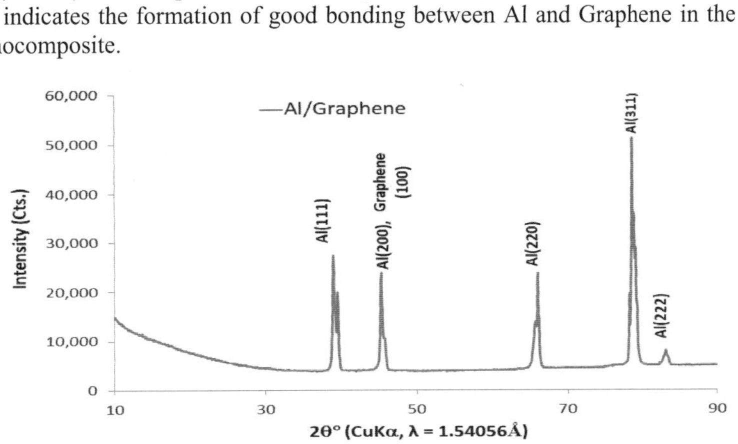

Figure 2: XRD pattern of an Al-Graphene (2 wt %) nanocomposite

IOP Conf. Series: Materials Science and Engineering **653** (2019) 012021 doi:10.1088/1757-899X/653/1/012021

3.3. Tensile testing

The tensile test results are given in **Table 1**. As the specimens were externally loaded, the presence of Graphene nanoparticles at the grain boundaries provides the resistance and this is attributed as increase in the tensile strength. However, increasing Graphene percentage (more than 2%) decreased the strength again. The main reason for such trend is due to the agglomeration of nanoparticles which generally forms a fractal surface because of attractive Van der Waals force between them [15]. Ag-glomeration of Graphene in form of sheets introduce micro-cracks in the composite, which coalesce under tensile loads to propagate inter-granular fracture, leading to brittle fracture thus decreasing tensile strength of the composite at higher concentration of Graphene i.e. (3 wt%).

Table 1: Tensile test results of Al-Graphene nanocomposites

Percentage Graphene	Yield Strength (MPa)	Ultimate tensile strength (MPa)
1%	90.591	113.446
2%	111.413	139.953
3%	97.113	121.905

3.4. Fracture toughness testing

The fracture toughness of Al/Graphene nanocomposite C (T) specimens for a/W ratio of 0.5 was experimentally obtained in mode –I. Under plane strain condition the fracture toughness reduces with increasing Graphene concentration as it becomes brittle in nature thus avoiding plastic deformation at the crack tip. Plastic deformation leads to strain hardening and also reduces stress concentration at the crack tip, thus results in increasing the fracture toughness. The fracture toughness results of Al-Graphene nanocomposites are given in **Table 2**.

Table 2: Fracture toughness results of Al-Graphene nanocomposites

Percentage Graphene	Fracture Toughness (K_Q) (MPa/m)
1wt%	9.20
2wt%	8.93
3wt%	8.41

4. Conclusions

Al-Graphene nanocomposites were fabricated by stir casting because stir casting is an economic method for manufacturing such composites. FESEM images show the presence of Graphene on the surface of Al, where Graphene is deposited in the form of sheets at the grain boundaries. The FESEM images of fractured surfaces of tensile specimens also show relatively brittle fracture of 3wt% Al/Graphene nanocomposite as compared to 1wt% Al/Graphene nanocomposite. This is due to high porosity of the composite as Graphene has less wettability with Al and also due to agglomeration of nanoparticles leading to larger fractal surfaces. XRD shows 5 identical peaks for pure Al. Due to presence of Graphene in the composite, a sharp XRD peak near $2\theta=46°$is observed which merges with the peak of Aluminium. This indicates formation of good bonding between Aluminium and Graphene for (2 wt %) Al-Gr nano-composite. Tensile strength of the specimens is found to increase with increasing Graphene content from 1wt% to 2wt%, however it decreases with Graphene content above 2wt% i.e., at 3wt% Graphene. Since the material tends to become more brittle with increase in Graphene percentages, it avoids plastic deformation at the crack tip. This decreases the fracture toughness of the composite with increase in Graphene percentages. Analysis of different test results suggest, addition of (2 wt %) Graphene is the saturation point in the nanocomposite. Further addition of Graphene leads to degradation in the quality of the composite.

References

1. Rambabu P., Eswara Prasad N., Kutumbarao V.V., Wanhill R.J.H. 2017, Aluminium Alloys for Aerospace Applications, *Aerospace Materials and Material Technologies, Indian Institute of Metals Series*, **DOI 10.1007/978-981-10-2134-3_2**.

2. Starke E. A., Staley J. T. 1996 Application of modern Aluminium alloys to aircraft. Pro 9. *Aero-space Science*, **32**, 131-172.

3. Yolshina L.A., Muradymov R.V, Vichuzhanin D.I, Smirnova E.O. 2016 Enhancement of the Mechanical Properties of Aluminium-Graphene Composites',Mechanics, *Resource and Diagnos-tics of Materials and Structures , AIP Conference Proceeding.*

4. Czeppe T., Korzokikova E., Robert L. 2015 Application of the high pressure torsion supported by mechanical alloying for metal-Graphene composites preparation. *Innovative Manufacturing Technology*, 2 (15).

5. Konakov V.G., Kurapova O.Yu., Lomakin I.V., Archakov I.Yu., Solovyeva E.N. 2016 Fabrication of Aluminium-Graphene and metal-ceramic nanocomposites. *Review of Advance Material Science*, 44, 361-369.

6. Prashantha K. H. Ga, Anthony X. M.: Graphene Reinforced Metal Matrix Composite (GRMMC) 2014 A Review. *12th global congress on manufacturing and management.* GCMM.

7. Yan S.J., Dai S.L., Zhang X.Y., Yang C., Hong Q.H., Chen J.Z., Lin Z.M. 2014 Investigating Aluminium alloy reinforced By Graphene nanoflake. *Materials Science & Engineering A, 612, 440-444.*

8. Sai Jagadish B. 2015 Synthesis and characterization of Aluminium2024 and Graphene metal matrix composites by powder metallurgy means. *SSRG International Journal of Mechanical Engineering (SSRG - IJME)*, 2, 14-18.

9. Sahoo B., Joseph J., Sharma A., Paul J. 2018 Surface modification of Aluminium by Graphene impregnation. *Materials and Design*, 116, 51-64.

10. Lokesh K., Chetan I, Thomas P., Umesh C 2018 Influence-of-Graphene-on-Hardness-Number-of-aluminm-7075-Based-Metal -Matrix-Composites. *International Journal of Mechanical Engineering and Technology (IJMET)*, 9 (1), 327–335.

11. Lokesh K S, Chethan I C, Naveen K R, Vinayaka K. 2017 Determination of compressive strength of Graphene reinforced with Aluminium -7075 metal matrix composites, *International Journal of Innovative Science and Research Technology*, 2, 285 -289.

12. Rutkowski P., Jaworska P. K. L., Stobierski L., Dubiel A. 2015 Thermal properties of pressure sintered alumina–Graphene. *Journal of Thermal Analysis and Calorimetry*, 122 (1), 105-114.

13. Li M., Gaoa H., Lianga J., Gub S., Youb W., Shua Da, Wanga J., Suna B. 2018 Micro struc-ture evolution and properties of Graphene nano platelets reinforced Aluminium matrix composite. *Material Characterization*, 140, 172-178.

14. Sun Y., Zhang Chi, Liu B., Meng Q., Ma S. and Dai W. 2017 Reduced Graphene Oxide Rein-forced 7075 Al Matrix Composites: Powder Synthesis and Mechanical Properties. *MDPI Metals*, 7, 499.

15. Deng X., Dave R. N. 2016 Investigation of nanoparticle agglomerates properties using Monte Carlo simulations. *Advanced Powder Technology*, 27, 1971-1979.

16. Chandla N. K., Yashpal C. S, Jawalkar N., Suri M 2017 Effect of Different Weight Percentage on Mechanical Properties of Aluminium Metal Matrix Composites through Hybrid Reinforcement – A Review. IJSRD - *International Journal for Scientific Research & Development*, 5, 91-96.

17. Yue H., Yao L., Gao X, Zhang S., Guo E., Zhang H, Lin X., Wang B. 2017 Effect of ball-milling and Graphene contents on the mechanical properties and fracture mechanisms of Graphene nanosheets reinforced copper matrix composites. *Journal of Alloys and Compounds*, 691, 755-762.

18. ASTM E8- Standard test methods for tension testing of metallic materials.

19. ASTM E1820 Standard Test Method for Measurement of Fracture Toughness.

20. ASTM E-399 Standard Test Method for Plane-Strain Fracture Toughness of Metallic Materials.

21. ASTM B-645 Standard Test Method for Plane-Strain Fracture Toughness of Aluminium alloys.

22. Cambel P., Martins R. F. 2016 Research on fatigue crack propagation in CT specimen's subjected to loading in modes I, II, III. *Procedia Structural Integrity*, 1, 134–141.
23. Ulukoy A., Topcu M., Tasgetiren S. 2009 An Experimental Crack Propagation Analysis of Aluminium Matrix Functionally Graded Material. *Advances in Functionally Graded Materials and Structure*, 74-95.
24. EL-Desouky A. R., EL-Wazer M. S. 2013 Mixed Mode Crack Propagation of Zirconia/Nickel Functionally Graded Materials. *International journal of engineering*, 26, 885-894.
25. Hill M. R., Doug Carpenter R., Paulino G. H., Munir Z. A., and Jeffery C. 2002 Fracture testing of a layered functionally graded materials. *Fracture Resistance Testing of Monolithic and Composite Brittle Materials, ASTM STP* 1409.

Numerical Analysis of Composite Coating(Al_2O_3+ZrO_2·5CaO) applied on Cast Iron

Abhinav[1], Bal Sasmita[2] and K Kapilesh[3]

Alliance College of Engineering and Design, Alliance University, Bangalore-106

Email: abhinavtechno5@gmail.com & sasmitabal@gmail.com

Abstract. A new method, the rule of the mixture is developed for the determination of thermal conductivity (K), allied properties viz.density and the volume of the individual composite coating powders. Optimum coating thickness through steady-state thermal analysis of composite coating (Al_2O_3+ZrO_2.5CaO) applied on cast iron surface is carried out. Also, the thermodynamic behaviour of TGO and bond coat were simulated .The thickness of the topcoat, TGO, and the bond coat was varied independently in the range of 0.1 to 0.04 mm. Results show that topcoat thickness doesn't have much influence on the heat transfer. However, TGO and the bond coat plays a crucial role in heat transfer study. Maximum heat resistance ,171.47 ºC observed at 1 mm TGO thickness, compared to 88.94 ºC in case of bond coat for the same coating thickness. It is also understood that 1 mm of optimal TGO thickness may be beneficial in the design of thermal barrier composite coating systems.

1. Introduction

Thermomechanical analysis is a well-known finite element technique to predict the residual stresses in the coating systems [1].The residual stress induced in the thermally grown oxide (TGO) may lead to early spallation of the topcoat. Studies have been carried out to examine the stress built up at the interface of the topcoat and TGO and also between TGO and bond coat and found that TGO thickness and radii of curvature significantly affects the life of the coating systems. It has been also, found that at a higher thickness of TGO yields a high concentration of residual stress that may lead to early spallation of the topcoat [2]. A dual model based on a 2D wave profile was developed to visualize the attributes in the interface. In his study diffusion model was utilized to determine TGO growth [3, 4]. TGO growth behavior found to be one of the vital parameter affecting stress development. Extensive work has been carried out using a 2-D profile in this area [5]. Attempts have been made by many researchers to detect TGO profile but it was found that experimental methods for this purpose are very challenging [6, 7]. For this reason modeling of TGO will be more effective than conducting an experiment. A model based on the Finite element method (FEM) was proposed to find conductivity of Thermal barrier coating (TBC) made by EB-PVD [8]. A numerical model was developed to determine TGO growth and to study the strength of the topcoat which affects the life span of a TBC system [9]. In another investigation carried out on graded coating system and found that that presence of TGO with a thickness of 2-3 μm has a positive effect on the resistance against thermal shock [10]. In the present work mathematical model has been developed for the determination of thermal conductivity (K) using the rule of the mixture and also Optimum coating thickness through steady-state thermal analysis is carried out for the design of improved thermal barrier coating systems. The thickness of the topcoat, TGO, and the bond coat was

Published under licence by IOP Publishing Ltd

varied independently in the range of 0.1 to 0.04 mm and results are presented in terms of temperature drop and heat flux.

2. Mathematical Formulation-Rule of Mixture

A new mathematical model has been developed for the determination of thermal conductivity (K) using the rule of mixture. The conductivity "K" considered an inertial property. The effective conductivity, effective density and the total mass of the composite coatings include a topcoat (Al_2O_3+ZrO_2·5CaO) and bond coat ($Fe_3$8Ni10Al) were calculated using the empirical relation shown in Equation 1, 2 and 3.The topcoat comprises of 50 wt. % Alumina, 45 wt. % Zirconia dioxide and 5 wt. % Calcia Oxide, similarly the bond coat contains 52 wt. % Fe 38 wt. % Ni and 10 wt. % Al. Their masses were calculated along with the density shown in Table 1. The K_{eff} of the topcoat and bond coat were found to be 6.8329 W/m^{2}k and 69.99 W/m^{2}k respectively at ambient temperature. The total mass of the coating can be calculated by multiplying effective density (ρ_{eff}) with total volume ($\forall_T$) and the total volume is based on the available area and thickness of the coatings which is in the present case (10mm×10mm×5mm). The K_{eff} of TGO (7 W/m^{2}k) obtained from the previously published literature [11].

$$K_{eff}= (m_1/M_c)\,K_1 + (m_2/M_c)\,K_2 + (m_3/M_c)\,K_3 \quad (1)$$

$$\rho_{eff}= 0.5\rho_1+0.45\rho_2+0.05\rho_3 \quad (2)$$

$$M_c=\rho_{eff} \times \forall_T \quad (3)$$

where m_1, m_2, m_3 are Individual mass, ρ_1, ρ_2, ρ_3 are individual density, K_1, K_2, K_3 are Individual thermal conductivity of the powders respectively and M_c is the total mass of the composite powder ,ρ_{eff}, K_{eff}, & $\forall_T$, Total Effective density, thermal conductivity and Volume of the coatings respectively.

Table 1. Individual powders mass and density values obtained from the rule of mixture.

Topcoat powders	Al_2O_3	ZrO_2	CaO
Mass, (m= wt.% ×M_c), (g)	117.46	105.71	11.74
Density(g/cm^3)	3.95	5.68	3.35
Bond Coat Powders	Fe	Ni	Al
Mass (wt.% ×M_c), (g)	201.29	147.09	38.71
Density(g/cm^3)	4.09	3.38	0.27

3. Methodology

A two-dimensional a physical model developed using commercial software (Ansys 15). A steady-state heat transfer analysis carried on the topcoat, TGO, BC whose thickness individually varies from 1mm to 0.04 mm. The heat is applied from the topcoat side (Al_2O_3+ZrO_2·5CaO) considered as a heat source and a substrate end considered as a sink in the present model. The lateral side of the topcoat subjected to 375 ºC and the remaining sides were kept at the ambient temperature of 22 ºC .The selection of the temperature is based on approximate temperature commonly observed in the case of internal combustion engines (diesel engines). A perfect bonding established at the interfaces of the graded coatings. The physical model along with boundary conditions is shown in Fig. 1(a) & (b). The thermal conductivity of the topcoat and bond coat calculated using the rule of mixture discussed and obtained in section 2. A quadratic element was used to mesh the model. The total number of nodes and elements developed in the 2D model is approximately 17980 and 17280 respectively.

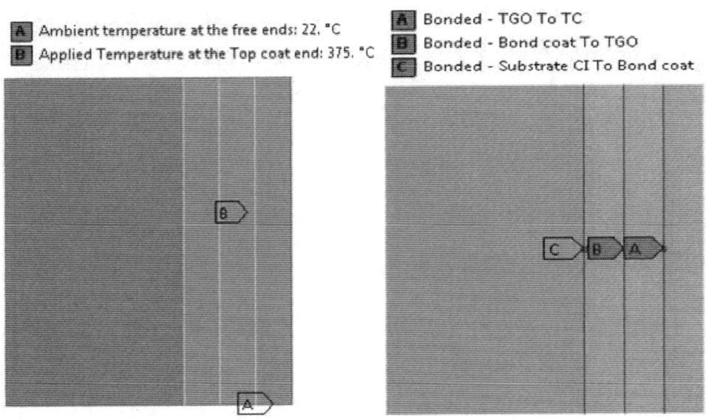

Figure 1. 2D physical model with (a) applied boundary condition (b) applied bonded contact relation at the interfaces

4. Results and Discussion

The thermal conductivity (K_{eff}) of Fe38Ni10Al (BC) and Al_2O_3+ZrO_2·5CaO (TC) was found to be 69.99 W/m^{2}k and 6.8329 W/m^{2}k respectively. Figures 2(a),(b),(c),(d),(e) & 3(a),(b),& (c) respectively show temperature and heat flux variation in topcoat, thermally grown oxide, bond coat, the substrate (CI) ,and composite systems. In the topcoat, no significant change in temperature drop noticed when the coating thickness varies from 1 to 0.04 mm refer Fig 2(a). It is understood that if the coating thickness varies in this range, temperature drop can be arrested. Also, one of the apparent reasons may be attributed to a very low thermal conductivity value, 6.893 W/m^{2}k obtained for the topcoat (Al_2O_3+ZrO_2·5CaO) mixture. However, temperature gradient (temperature drop) observed at all the thickness in the graded coating systems. In the case of TGO, BC, and substrate as the coating thickness change from 1 to 0.04 mm, a significant amount of temperature drop observed refer Fig.2 & 4. The reason for significant change attributed to the comparatively large magnitude of K_{eff} in case of BC (69.99 W/m^{2}k) $Fe_3$8Ni10Al whereas in case of TGO (K_{eff}=7 W/m^{2}k) drop in temperature observed less compared to BC. The temperature distribution at TC, TGO, BC, substrate & on overall coating systems is shown in Fig.2 (a), (b), (c), (d) and (e) respectively. From the Fig.4 a constant decrease in temperature understood when the thickness of the coatings (BC, TGO & Substrate) varied from 0.04 to 0.09 and the rectangular parabolic trend observed when the thickness changes from 0.1 to 1 mm. The effect of thickness of TGO has an important significance in the design of graded coating systems [12]. In the heat transfer analysis, the heat flux behavior was also examined and found that the concentration of heat flux is more at the free edge compared to the coatings in contact. Also, an abrupt change in heat flux noticed when thickness varies in the range of 0.04 to 0.1 mm refer Fig.5. The reason attributed to an abrupt change in temperature over a small area where the heat couldn't able to transfer the heat due to direct exposition to ambient temperature, however, the dissemination of heat flow found easily as heat flows from the free edge refer to Fig.3.

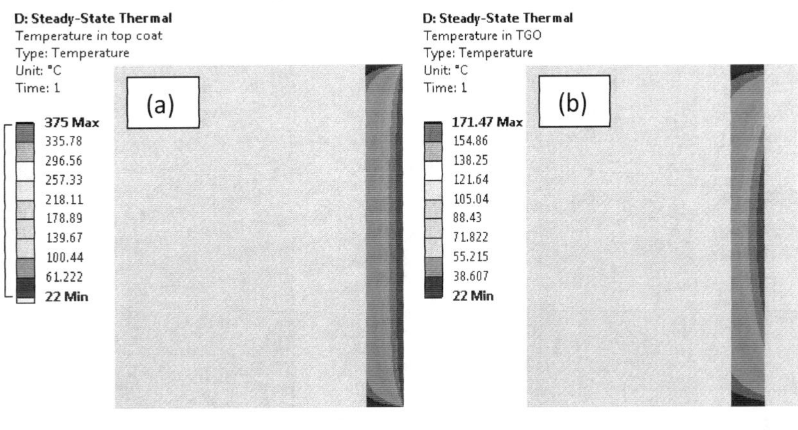

IOP Conf. Series: Materials Science and Engineering **653** (2019) 012022 doi:10.1088/1757-899X/653/1/012022

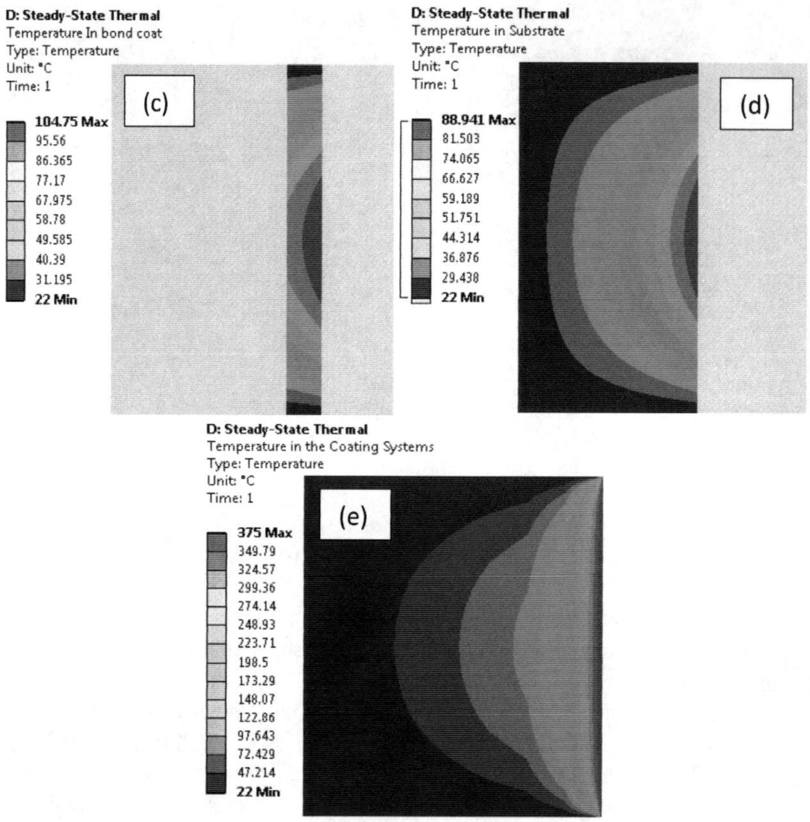

Figure 2. Schematic of temperature drop in (a) TC, (b) TGO, (c) BC, (d) Substrate(CI), (e) In composite systems

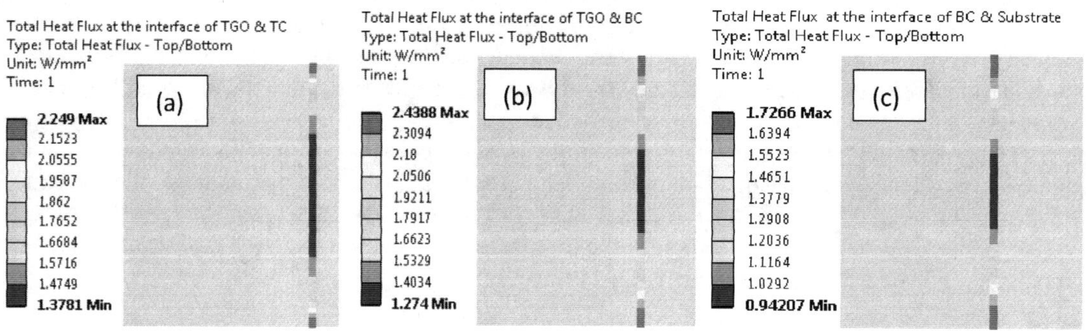

Figure 3. Schematic of change in Heat flux at the interfaces of (a) TGO and TC, (b) TGO and BC, (c) BC and substrate

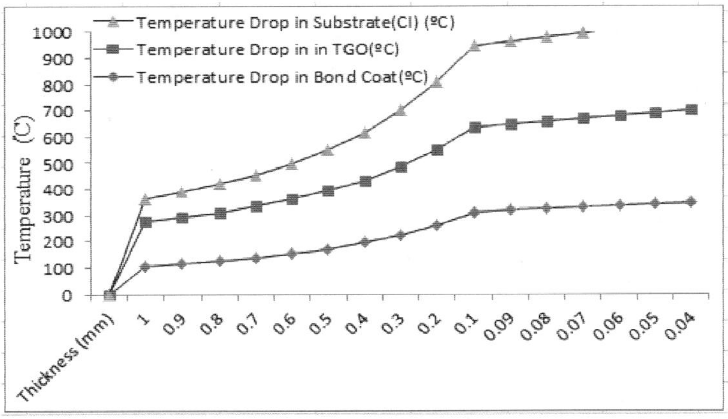

Figure 4. Temperature variation in substrate ,thermally grown oxide and bond coat.

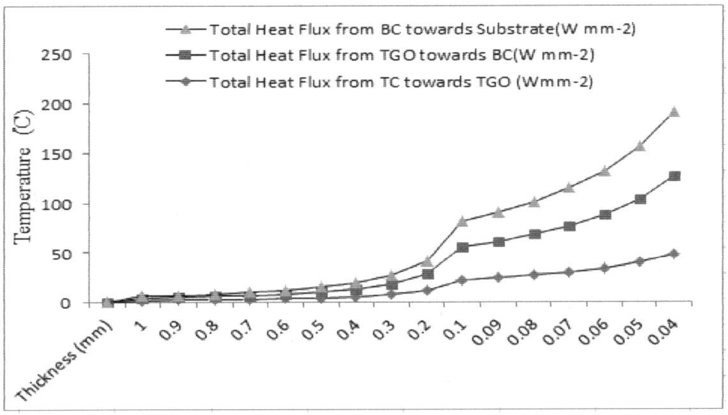

Figure 5. Heat flux variation in substrate ,thermally grown oxide and bond coat.

5. Conclusions

1. In the present graded coating system (Al_2O_3+ZrO_2·5CaO) an optimal coating thickness below 1 mm is recommended for extended life span.
2. Linear and Non-linear parabolic temperature signature examined at the thickness of 0.04 to 0.09 mm and 0.1 to 1 mm respectively irrespective of TC, BC, and TGO. However, change in heat flux shows a parabolic signature at above thicknesses.
3. Optimization results revealed that the concentration of heat flux was found to be more at the free end compared to the coatings in contact.
4. Since the conductivity of the bond coat is more compared to TGO, significant temperature drop observed in the bond coat compared to TGO. Therefore, from the above studies, it can be anticipated higher the bond coat thickness lesser will be the temperature drop.

References

[1] Ksi K., Arthur, J., 2013, Predicting failure within TBC system: Finite element simulation of Stress within TBC system as affected by sintering of APS TBC, geometry of substrate and creep of TGO. *Engineering Failure Analysis*,Vol. 27, pp 150-164.

[2] T.-J. Chuang, E. R. Fulle Jr.,2002,Analysis of Residual Stress State in Thermal Barrier Coatings,*Fracture Mechanics of Ceramics*,Vol. 13, pp 169-178 .

[3] Hermosilla, U., Karunaratne, M. S. A., Jones, I. A., Hyde, T. H., & Thomson, R. C., 2009, Modelling the high temperature behaviour of TBCs using sequentially coupled microstructural–mechanical FE analyses, *Materials Science and Engineering: A*, Vol.513, pp 302-310.

[4] Busso, E. P., Lin, J., Sakurai, S., & Nakayama, M. 2001, A mechanistic study of oxidation-induced degradation in a plasma-sprayed thermal barrier coating system: Part I: model formulation. *Acta materialia*, Vol. 49(9), pp 1515-1528.

[5] Bednarz, P, 2006, Finite element simulation of stress evolution in thermal barrier coating systems *Ph.D. thesis*, Forschungszentrum Julich GmbH, Julich.

[6] Schweda, M., Beck, T., & Singheiser, L., 2012, Thermal cycling damage evolution of a thermal barrier coating and the influence of substrate creep, interface roughness and pre-oxidation *International Journal of Materials Research*, Vol. 103(1), pp 40-49.

[7] Czech, N., Juez-Lorenzo, M., Kolarik, V., & Stamm, W., 1998, Influence of the surface roughness on the oxide scale formation on MCrAlY coatings studied in situ by high temperature X-ray diffraction, *Surface and coatings Technology*, Vol.108, pp 36-42.

[8] Lu, T. J., Levi, C. G., Wadley, H. N., & Evans, A. G., 2001, Distributed porosity as a control parameter for oxide thermal barriers made by physical vapor deposition, *Journal of the American Ceramic Society*, Vol.84(12), pp 2937-2946 .

[9] Hille, T. S., Turteltaub, S., & Suiker, A. S. J., 2011, Oxide growth and damage evolution in thermal barrier coatings. Engineering Fracture Mechanics,Vol.78(10),pp 2139-2152.

[10] Kaveh Torkashvand et al.,June 2018, Effect of TGO thickness on the Thermal Barrier Coatings Life under Thermal Shock and Thermal Cycle Loading, *Ceramic International*, Vol.44 (8),pp 9283-9293.

[11] Vaben, R., Giesen, S. and Stöver, D., 2009, Lifetime of Plasma-Sprayed Thermal Barrier Coatings: Comparison of Numerical and Experimental Results. *J. Therm. Spray Technol.,*Vol. 18(5-6),pp 835-845.

[12] Kaveh Torkashvand et al., September 2018, Effect of temperature and ceramic bonding on BC oxidation behavior in plasma-sprayed thermal barrier coatings, *Surface & Coatings Technology*, Vol. 349, pp 177-185.

IOP Conf. Series: Materials Science and Engineering **653** (2019) 012023 doi:10.1088/1757-899X/653/1/012023

Study of the output responses using Utility concept coupled with Taguchi Philosophy during EDM machining of nickel-based superalloy Inconel 718

Dipraj Banik[1], *Rahul[2], Subhankar Bhattacharjee[3], Biswajit Debnath[4], Hu Huang[5], B.C Routara[6]

[1,3,4,6]KIIT Deemed to be University, Bhubaneswar, Odisha, India.
[2,5]School of Mechanical and Aerospace Engineering, Jilin University, Changchun, 130025, China

*Corresponding author: rahulkumar589@gmail.com

Abstract. The primary objective of this research paper is to get the optimum combination of input process parameters such as peak current (IP), open voltage (Vg), pulse-on-time(Ton) & duty factor(τ) for EDM machining of Inconel 718. The output process parameters that we got and studied after experiments were materials removal rate (MRR), electrode wear rate (EWR), and surface roughness (SR). An orthogonal array of process parameters was created, using Minitab software as the design of experiments (DOE). For a single domain of multiple output response parameters, we used the utility method. After that, we used the Taguchi optimization method to find out the optimal parameter setting for higher materials removal rate (MRR), lower electrode wear rate (EWR), and lower surface roughness (SR).

Keywords: EDM, Inconel 718 Super Alloy, Copper tool, Utility Method, Taguchi Philosophy.

1. Introduction: Background and Motivation

This experiment proved a challenge to the modern machining process and led to the Advanced Machining Process in the field of Technology of Manufacture. Inconel 718 is a nickel-based alloy, which is a relevant material. The key benefits of using this Inconel 718 material are improved yield strength, tensile strength, and high-temperature resistance capacity. This Inconel 718 alloy is used during the manufacturing of jet engines and various airplane parts such as blades, discs, etc. A most significant and remarkable technology Electro Discharge Machine (EDM) is used in doing this experiment with Inconel 718. Here, in this technology, place both the tool and the electrolyte in the dielectric and then apply a high potential, thereby producing a high pulse spark. A large amount of heat will generate and dissipates, which, in turn, melt the workpiece. We use EDM for doing this experiment because Inconel 718 is a tough material, and it cannot be melt by another machining process.

Rahul et al. [1] have studied the surface roughness and metallic properties of Inconel 825 during EDM machining using CTT and NTT. He found a set of parameters for which surface crack density is less while EDM machining using CTT concerning NTT. Further, he has studied the factors affecting the tool life. Singh et al. [2] has done the experimental study on Inconel 601 in EDM using RSM and studied that the increase in current MRR increases, and surface finish varies directly with the current. Vishnu et al. [3] have presented an experiment with Inconel 718 using EDM for predicting the MRR, SR, and EWR. Jeykrishnan et al. [4] studied on Inconel 625 using Electro Discharge Machine to find the MRR and studied that surface finish is directly proportional to the current. Kumar et al. [5] studied on Inconel 825 using Electro Discharge Machine and studied the machinability of Inconel 825 using Graphene Nanopowder and predict the results of MRR, EWR, and R_a. Raju et al. [6] have presented an experimental study on Inconel 600 using micro-electro Discharge Machine (EDM) to find the optical parameter for the EDM Drilling process. Tanjilul et al. [7] have presented an experimental study on Inconel 718 using Electro Discharge Machine (EDM) for drilling holes on Inconel 718 through Spark erosion. Basha et al. [8] have presented an experimental study on Inconel X-750 using Electro Discharge Machine (EDM) to investigate the MRR, TWR, SR and also studied the influence of this parameter on the tungsten-copper electrode. Rahul et al. [9] also studied the optimization of Inconel 601, 625, 718, and 825 superalloy parameters were carried out by a 5 factor 4 L16 (OA) based EDM machine, and the best results were calculated using the Topsis and Taguchi methods. Rahul et al. [10] did experimental study on the processability of processing Inconel 718 with copper tools. They conducted research based on L25 OA by adopting various parameters such as Vg. Ip, Ton, τ, and Flush pressure and received various responses like TWR, Ra, SCD, and WLT.

The primary objective of this experimental work is to investigate the output responses like Material removal rate (MRR), Electrode Wear Rate (EWR), Surface Roughness (Ra) on a nickel-based alloy Inconel 718 using Electro Discharge Machine (EDM) by giving input process parameters like Peak Current (IP), Open Voltage (Vg), Pulse-on time (Ton) & Duty Factor (τ).

2. Experimental details and data collection

Inconel 718, which is a super hard nickel-based alloy, the square workpiece having a dimension (50 mm×5 mm) is used as a workpiece(Fig.1). The tool electrode of Copper (20mm diameter) has been used(Fig.2). An experimental study has been done on EDM (Model: ELEKTRA EMS 5535 Machine, Pune, India) setup(Fig.3). For dielectric fluid, EDM oil 30 is used. The EDM fluid has the viscosity of 36 ssu at a temperature of 38°C. The polarity of the workpiece has been chosen as a positive. The experimental design is taken as per 4-factor-4-level L_{16} Orthogonal Array (OA). Here peak current, voltage, pulse on time, and duty factor are considered as the input process parameters where each of these varied at four different levels of affecting experiments (Table.1). For every machining process, time taken as 30 minutes. Then the output parameters like MRR, EWR, SR are studied and recorded for every experiment.

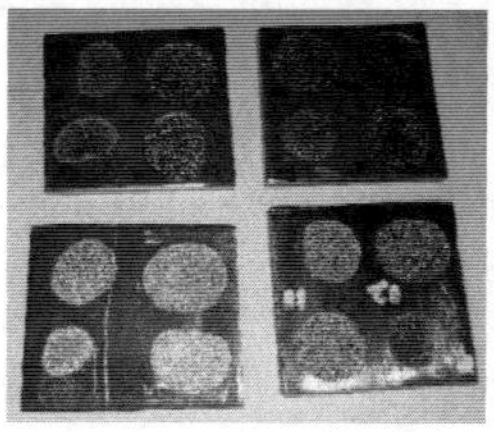

Fig. 1. EDMed Inconel

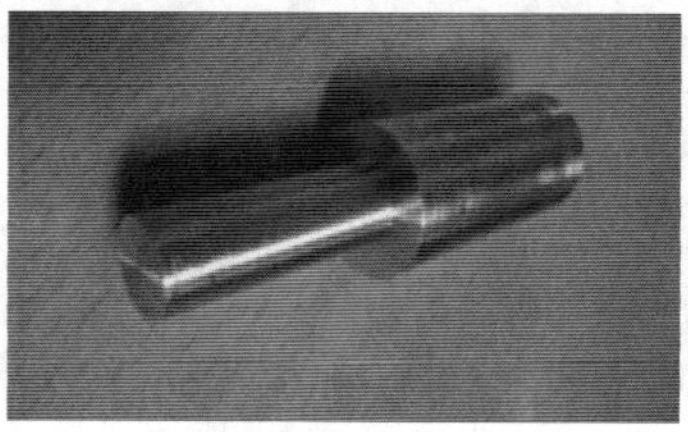

Fig.2. Copper tool

Fig.3. EDM Machine

Table 1. The domain of experiments

Input Parameters	Unit	Symbol	Levels of Variation			
			1	2	3	4
Peak Current (I_P)	Amperes	A	8.5	10.5	12.5	14.5
Open Voltage(V_g)	Volts	B	16	21	26	31
Pulse-On-Time(T_{on})	μ secs	C	52	102	152	202
Duty Factor(τ)	%	D	49	66	83	100

3. Methodology

3.1. Utility Method

The definition of utility is the aptness of any job or any procedure concerning the consumer's level of expectancy. The appraisement and calculation of any machining process lie upon various characteristics of output. Hence, an integrated measure is required to collect its overall execution that makes sure that the correlative performance of all quality features should take into account. We have to first calculate the individual utility of a product; then, we have to calculate the overall utility of a process, which is represented by the composite index. This utility method gives a method to construct a frame for the calculation of substitute characteristics, which is made by any person, firm, and organization. The actual meaning of utility is the satisfaction provided by all attributes to a decision-maker. Hence, utility theory takes an assumption that every decision should be formed based on the principle of maximization of utility, which means that the ideal option should be made that gives the best satisfaction to any judgment maker. So, utility theory says the best choice is such, which provides the best satisfaction to the decision-maker.

The preference number can be expressed as:

$$P = A \times \log\left(\frac{X_i}{X_i^*}\right)$$

Here, P is performance number; A is constant, X_i =Measurement of any quality characteristic or attribute i and X_i^* is a receivable value of quality characteristic or attribute i.

$$A = \frac{9}{\log\left(\frac{X^*}{X_i^*}\right)}$$

Here, X^* is the optimum value.

Now, overall utility function can be written as:

$$U = \sum_{i=1}^{n} W_i P_i$$

Here, P_i is the performance number for each attribute, and W_i is weightage of each attribute.

4. Data Analysis: Results and Discussions

The main purpose of this work is to find the optimal set of parameters for high metal removal rates (i.e., peak current, gap voltage, pulse on time, and duty cycle), but electrode wear rate and surface roughness are low. In this utility method, an individual utility for each response, such as MRR, EWR, and SR is calculated. Then overall utility is calculated by taking equal weightage of each response (Table.2). Here we have used the Taguchi method to find the optimum values of the overall utility function ($U_{overall}$) (Table.3). As the $U_{overall}$ is required to be the highest, so, Higher-is better (HB) property is used for S/N ratio calculation (Fig.4).

Table 2. Calculations of utility value corresponding to each response

Sl. no	Individual Utility			Overall Utility	Corresponding S/N ratio	Predicted S/N ratio
	U_{MRR}	U_{EWR}	U_{SR}	$U_{Overall}$	dB	dB
1	6.492986	0.000000	7.214865	4.523591	4.523590	7.793980
2	7.962413	3.890880	5.781202	5.819383	5.819380	
3	5.832318	7.756115	7.238679	6.872947	6.872950	
4	5.832436	8.062941	6.084117	6.593233	6.593230	
5	9.004931	3.890880	4.228906	5.651157	5.651160	

6	7.051666	6.583450	0.000000	4.499588	4.499590
7	8.962878	7.387633	5.254880	7.129779	7.129780
8	7.793980	8.999866	7.062117	7.443140	7.443140
9	6.482978	3.890880	3.652765	4.628786	4.628790
10	7.051666	0.406891	4.548048	3.962180	3.962180
11	0.000000	3.890880	2.296734	2.041913	2.041910
12	2.511648	7.930026	2.754982	4.354897	4.354900
13	8.926505	0.274554	5.495288	4.849794	4.849790
14	5.832318	3.890880	8.003187	5.849707	5.849710
15	6.492986	5.692971	8.999993	6.991363	6.991360
16	5.023593	6.153977	6.544025	5.848126	5.848130

Table 3. Response table for means of $U_{overall}$

Level	A	B	C	D
1	5.952	4.913	4.228	5.464
2	6.181	5.033	5.704	5.039
3	3.747	5.759	6.199	5.584
4	5.885	6.060	5.634	5.678
Delta	2.434	1.147	1.970	0.640
Rank	1	3	2	4

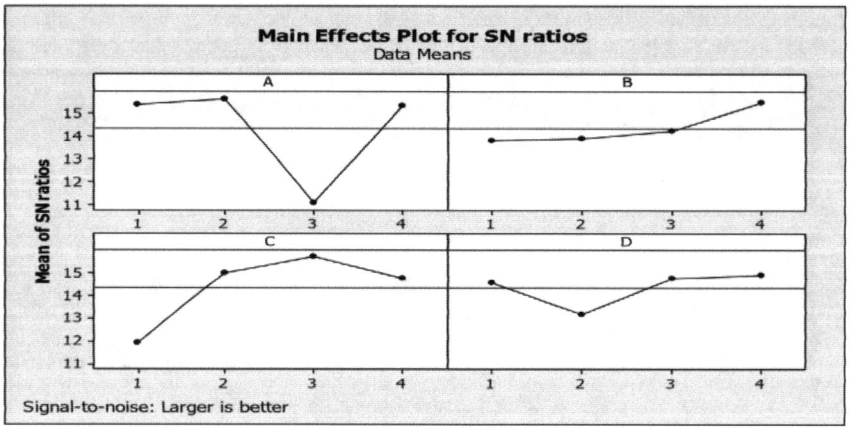

Fig.4. S/N Ratio Plot: Evaluation of Optimal Setting (Optimization of $U_{Overall}$) [Optimal Setting: $A_2B_4C_3D_4$] (Inconel 718)

5. Conclusion

The key points of this research work have been summarized below.

- In this research work, utility function integrated with Taguchi's philosophy used to determine an optimum set of EDM parameters (i.e., gap voltage, peak current, pulse-on-time, and duty factor) such that MRR will be highest, but EWR and SR will be lowest simultaneously. The proposed utility function was instrumental in integrating multiple responses into an equivalent single index (i.e., $U_{overall}$ in the present case), which was optimized finally by the Taguchi method. ➢ The optimal process parameters setting appears as $(A_2B_4C_3D_4)$ i.e. I_P=10.5 A, V_g=31 V, T_{on}=152 μs and τ=100%.
- The mean response table (mean S/N ratio of $U_{overall}$) shows that the peak current (I_p) is the most significant parameter to influence machining performances.

6. References

[1] Rahul, Datta S., Biswal B. B., Mahapatra S.S 2017 Electrical discharge machining of Inconel 825 using cryogenically treated copper electrode: Emphasis on surface integrity and metallurgical characteristics. *J. Manuf. Processes* **26** pp 188–202

[2] Singh N., Routara B.C, Das D 2018 Study of machining characteristics of Inconel 601in EDM using RSM. *Mater. Today: Proc. 5* pp 3438–3449

[3] Vishnu P, Kumar N.S, Manohar M 2018 Performance Prediction of Electric Discharge Machining of Inconel-718 using Artificial Neural Network. *Mater. Today: Proc. 5* pp 3770-3780.

[4] Jeykrishnan J, Ramnath B.V, Ram N.C, Nikhil K, Babu R.N, Naveen D 2018 Investigation on the Influence of Performance Characteristics on Machining Inconel 625 alloy in Electro-Discharge Machining (EDM). *Mater. Today: Proc. 5* pp 20449-20454.

[5] Kumar A, Kumar S, Mandal A, Dixit A.R 2018 Investigation of powder mixed EDM process parameters for machining Inconel alloy using response surface methodology. *Mater. Today: Proc. 5* pp 6183-6188.

[6] Bhosle R. B, Sharma S.B 2017 Multi-performance optimization of micro-EDM drilling process of Inconel 600 alloy. *Mater. Today: Proc. 4* pp. 1988-1997.

[7] Ahmed M.T.A, Kumar A.S, Rahman M 2018 A study on EDM debris particle size and flushing mechanism for efficient debris removal in EDM-drilling of Inconel 718. *J Mater Process Tech*, 255 pp. 263–274.

[8] Bashaa S.K, Raju M.V.J, Kolli M 2018 Experimental Study of Electrical Discharge Machining of Inconel X-750 Using Tungsten-Copper Electrode. *Mater. Today: Proc. 5* pp. 11622-11627

[9] Rahul, Abhishek K., Datta S., Biswal B. B., Mahapatra S. S. Machining performance optimization for electro-discharge machining of Inconel 601, 625, 718 and 825: an integrated optimization route combining satisfaction function, fuzzy inference system and Taguchi approach. *J. Braz. Soc. Mech. Sci. Eng.*, **Vol. 39**, Issue 9, pp. 3499–3527

[10] Rahul, Datta S., Biswal B. B., Mahapatra S.S 2018 Optimization of Electro-Discharge Machining Responses of Super Alloy Inconel 718: Use of Satisfaction Function Approach Combined with Taguchi Philosophy. *Mater. Today: Proc. 5* pp 4376-4383

A review on Manufacturing, Machining, and recycling of 3D printed composite materials.

P.S.Ujeniya[1] and N.V.Rachchh[2]

[1]Assistant Professor [2]Associate Professor and Head, Rajkot, India

prashantujeniya@gmail.com

Abstract. This paper includes a comprehensive review of the literature on the latest research and development in 3D printed composite materials and its behavior with several machining processes. Machining parameters need to be considered while performing an operation like drilling, turning, shaping, and surface finishing on 3D printed composite material for providing some useful new application with excellent benefits. Also, recycling of plastic waste which is the most challenging hurdle of society nowadays is discussed, and the scope of its real-life implementation is observed.

Keywords: Additive manufacturing, Machining of composites, recycling of plastic waste.

1. Introduction

It's a very well known that composite provides superior strength to weight ratio, excellent fatigue, and corrosion resistance. Apart from all these qualities, composite material provides long life durability and light weight structure [1]. A composite is considered as a structural material in which 2 or more than 2 materials that are united at a macroscopic level. Any one constituent is reinforcing phase, and the second is matrix phase. Reinforcing material may be in form of small scrappy, flakes or fibers and the matrix phase constituents are generally uniform [2].

Additive manufacturing has become ordinary for the moulding parts and objects for uneven geometries and structures. Latest innovations that validate formation of composite materials shows that the additive manufacturing technology or commonly known as 3D printing can offer a much broad design space besides unusual forming [3].

3D printing or additive manufacturing is emerging as an enabling technology in recent years for a wide range of applications. It can be stated as the "joining process for materials to manufacture parts or model from 3D modeling data, by applying layer upon layer, as opposed to subtractive and formative manufacturing technologies" Additive manufacturing is versatile regarding materials. The versatility of 3D printing material comes from the huge system variety, For all application printing materials are still having shortage and further research and development of materials still in progress. There are two main challenges in additive manufacturing technologies: (1) Improvement in resolution and velocity of the printing process with lower power consumption. (2) It requires developing new printing materials with more effective physical, chemical and mechanical properties [4].

Carbon fiber reinforced polymers/plastics (CFRP) offers excellent mechanical strength. Post machining of CFRP is a mandatory step that assures that the fabricated parts form their measurements, shape and size and also meet their surface quality tolerances along with other functional requirements. Also the machining of CFRPs now a day's considered an very difficult process due to the non homogeneous structure, abrasive nature and highly nonlinear. Processes in CFRP machining, unconventional hybrid processes, and conventional processes, were invented based on the requirements of market [5].

2. Additive Manufacturing (AM)

3D printing technique or Additive manufacturing that takes the digital data from a CAD file that is later converted to STL file and produce a model, sculpture or parts as shown in figure 1. 3D printing is easier for architects, automobiles, mechanical components, and many more applications. Still, the precision requires improvement to reduce the requisite of a finishing process and to become able to manufacture models that demands the high level of accuracy [6]. From the dependency on the material, many problems occur during the manufacturing process, and the manufactured part gives many varieties in mechanical properties. By experimenting on research for reproducing textile structures, it is found that textile-based structures are 3D printed using SLS (selective laser sintering) and FDM (fused deposition modeling method), While Acrylonitrile butadiene styrene (ABS) material has turned out to be often brittle for the desired structures [7].

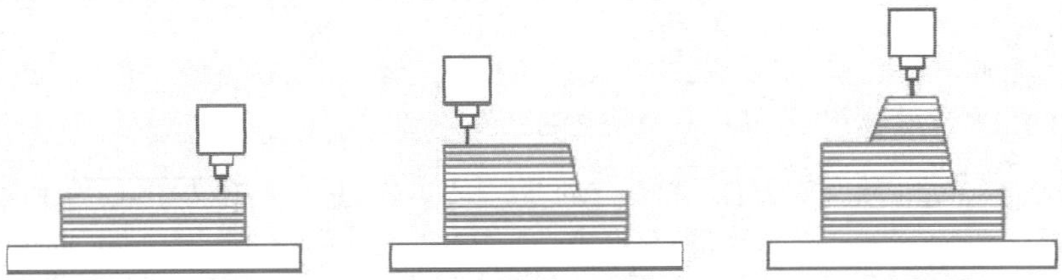

Figure 1. 3D printing technology

Additive manufacturing application is limited by printing materials available. Currently, thermoplastic polymer, materials formed in powder and photopolymers are possible to use in 3D printing. Moreover, this limitation could not fulfill the requirement of additive manufactured parts with huge verities in industries [8]. 3D printing is technology which is rapidly expanding in industrial sectors. However, it's growth is still hampered by lower mechanical properties, low productivity, poor quality and uncertainty of final product. The most commonly modeled additive manufacturing process is SLA, followed by FDM and SLS/SLM (selective laser melting) [9].

The ability to produce complete microfluidic parts or model from a CAD model has an obvious attraction. Current states for 3D printing of microfluidics, specifically focus on four widely used 3D printing methods are Inkjet, FDM, 2PP, and SLA. Fused deposition modeling printer allows for the use of a wide range of thermopolymers and gives material compatibility with mass production. The resolution of rough channel surfaces that can be obtained in FDM is 100 μm. in the future fabrication of micro channels needs to be lowered to 10 μm, in a range of different materials with variation in physical and chemical properties [10]. FDM is limited by the particular operational materials for respective application. Printing of non-conventional polymer Carbon nanotubes (CNT) and grapheme based polybutylene terephthalate (PBT) nanocomposites on a currently available 3D printer are illustrated by K. Gnanasekaran et. Al, and evaluation is done for conductivity of electricity, printability of material and stability in structure at microlevel of the composites before & after additive manufacturing. 3D printed parts with PBT/CNT gives higher mechanical properties, and electrical conductivity and overall better performance compared to PBT/grapheme models fabricated by 3D printing technology. Additionally to printing of multi-materials, problems faced in abrasive fillers are also discussed. The most favorable printing conditions are not changing for the PBT/G and PBT/CNT [11].

3. Manufacturing of composite parts using 3D printing technology

Recent additive manufacturing process presents the capacity to create a fine shaped structure with difficult shapes and size, thus eliminating the need for joints/fasteners and multi-step processing. AM provides unique chance for the manufacture of multi-directional performs [12]. Fused deposition modeling additive manufacturing technique is used to manufacture continuous carbon, glass and Kevlar fiber reinforced composites in Markforged Mark One 3D printer. The investigation made on the influence of volume fraction, fiber orientation and fiber type; tensile strength values ware up to 6.3 times that obtained with the non-reinforced polymer [13].

The AM technique known as 3-dimensional printing technology has attracted much more attention in industry and academics in recent years [14]. Poly lactic acid (PLA) and Carbon fiber filament were used as matrix phase and reinforcing phase respectively, and supply into the FDM (fused deposition modeling) 3D printing process for manufacturing and forming CFRTPCs (Continuous Fiber Reinforced Thermoplastic Composites) [15]. The development of design to get benefit of the possibilities given by additive manufacturing and to eliminate the constraints connected with the technology has lagged [16]. AM holds tremendous promise regarding revolutionizing latest manufacturing processes, but the fundamental hurdles limit the use of this technology like lower production rate, physical size, poor mechanical properties [17].

For manufacturing continuous carbon fiber thermosetting composite 3D printing process was created to design and fabricate the composites on FDM. A 3D printed thermosetting composite material gives the superior mechanical performance compare to 3D printed thermoplastic composites [18]. Thermo-mechanical properties of new materials can be observed by manufacturing new polymer composite filament. The acrylonitrile butadiene styrene (ABS) thermoplastic was mixed with iron and copper particles in various proportions to find the effect of metal particles on various mechanical properties. It was observed that tensile strength decreased by increasing metal percentage and thermal conductivity increased by increasing the metal particles percentage [19].

Bio-inspired composite materials were manufactured by computationally designed 3D printing technology with predicted fracture response and compared to synthetic composites with having similar mechanical properties. The results point out that CAD models can be used to create new materials for printing to reveal fracture properties [20]. FDM technology has become one of the most used rapid prototyping methods because it offers the opportunity to design and manufacture new materials [21].

Additive manufacturing technology is now repeatedly spreads in many fields of research and development. A comparatively narrow range if 3D printing materials with a limited spectrum of physical, chemical and mechanical properties still constrain the competence of this undisciplined additive manufacturing technology [22].

4. Machining of 3D printed composite parts

Mechanical machining has been repetitively used for the fabrication of hybrid CFRP (carbon fiber reinforced polymer) stacks in order to ensure suitable applications. Due to inhomogeneous and poor machinability, CFRP machining has bought difficult challenges to the latest designing and fabrication industries. Drilling experiment of CFRP/Ti stacks were executed on a 5-axis CNC (Computerized Numerical Control) machining center in dry cutting condition. The interface machining is usually most challenging cutting stage in drilling due to multi-too-work interaction existence, in which undesirable transformation of physical/mechanical reactions dominate and chip formation modes coupled [23]. Aerospace composite and alloys need high-performance cutting performance. Research in tool development and its impression on surface integrity characteristics & tool wear were studied for titanium, nickel-based alloys, and composites. Recent machining problems have now come out with the research of even more latest aerospace alloys and composites [24].

The combination of glass and carbon fibers in composite materials creates the machining of these excellent materials challenging. Machining of CFRP/GFRP (Glass fibre reinforced plastic) material gives poor hole qualities, experiments shows that CVD diamond coating cannot achieve any advantages [25]. Turning is a 3D process through the primary and secondary cutting edges of part; it can be make process easier into an orthogonal cutting. The emerging and advanced field of multi-scale 3D printing is playing a principle part in the machining for composite laminates. Powder metallurgy were used to manufacture Metal matrix composites, and the effect of milling characteristics on cutting force and surface roughness with the use of uncoated carbide tool were investigated. ANOVA (Analysis of variance) was used to describe the impression of the cutting characteristics. The result determines that the structure of material was the highly effect full factor on roughness of surface and the force of cutting is affected by feed rate [26].

The effect of cutting parameters like feed rate, cutting speed and angle of approach was investigated on roughness in turning process of Al 7050 composite and Al 7050 hybrid composite by using the polycrystalline diamond tool. ANN (Artificial neural networking) and response surface methodology (RSM) is applied for comparison of the experimental results. Cutting speed and interaction of cutting speed and feed rate affect surface roughness. Involvement Graphite particle improves the surface roughness [27].

5. Recycling of composite materials by 3D printing

CFRTPCs (Carbon fiber reinforced thermoplastic composites) with comparatively very high performance have been implied at the ground level for recycling of 3D printing of continuous carbon reinforced poly lactic acid composites. The tensile strength of recycled fiber filament observed during experimentation was much more than originally printed without CFRTP composites, and in bending strength, 25% of improvement was observed. Recovery rate of material was 100% for CF (carbon fiber) and 73% for PLA (poly lactic acid) were gained for a higher environmental and social impact. Use of energy for recycling was near about 67 MJ/Kg, comparison was made of this recycled composite with previous composite which made up with conventional method [28].

Many researches in 3D printing technology give opening to recycle thermoplastics. Using wastage materials to manufacture composites in this manner enhance value to the polymer by developing excellent aesthetic and mechanical properties. Composite filament for 3D printing with differing hemp fiber, harakeke and recycled gypsum contents in recycled polypropylene were assessed and produced for mechanical properties. The highest tensile property observed in 30% harakeke fiber filament [29].

IOP Conf. Series: Materials Science and Engineering **653** (2019) 012024 doi:10.1088/1757-899X/653/1/012024

6. Conclusion

Additive manufacturing/3D printing technology is lagging due to higher production time and secondly limited availability of printing materials variety. A composite material offers superior mechanical, chemical and physical properties with higher durability. Combination of two or more materials successfully adopted as 3D printing materials but still much more research and development are required to fulfill the demand of society. After that machining of 3D printed part is more complicated compared to traditional metals because of unavailability in the design of machining especially for plastic parts. Also, the plastic waste is a huge problem now a day for the environment so, significant research and development is needed in a field of additive manufacturing to recycling plastic waste. Main objective of this work is to develop a novel composite filler material for additive manufacturing by recycling the Plastic waste and manufacture a test samples from that developed material. Mechanical characterization and Validation is planned to evaluate from the manufactured composite material for some useful application.

References

[1] C. R. Dandekar and Y. C. Shin, "Modeling of machining of composite materials : A review," Int. J. Mach. Tools Manuf., vol. 57, pp. 102–121, 2012.

[2] N. V Rachchh, P. S. Ujeniya, and R. K. Misra, "Mechanical characterisation of rattan fibre polyester composite," Procedia Mater. Sci., vol. 6, no. Icmpc, pp. 1396–1404, 2014.

[3] D. Kokkinis and M. Schaffner, "Multimaterial magnetically assisted 3D printing of composite materials,"Nature communications, 2015.

[4] J. Lee, J. An, and C. K. Chua, "Fundamentals and applications of 3D printing for novel materials," Appl. Mater. Today, vol. 7, pp. 120–133, 2017.

[5] D. Che, "Machining of Carbon Fiber Reinforced Plastics / Polymers : A Literature Review," Journal of Manufacturing Sci. and Engg., vol. 136, 2014.

[6] K. V Wong and A. Hernandez, "A Review of Additive Manufacturing," ISRN Mech. Engg. vol. 2012, 2012.

[7] R.Melnical et. al., "3D printing of textile-based structures by Fused Deposition Modelling (FDM) with different polymer materials," I.O.P. Science. vol. 012018, 2014.

[8] X. Wang, M. Jiang, Z. Zhou, J. Gou, and D. Hui, "3D printing of polymer matrix composites : A review and prospective," Compos. Part B, vol. 110, pp. 442–458, 2017.

[9] H.Bikas et. al., "Additive manufacturing methods and modelling approaches a critical review" Springer, 2015.

[10] Sidra Waheed et. al., "3D Printed Microfuidic devices: Enablers and Barriers"Lab on Chip, Springer, 2016.

[11] K. Gnanasekaran, T. Heijmans, S. Van Bennekom, H. Woldhuis, and S. Wijnia, "3D printing of CNT- and graphene-based conductive polymer nanocomposites by fused deposition modeling," Appl. Mater. Today, vol. 9, pp. 21–28, 2017.

[12] Z. Quan et al., "Additive manufacturing of multi- directional preforms for composites : opportunities and challenges," Biochem. Pharmacol., vol. 18, no. 9, pp. 503–512, 2015.

[13] A. N. Dickson, J. N. Barry, K. A. Mcdonnell, and D. P. Dowling, "Fabrication of continuous carbon , glass and Kevlar fibre reinforced polymer composites using additive manufacturing," Addit. Manuf., vol. 16, pp. 146–152, 2017.

[14] Kun Fu et. al., "Progress in 3D Printing of Carbon Materials for Energy Related Applications.", Advanced Materials, 2017.

[15] X. Tian, T. Liu, C. Yang, Q. Wang, and D. Li, "Composites : Part A Interface and performance of 3D printed continuous carbon fiber reinforced PLA composites," Compos. Part A, vol. 88, pp. 198–205, 2016.
[16] M. Kathryn et al., "Design for Additive Manufacturing : Trends , opportunities , considerations , and constraints," CIRP Ann. - Manuf. Technol., vol. 65, no. 2, pp. 737–760, 2016.
[17] L. J. Love, A. M. Elliott, B. K. Post, R. J. Smith, and C. A. Blue, "The importance of carbon fi ber to polymer additive manufacturing," Cambridge Journal, pp. 1893–1898, 2014.
[18] W. Hao, Y. Liu, H. Zhou, H. Chen, and D. Fang, "Preparation and characterization of 3D printed continuous carbon fi ber reinforced thermosetting composites," Polym. Test., vol. 65, no. September 2017, pp. 29–34, 2018.
[19] S. Hwang, E. I. Reyes, K. Moon, R. C. Rumpf, and N. A. M. S. O. O. Kim, "Thermo-mechanical Characterization of Metal / Polymer Composite Filaments and Printing Parameter Study for Fused Deposition Modeling in the 3D Printing Process," Journal of Electronic Materials, vol. 44, no. 3, pp. 771–777, 2015.
[20] L. S. Dimas, G. H. Bratzel, I. Eylon, and M. J. Buehler, "Tough Composites Inspired by Mineralized Natural Materials : Computation , 3D printing , and Testing," Advanced Fuctional Materials, pp. 1–10, 2013.
[21] P. Dudek,"FDM 3d printing technology in manufacturing composite elements," Archive of Metallurgy and Materials, Vol.58, Issue 4, 2013.
[22] U. Kalsoom, N. Nesterenko, and B. Paull, "Advances Recent developments in 3D printable composite materials," RSC Adv., vol. 6, pp. 60355–60371, 2016.
[23] J. Xu and M. El, "Experimental study on drilling mechanisms and strategies of hybrid CFRP / Ti stacks," Compos. Struct., vol. 157, pp. 461–482, 2016.
[24] R. M. Saoubi et al., "High performance cutting of advanced aerospace alloys and composite materials," CIRP Ann. - Manuf. Technol., vol. 64, no. 2, pp. 557–580, 2015.
[25] E. Uhlmann, F. Sammler, S. Richarz, G. Reucher, R. Hufschmied, and A. Frank, "Machining of carbon and glass fibre reinforced composites," vol. 46, pp. 63–66, 2016.
[26] M. Program, "Optimization of surface roughness and cutting force during AA7039 / Al 2 O 3 metal matrix composites milling using neural networks and Taguchi method," Mesurement, vol. 66, pp. 139–149, 2015.
[27] R. Kumar and S. Chauhan, "Study on surface roughness measurement for turning of Al 7075 / 10 / SiCp and Al 7075 hybrid composites by using response surface methodology and artificial neural networking," MEASUREMENT, vol. 65, pp. 166–180, 2015.
[28] X. Tian, T. Liu, Q. Wang, A. Dilmurat, D. Li, and G. Ziegmann, "Recycling and remanufacturing of 3D printed continuous carbon fi ber reinforced PLA composites," J. Clean. Prod., pp. 1–10, 2016.
[29] D. Stoof and K. Pickering, "Sustainable composite fused deposition modelling filament using recycled pre-consumer polypropylene," Compos. Part B Eng., vol. 135, pp. 110–118, 2018.

IOP Conf. Series: Materials Science and Engineering **653** (2019) 012025 doi:10.1088/1757-899X/653/1/012025

Role of microstructure on the degradation behaviour of friction stir processed AZ series Mg alloys assessed in simulated physiological solutions

V V Kondaiah [1], Ch Vasu [2], Ravikumar Dumpala [3] and B Ratna Sunil [4,*]

[1] Center for Design and Analysis, Department of Mechanical Engineering, CMR College of Engineering & Technology, Kandlakoya, Hyderabad 501401, India
[2] Department of Mechanical Engineering, Rajiv Gandhi University of Knowledge Technologies, Nuzvid 521201, India
[3] Department of Mechanical Engineering, Visvesvaraya National Institute of Technology, Nagpur 440010, India
[4] Department of Mechanical Engineering, Bapatla Engineering College, Bapatla 522101, India.

[*] corresponding author's e-mail address: bratnasunil@gmail.com

Abstract. In the current work, pure Mg and two Mg alloys (AZ31 and AZ91) were selected and processed by friction stir processing (FSP). After FSP, the effect of microstructure and the corroding environment on the degradation behaviour of processed and unprocessed materials in two different solutions (0.9M NaCl and Ringer's solution) by immersion method. The samples after immersion were characterized by X-Ray diffraction (XRD) method and electron microscopy. Interestingly, corrosion rate of FSPed samples was observed as decreased compared with unprocessed condition when immersed in Ringer's solu-tion. On contrary, the corrosion rate of FSPed samples was noticed as increased compared with unprocessed condition when immersed in 0.9M NaCl solution. The results strongly suggest that the effect of smaller grain size and the ion concentration of the simulated physiological solution have a significant influence on the degradation behaviour of Mg alloys which must be considered while developing Mg based biodegradable implant materials

1. Introduction

In recent years, magnesium (Mg) based materials attracted the attention of materials researchers to investigate as promising candidates for degradable implant applications. Bio-compatibility, nontoxicity and mechanical properties of Mg similar to human bone made selection of Mg based materials as promising for biomedical applications [1]. Mg corrodes in biological environment safely without leaving any abnormal health issues. However, the degradation rate is uncontrolled and rapid that invites special attention in producing Mg based implants [2]. Several new alloys and composites were developed in addition to surface coatings and microstructural modifications as the important strategies to tailor the bio-corrosion properties of Mg alloys suitable to produce temporary implants for bone fixing plates, nails and screws [3, 4, 5, 6, 7]. Altering the microstructure is one of promising strategies reported in the literature to develop Mg based biomaterials [8, 9, 10]. Friction stir processing (FSP) is a new method, recently developed to process metals and alloys to alter the microstructure

[11]. Without melting the substrate, FSP results fine grain structure. AZ series Mg alloys are widely used commercial alloys available with good mechanical proper-ties. Therefore, in the current work, AZ series Mg alloys have been selected and processed by friction stir processing (FSP) to alter the microstructure. Further, the role of smaller grains on degradation rate was assessed by using two different electrolytes (0.9M NaCl solution and Ringer's solution) to understand the role of corroding environment on the degradation properties.

2. Experimental methods

Commercial grade pure Mg (99% purity), AZ31 Mg alloy in the form of sheets (2.75%Al, 0.001%Fe, 0.91%Zn, 0.01% Mn and balance with Mg by wt.%) and AZ91 die cast Mg alloy in the form of billet (8.67%Al, 0.002%Fe, 0.85%Zn, 0.03% Mn and balance with Mg by wt.%) were purchased from M/S Exclusive Magnesium (Balanagar, Telangana, India) and workpieces (100 × 100 × 4 mm^3 size) were prepared and FSP experiments were done on a vertical milling machine having automated controls (Bharath Fridge Werner, Hyderabad, India). The elemental composition of the workpieces is listed in Table 1. FSP tool was fabricated with tool steel (H13) having 15 mm diameter of shoulder. Circular pin with tapered profile was selected. Pin diameter varies from 2 mm to 5 mm and length of the pin was 4 mm. Process parameters were selected based on the earlier experiments (1400 rpm and 25 mm/min feed) [12]. The sample were subjected to X-ray diffraction (XRD) analysis (D8, Bruker, USA) using Cu as source for radiation before and after the immersion test between 20 to 80° 2θ range. Samples from FSPed region and unprocessed materials were collected and microstructural studies (Leica, Germany) have been done as per the standard metallurgical procedure. Microhardness measurements (Omnitech, India) have been done on the FSPed samples (100 g load for 15 min). For degradation studies, samples (10×10×2 mm^3 size) were prepared from unprocessed and FSPed materials. Two solutions i) 0.9M NaCl solution and ii) Ringer's solution were prepared and the samples were immersed for 3 days. 0.9M NaCl solution was prepared by dissolving 52.605 grams of NaCl in 1 liter of de-ionized water. Ringer's solution was prepared by adding 7.2 g of NaCl, 0.17 g $CaCl_2$ and 0.37 g of KCl in one liter of water (de-ionized). pH of Ringer's solution was maintained at 7.2-7.3 using HCl and $NaHCO_3$ as buffer. The samples were immersed in 35 ml of NaCl solution and Ringer's solution. The containers with solution and samples were stored in a water bath maintained at 37°C. After 24 h, the samples were collected from the containers and gently cleaned and the corrosion products were removed by immersing in an aqueous solution of CrO_3. Then the corrosion rate of the samples was calculated from the weight loss (W) measurements as per procedure explained elsewhere [13].

3. Results and discussion

FSP led to develop smaller grains in pure Mg, AZ31 and AZ91 Mg alloys as compared with the starting size of the grains as shown in Fig. 1. From the microstructural observations, it is evident that the smaller grains appeared after FSP compared with unprocessed condition measured as 2000 μm, 5-25 μm and 160 μm to 7 μm, 4 μm and 3 μm for pure Mg, AZ31 and AZ91 respectively. Additionally, the amount of intermetallic phase also influences the corrosion behaviour of Mg based alloys. From Fig. 1 (c), appearance of a compound made of Mg-Al (β, $Mg_{17}Al_{12}$) can be seen at the grain boundaries of α-Mg as shown by black arrow. Furthermore, presence of two phase regions (α+β) as indicated by white arrow adjacent to β also can be seen from Fig. 1 (c). Due to FSP, grain size was observed as decreased in Mg, AZ31 and AZ91 (Fig. 1(d), (e) and (f)). Interestingly, the fraction of β phase has been decreased in AZ91 after FSP. This kind of microstructure certainly influences mechanical and corrosion behaviour. Fig. 2 shows the XRD plots of the workpieces. Interestingly, the peak intensities (basal plane) in pure Mg, AZ31 and AZ91 were observed as increased after FSP which indicates the texture effect. Furthermore, in AZ91 Mg alloys, the intensities of peaks belong to $Mg_{17}Al_{12}$ phase was observed as greatly decreased. This observation strongly suggests that the secondary phase has been decreased in AZ91 due to FSP. This observation is in good agreement with the microstructural observations where, the amount of intermetallic phase at grain boundary was noticed as disappeared as shown in Fig. 1 (f).

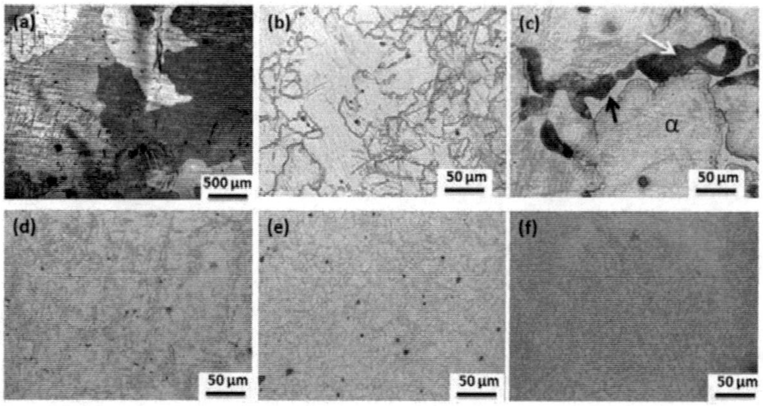

Figure 1. Optical microscope images of processed and unprocessed samples: a) pure Mg, b) AZ31, c) AZ91, d) FSPed pure Mg, e) FSPed AZ31 and f) FSPed AZ91 Mg alloys.

Microhardness measurements showed increased hardness for all the samples after FSP. This is attributed to the grain refinement effect. Fig. 3 shows the microhardness comparison of the samples. Along with the reduced grain size; reduced β phase and deceased area of eutectic regions as shown in FSPed AZ91 (Fig. 2 (f) play very important role on hardness. This is similar to our earlier observation [11]. The variations in microhardness within the AZ91 sample were observed as more before FSP. This can be explained by understanding the phases and their distribution in the AZ91 Mg alloy. When the hardness is measured on the regions containing hard and brittle β phase surrounded with eutectic zones, value is obviously higher compared with the α-grains. Therefore, the standard deviation for unprocessed AZ91 was observed as higher. However, after FSP, uniform grains with decreased β phase lead to show lower variations within the FSPed AZ91.

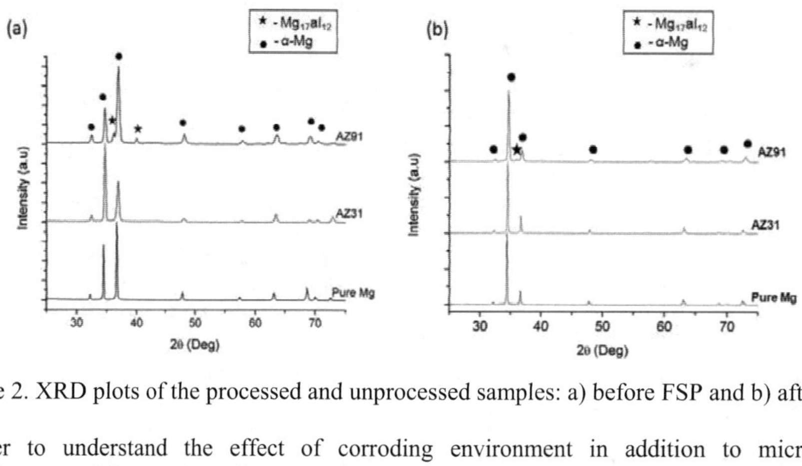

Figure 2. XRD plots of the processed and unprocessed samples: a) before FSP and b) after FSP

In order to understand the effect of corroding environment in addition to microstructural modification, two different physiological solutions were considered, one with rich chloride ions (0.9M NaCl) and other solution with additional ions which present in human blood plasma (Ringer's solution).

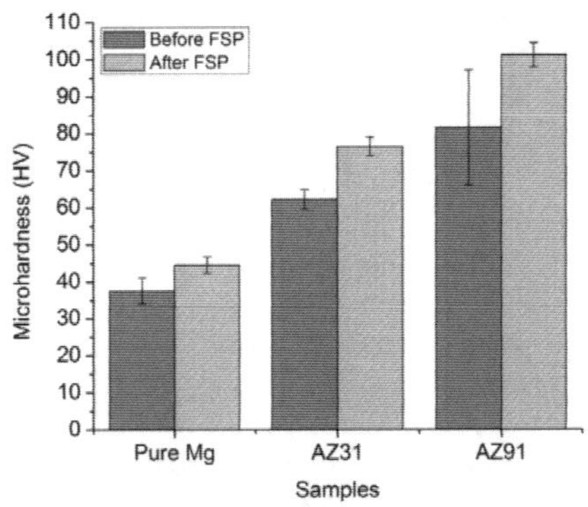

Figure 3. Microhardness comparison of the samples before and after FSP

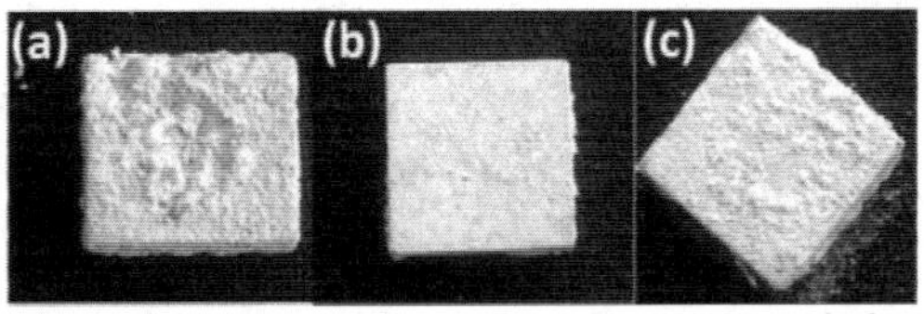

Figure 4. Photograph showing specimens after immersion tests carried out in Ringer's solution

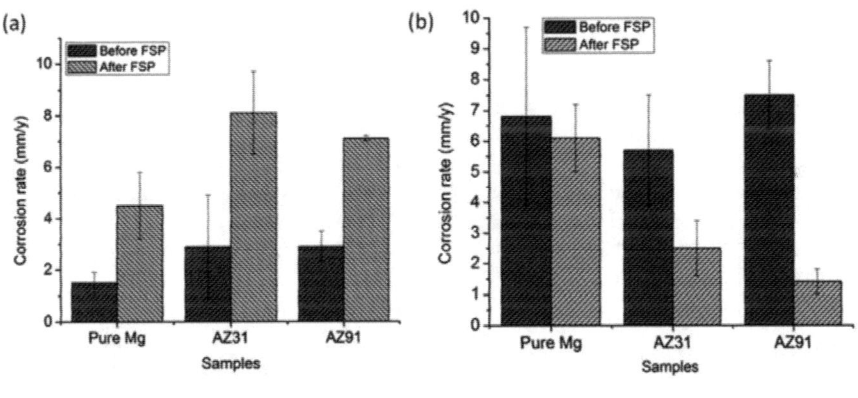

Figure 5. Corrosion tests results from the immersion studies carried out in a) 0.9M NaCl solution and b) Ringer's solution.

Fig. 4 presents the samples morphologies after immersion tests in Ringer's solution. A white color phase appeared on all of the samples which is usually characterized as magnesium hydroxide (Mg(OH)2) generally formed in aqueous solutions [1, 4]. Fig. 5 presents the corrosion data of the samples obtained from the immersion studies after 24 h. Fig. 5 (a) shows the corrosion rate of the samples before and after FSP, done in 0.9M NaCl environment. From the results, increased degradation can be seen after FSP for all the samples. After 1 day of immersion, the corrosion rate was calculated as 1.5 ±0.4, 2.9 ±2 and 2.9 ±0.6 mm/y for pure Mg, AZ31 and AZ91 Mg alloys respectively. Interestingly, after FSP, the corrosion rate of the samples was increased to 4.5 ± 1.3, 8.1 ±1.6 and 7.1±0.1 mm/y for pure Mg, AZ31 and AZ91 Mg alloys respectively. This behaviour can be explained by counting the effect of increased grain boundary into account due to smaller grains. With the higher fraction of grain boundary, galvanic corrosion is increased in metals particularly reactive metals such as Mg alloys.

However, from the immersion data obtained using Ringer's solution, corrosion rates were calculated as 6.8 ± 2.9, 5.7 ± 1.8 and 7.5 ± 1.1 mm/y for pure Mg, AZ31 and AZ91 respectively before FSP. But the corrosion rates were calculated as 6.1 ± 1.1, 2.5 ± 0.9, 1.4 ± 0.4 mm/y for pure Mg, AZ31 and AZ91 respectively after FSP. The results show that the formation of other mineral phases as observed on the surface of the samples when immersed in Ringer's solution which further decreased the corrosion rate compared with the samples immersed in 0.9M NaCl solution. Hence, from the results, it can be understood that the microstructure modified by FSP and the type of corroding medium has a significant effect on corrosion behaviour of Mg based alloys. Biological environment contains several ions and also biomineralization process influences the degradation rate during the implant tissue interaction. Therefore, assessing the corrosion rate in simulated physiological solutions is suggested compared with standard NaCl solution to precisely understand the degradation of Mg alloys in developing degradable implants.

4. Conclusions

Grain refined pure Mg and AZ series alloys (AZ31 and AZ91) were produced by FSP and increased hardness was measured due to smaller grains. Corrosion rate was observed as lower after FSP from immersion studies done in Ringer's solution but higher when tested in NaCl solution. Form the results, it is concluded that corroding medium also affects the degradation of Mg based alloys beside gain size.

References

[1] Witte F, Hort N, Vogt C, Cohen S, Kainer K U, Willumeit R and Feyerabend F 2008 *Curr. Opin. Solid. St. Mater. Sci.* **12** 63.

[2] Zeng R and Dietzel W 2008 *Adv. Eng. Mater.* **10(8)** B3.

[3] Shaylin S and George J D 2012 *Acta. Biomater.* **8** 20.

[4] Wang H, Estrin Y and Zúberová Z 2008 *Mater. Lett.* **62** 2476.

[5] Kirkland N T and Birbilis N, 2014 Magnesium Biomaterials Design, Testing, and Best Practice. Springer Science and Business Media, New York, USA, 1-131.

[6] Keerthi Soujanya G, Hanas T, Yogeshwar Chakrapani V, Ratna Sunil B and Sampath Kumar T S 2014 *Procedia. Mater. Sci.* **5** 817 – 823.

[7] Ratna Sunil B, Sampath Kumar T S and Uday Chakkingal 2012 *Mater. Sci. Forum.* 2012 **710** 264-269.

[8] Ratna Sunil B, Arun Anil Kumar, Sampath Kumar T S and Uday Chakkingal 2013 *Mater. Sci. Eng. C* **33** 1607–1615

[9] Ratna Sunil B, Sampath Kumar T S, Uday Chakkingal, Nandakumar V, Mukesh Doble, Devi Prasad V and Raghunath M 2016 *Mater. Sci. Eng. C* **59** 356–367.

[10] Saikrishna N, Pradeep Kumar Reddy G, Balakrishnan M and Ratna Sunil B, 2016, *J. Magnes. Alloys.* **4** 68–76.

[11] Mishra R S and Ma Z Y 2005 *Mat. Sci. Eng.* R **50** 1–78.

[12] Surya Kiran G V V, Hari Krishna K, Sameer Sk, Bhargavi M, Santosh Kumar B, Mohana Rao G, Naidubabu Y, Ravikumar Dumpala and Ratna Sunil B 2017 *Trans. Nonferrous. Met. Soc. China.* **27(4)** 804-811.

[13] ASTM Standard NACE TM0169/ G31 – 12a. 2012, Standard practice for laboratory immersion corrosion testing of metals, West Conshohocken, PA: ASTM International

IOP Conf. Series: Materials Science and Engineering **653** (2019) 012026 doi:10.1088/1757-899X/653/1/012026

A State of the art on Sequential Micro-Machining

R M Tayade*, B Doloi, B R Sarkar and B Bhattacharyya

Production Engineering Dept., Jadavpur University, Kolkata,

India *rmtayade@vjti.org.in

Abstract. The sequential micro-machining (SMM) is an emerging technique highly recommended for the manufacture of micro parts and also for generating micro features on parts. The review incorporates various traditional as well as non-traditional machining processes used in sequential manner. This paper focuses on recent trends in sequential micro machining combining diverse machining processes such as electrical discharge machining (EDM), laser beam machining (LBM), electrochemical machining (ECM), ultrasonic machining (USM), electrochemical discharge machining (ECDM), micro milling, micro grinding, electro-polishing, etc. The basic concepts, working principle, classification, mechanism of operation, the rationale behind the sequencing of individual machining processes etc. are described elaborately.

1. Introduction

In order to fulfill the difficult to meet requirements of tighter tolerances imposed on the parts to be produced, manufacturing industries are searching for different innovative methods and processes for the manufacture of micro-parts. Sequential machining being a promising futuristic technology; it is considered to be an important link in the chain of innovative and advanced micro-part manufacturing methods. The sequencing technique aims at machining in the same set up instead of relocating the workpiece in different machine tools for carrying out further operations on it. The multifunctional machine tool executing dissimilar machining processes in a sequential manner reduces effort, energy and machining time to a great extent. In spite of the several complimentary advantages, very few non-conventional machining processes are integrated for forming sequential machining combination. This methodology imparts a much higher level of accuracy to the micro-part and also improves the machining efficiency.

Basically, sequential machining is a plan of action where two or more machining processes are executed in sequence on the same or different machine tools. The strategy of implementation of different machining processes in a sequential manner, on the one hand, extracts the advantages from the individual process and minimizes the disadvantages. The trend of incorporating two or more micromachining systems at one workstation for performing various machining operations in a sequential manner has been increasing. It is well known that each and every discreet machining process has its own pros and cons. The μEDM cuts metal very fast but induces thermal stresses in the component, μECM offers stress-free machining but it is difficult to achieve perfect profile due to uncontrolled anodic dissolution. Frequent changing and setting of micro-tool is also a problematic area of micromachining which can be addressed by manufacturing tool online. Thus the methodology of applying these micro machining techniques in a sequence is adopted and it is termed as Sequential

IOP Conf. Series: Materials Science and Engineering **653** (2019) 012026 doi:10.1088/1757-899X/653/1/012026

Micro Machining (SMM). The sequential machining has the capability to fabricate complex micro features, channels, grooves; micro pyramids, etc on difficult to cut materials. It is more efficient, advantageous, stable, and economically feasible than other manufacturing techniques.

2. Basics of Sequential Micro-Machining

In sequential machining, two or more machining techniques are applied in a sequential manner on the same or different working systems. As all the required operations on the workpiece, right from initial cutting to final finishing are carried out without changing the tool and workpiece positions, the components with high geometric accuracy and better surface finish can be produced with high machining efficiency. The workpiece handling time, loading and unloading time reduces to a great extent. The efforts are being made to incorporate more and more machining processes in a single multifunctional machine system to avoid repositioning error and to achieve cost-effective manufacturing. The sequential micro-machining can be broadly classified into two major groups such as conventional sequential machining processes and non-conventional sequential machining processes. Figure 1 represents the broader classification of Sequential micromachining processes. The sequential machining processes are classified under five major heads based on the purpose of operation as; i).micro-tool making oriented processes like Turning and EDM, ii). Based on the increase in material removal rate, the machining combinations include; ECM and EDM, LBM and EDM, etc. iii).The processes enhancing surface quality such as EDM and ECM milling and EDM with electro-polishing, iv). The energy-efficient processes and v).The processes improving the microstructure such

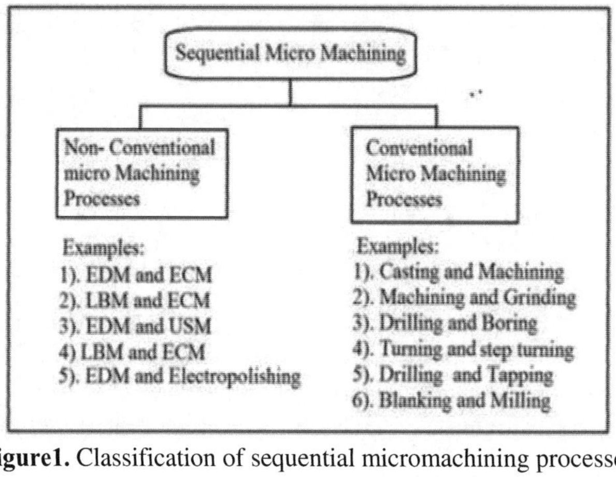

Figure1. Classification of sequential micromachining processes.

as surface defect machining (SDM) [1]. Most of the time the sequential machining and hybrid machining are considered as one and the same, but actually, both the machining techniques are totally different from each other. In hybrid machining, two or more micromachining processes occur simultaneously at one station. Hybrid manufacturing is defined as the simultaneous and controlled interaction of process mechanisms and/or energy sources or tools that have a significant effect on the process performance. Generally assisted or combined machining processes come under this category. For example, laser-assisted turning; where the workpiece materials which are hard are preheated and soften by the focused stream of the laser beam before it reaches to the tip of the cutting tool. In the case of sequential machining, the subsequent machining process is applied after the completion of the first process.

3. Development of sequential micro machining systems

The intention behind writing the review was to bring the area of sequential machining into the limelight by understanding the methodology of application, a strategy of sequencing, to focus on developing sequential machining processes and to highlight the most popular sequential machining combinations. The various non-traditional micromachining processes applied in a sequential manner are described elaborately.

3.1. Sequencing Laser Beam Machining (LBM) and μEDM

A nanosecond pulsed laser was used for micro hole drilling, as drilling by using μEDM is a time-consuming process and wear of tool during machining also takes place [2]. Therefore, initially, the pilot hole of 1mm depth was drilled by focusing a laser beam on the workpiece material. The LBM and the μEDM combination were used as shown in figure 2 for drilling holes in fuel injector [3]. Firstly the pilot holes in the injection nozzle of the Diesel engine were drilled by using laser ablation. The removal of the recast layer and finishing of the hole by reaming operation was conducted by

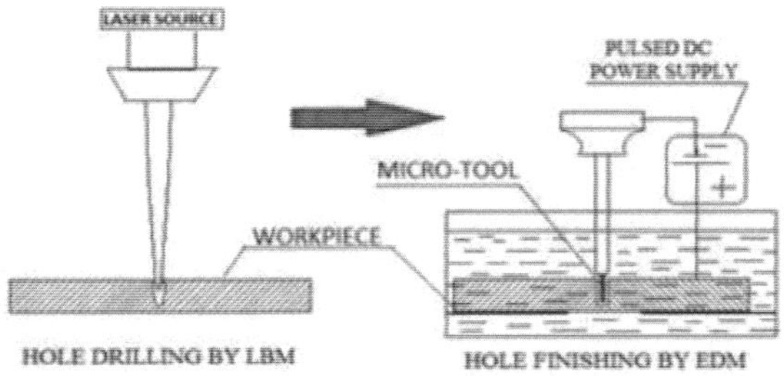

Figure 2. (a) LBM applied for hole drilling, (b) Hole finishing by μEDM.

μEDM drilling. This sequential machining allows 70% reduction in total drilling time in conventional EDM. The sequential combination of die sinking EDM and LBM was used for inscribing gear wheel in hardened steel [4]. Initially, the major portion of the gear i.e. gear blank was machined by using Die sinking EDM. The important gear tooth profile in detail then carved by using Nd: YAG laser.

3.2. Sequencing of LBM and micro Milling

The hot embossing dies used for micro-fluidic application were produced by implementing LBM process in sequential combination with micro- milling [5]. Both the machining processes were carried out on separate machining setups. Initially, the micro-fluidic channels were structured by using a laser ablation process. As such dies cannot be manufactured either by micro milling or simply by using LBM. The strategy of sequencing LBM and micro milling offers the advantages of an individual machining process.

3.3. Sequencing of Block μEDM and μEDM

This sequencing technique is generally implemented for the production of high aspect ratio microelectrodes with higher dimensional accuracy. The multifunctional machine tool used for manufacturing micro tool of difficult to cut metal. In Block μEDM process, a rectangular sacrificial block of tungsten carbide (WC) acts as a cutting tool. The cylindrical rod which is required to be reduced to a tool electrode acts as a workpiece. The electrode rod having a rotary motion is fed against the block surface during machining as shown in figure 3. The sacrificial block or the tool electrode is imparted to and fro scanning movement to avoid taper on the tool electrode. This process permits to fabricate tool electrodes of less than 40μm diameters [6].

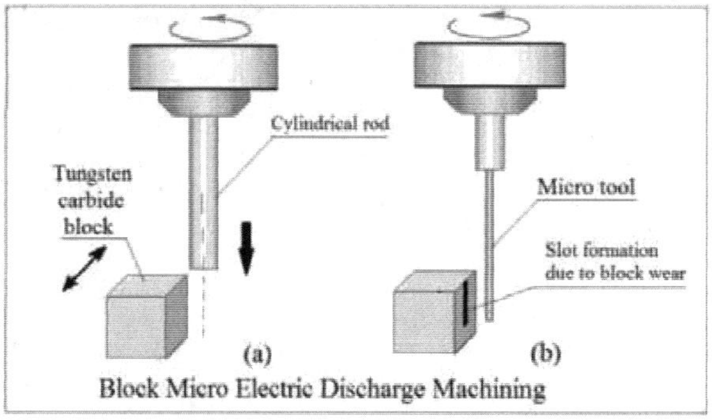

Figure 3. Schematic presentation of Block μEDM (a) Inception of the process And (b) Final step of micro tool generation.

3.4. Sequencing of μEDM and micro grinding processes

Polycrystalline diamond (PCD) is considered as on one of the hardest and toughest material usually used for cutting difficult to machine metals. The effect of the PCD tool on the glass during micro grinding has been investigated. The PCD micro tool was first made online by applying block μEDM as well as moving block μEDM process. The same PCD tool was then used for micro grinding the glass. Both the machining processes; block μEDM micro tool making and glass grinding were performed in the same set up without changing the working fluid [7]. Thus the fine finished micro tool imparts high surface quality and enhances dimensional accuracy of the hole.

3.5. Application of μECDM and μECM in sequential manner

A novel combination of micro ECDM (μECDM) shaping and μECM finishing process in a sequential manner for micro hole drilling have been presented [8]. Initially using μECDM process, a hole was

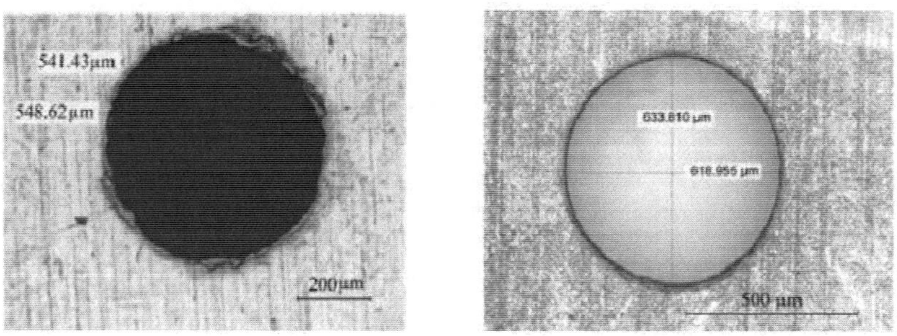

(a) (b)

Figure 4. (a) hole drilled by μECDM, (b) hole finished by μECM.

drilled on 500μm thick SS-304 stainless steel plate. The hole formed by μECDM process consists of the recast layer which was removed by subsequent application of the μECM process. Figure 4 (a) shows a hole drilled using the μECDM process and finished by a μECM process as shown in figure 4(b).

3.6. Sequential combination of block μEDM, μEDM and μECM

Three micromachining processes such as block μEDM, μEDM and μECM conducted in sequence on the same machining system without changing the tool and workpiece positions [9]. First of all, a tool diameter was reduced by Block Electric Discharge Grinding (BEDG) method. The micro -hole then drilled with the same tool electrode using μEDM process. The micro-cracks, micro-craters generated on the hole surface then finished by applying μECM process.

3.7. Sequential combination of μECM and μEDM

A prototype machine tool was designed and developed to conduct μEDM and μECM processes sequentially to investigate the effects of the application of μECM sinking and μEDM sinking for making star-shaped cavity [10]. The applied sequential machining strategy showed a double decrease in the machining time required for milling a cavity in comparison with the μEDM process.

3.8. Sequential combination of μEDM and μUSM

A sequential combination of μEDM tool shaping and micro ultrasonic machining (μUSM) for hole drilling on borosilicate glass was presented. Initially, tool diameter reduced to 150μm by using μEDM process. The ultrasonic vibrations of 30 kHz frequency transmitted to borosilicate glass. The same micro tool then used to drill high aspect ratio micro hole in borosilicate glass [11].

4. Conclusions

The review enumerated in this paper has tried to encompass various sequential combinations of different micromachining processes. A broader classification of sequential micromachining processes has been presented by taking into account a wide scope of conventional and non-conventional machining domain. The integrated multipurpose system set up of sequential micro machining is the current trend to accommodate various machining processes such as EDM, LBM, ECDM, USM, ECM, BEDG etc. to overcome the problems of an individual process. The sequential machining strategy offers various advantages such as improved surface quality, high dimensional accuracy and enhanced machining efficiency. The sequential micromachining is in rudimentary form, as very few sequential machining combinations are developed yet. The system set up accommodating EDM, ECM and ECDM can be developed. The SMM can be combined with hybrid machining processes.

References

[1] Chavoshi S Z, Goel S and Morantz P 2017 Current trends and future of sequential micro-machining processes on a single machine tool *Materials & Design* vol 127 pp 37-53

[2] Kim S 2009 Hybrid micromachining using a nanosecond pulsed laser and micro EDM *J. Micromech. Microeng.* vol 20 (1) 015037

[3] Li L, Diver C and Atkinson J 2006 Sequential Laser and EDM Micro-drilling for Next Generation fuel Injection Nozzle Manufacture *CIRP, Annals* vol 55 (1) pp 179-182

[4] Fleischer J, Schmidt S and Haupt S Combination of electric discharge machining and laser ablation in microstructuring of hardened steel 2006 *Microsyst Technol.* vol 12 pp 697–701

[5] Schubert A, Groβ S, Schulz B and Eckert U 2011 Sequential combination of micro-milling and laser structuring for manufacturing of complex micro-fluidic structures *Physics Procedia* vol 12 pp 221– 229

[6] Jahan M P, Rahman M, Wong Y S and Fuhua L 2010 On-machine fabrication of high-aspect-ratio micro-electrodes and application in vibration-assisted micro-electrodischarge Proc. *Inst. Mech. Eng. B J. Eng. Manuf.* vol 224 (5) pp 795–814

[7] Perveen A , Jahan M P , Rahman M and Wong Y S 2012 A study on micro grinding of brittle and difficult-to-cut glasses using on-machine fabricated poly crystalline diamond tool

[8] Tayade R M, Doloi B, Sarkar B R and Bhattacharyya B 2017 Study of Sequential Electro Micro Machining (SEMM) System for Enhancing Machining Performances *Int. Conf.(COPEN, IIT Madras)* ISBN 978-93- 80689-28-9 pp 120-124

[9] Xiaolong H, Yukui W, Zhenlong W and Zhaoqi Z 2013 Micro-hole drilled by EDM-ECM combined processing *Key Engineering Materials* Online ISSN: 1662-9795 Vol 562-565 pp 52- 56
[10] Skoczypiec S and Ruszaj A 2014 A sequential electrochemical electro-discharge process for micropart manufacturing *Precision Engineering* vol 38 pp 680-690
[11] Yan B H, Wang A C, Huang C Y and Huang F Y 2002 Study of precision micro holes borosi-licate glass using micro EDM combined with micro ultrasonic vibration machining *International Journal of Machine Tools & manufacture* vol 42 pp 1105–1112

IOP Conf. Series: Materials Science and Engineering **653** (2019) 012027 doi:10.1088/1757-899X/653/1/012027

Experimental investigation into Micro Ultrasonic Machining of Quartz

S Kumar*, B Doloi and B Bhattacharyya

Production Engineering Department, Jadavpur University, Kolkata, India

*santosh14fiem@gmail.com

Abstract. Quartz has been broadly applied in industries including electronics, optics, micro electromechanical system (MEMS), biomedical. Micro ultrasonic machining (Micro USM), is a non traditional micro machining technique, extensively used to produce micro features and parts on quartz, silicon, ceramics and glass. A micro tool of stainless steel having diameter 330μm has been develop for generating micro hole on quartz. Experiments have been done by using different types of abrasives like boron carbide, silicon carbide and aluminum oxide of average grain size 14 μm. Experiments have been performed under different parametric setting of tool feed rate and power rating. The range of tool feed rate is 0.8 to 1.1 mm/min. The range of power rating is 200 to 500 W. Micro hole has been generated on quartz. Overcut and taper angle has been considered as response of micro ultrasonic drilling. In the present research effect of tool feed rate and power rating on taper angle and overcut has been investigated while using different type of abrasives.

1. Introduction

Quartz is widely applied in industries including electronic, optics, micro electro mechanical system (MEMS) and biomedical etc. Due to hard and brittle nature, quartz is difficult to machine. This material is machined by non-traditional technique of machining. Due to non-conducting in nature, it is not machined by ECM, EDM etc. Micro ultrasonic machining (Micro USM) is a nontraditional micro machining process; employ to produce micro features on hard and brittle materials like quartz, silicon and glass and its parts [1]. The advantages of micro USM are no thermal injury to the workpiece and no limitation by the electrical or chemical aspects of workpiece materials [2]. A tool vibrates at ultrasonic frequency with low amplitude. In micro USM, the material is removed by the shock of abrasive grains with high kinetic energy generated by the vibration of an acoustic system. The objectives of this paper are to investigate micro hole characteristics such as overcut and taper angle produced by micro-USM using simple cylindrical shape micro-tool.

Some research work have been performed on ultrasonic machining on different materials but still more in depth studies are very much needed to explore the micro ultrasonic drilling on quartz materials. The machined surface roughness varies on depth of cut. Low surface roughness was obtained when low depth and the small size abrasive particles are used [3]. Micro drill produces on borosilicate glass by USM. The MRR mainly depends on static load while tool wear rate depends on abrasive size [4]. In micro-USM, material removal is decided by the workpiece crack propagation and abrasive particle wear. By using harder and spherical abrasive bigger chip removal occur [5]. During quartz machining by USMM process, slurry concentration is a great influenced compare to power

IOP Conf. Series: Materials Science and Engineering **653** (2019) 012027 doi:10.1088/1757-899X/653/1/012027

rating and tool feed rate during the micro hole generation on quartz [6].

The present research work is to carry out experimental investigation into micro ultrasonic machining on quartz during micro hole generation using different type of abrasives like boron carbide, aluminium oxide and silicon carbide. The influences of process parameters on responses during the micro-ultrasonic machining have been investigated.

2. Experimental Setup and planning

Figure 1 represents the micro USM system is used to conduct experiments. The ultrasonic frequency of micro USM system fixed at 20 kHz and amplitude of frequency fixed at 25 μm. The quartz material was set on the fixture made of conductive material put on magnetic base. Tool was fed in the longitudinal direction. Boron carbide, silicon carbide and aluminum oxide fine particles combined with water at normal temperature was selected for slurry. Stainless steel (SS304) cylindrical tools with tip diameter 330 μm were developed as per designed and fixed with the help of brazing to hexagonal head. Figure 2 shows the tool used to generate micro hole on quartz.

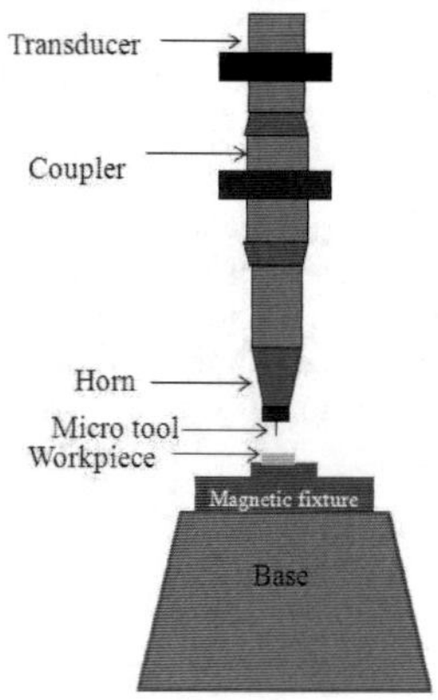

Figure 1. Representation of micro USM system

Figure 2. Micro tool used in USM

3. Experimental result and Discussion

Micro hole are generated in the quartz workpiece of thickness 1 mm. Experimental condition used in the investigation is listed in table 1.

3.1. Influence of process parameters on overcut

In micro ultrasonic machining process the power rating is one of the important parameters. The variation of power rating on overcut has been analyzed through the experimental result and graph. Figure 3 shows deviation of overcut with power rating graph. While 500 W power rating is set then the overcut is large. Generally low power rating offers the low overcut. As the power rating increases then overcut also in-creases because with more ultrasonic power, abrasive particles in slurry were hitting with additional impetus and driving force hence removing more material also from the side of the tool. And at low power rating abrasive particles are hitting through low driving force and hence, overcut becomes less. With the increment power rating after a certain value, crack formation started at surrounding of entrance of micro hole which is not desirable for accuracy aspects.

Table1. Experimental conditions

Experimental Conditions	Explanation
Workpiece	Quartz
Material of Tool	Stainless steel-304 (Diameter 330 μm)
Tool cross sectional	Circular
Abrasive material	Boron carbide, Silicon Carbide, Aluminum oxide
Average Abrasive size	14 μm
Frequency	20 kHz
Amplitude	25 μm
Concentration of Slurry	20% by weight
Power rating	200-500 W
Feed rate	0.8-1.1 mm/min
Slurry medium	Water

Figure 4 show the variation of overcut with the tool feed rate. It indicates that as tool feed rate is enhancing the overcut first rises gradually and then slowly while B_4C abrasive are used. When low tool feed rate is used, overcut is also small since abrasive particle hits through minute force on work piece. When tool feed rate is enhancing, the abrasive particle hits through additional force at work surface, hence overcut also enhancing. Using Al_2O_3 abrasive the overcut is small because of low hardness of abrasive and also low sharpness as compared to B_4C and SiC.

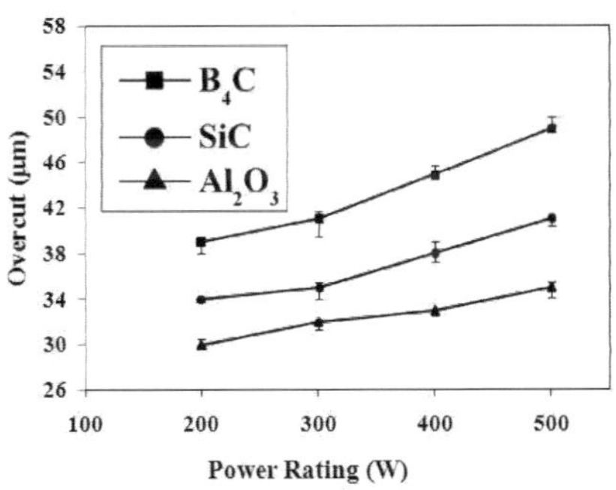

Figure 3. Effect of power rating on overcut

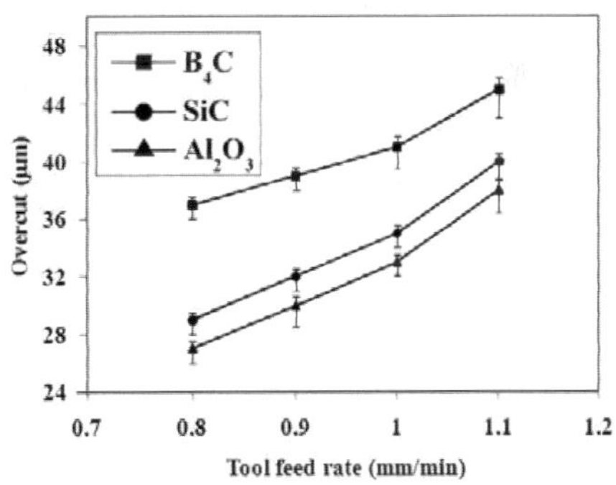

Figure 4. Effect of tool feed rate on overcut

3.2. Influence of process parameters on taper angle

Figure 5 shows that taper angle decreases as power rating increases. With increasing power rating, the shock power through abrasive particles boost that guide to a large amount of momentum transfer to the abrasives hence more material is removed. Material removals take place with uniform rate at the exit surface of hole. As a result, taper angle value is less. To attain less taper angle, large power rating is favoured.

Figure 6 show that variation of taper angle with tool feed rate. To investigate the effects of tool feed rate on taper angle, tool feed rate was varied along longitudinal direction. At less tool feed rate, taper angle is small. When tool feed rate is raising the value of taper angle also enhanced because with large tool feed rate, the material removal from the entrance surface of micro hole is more than exit surface of the micro hole. Hence value of taper angle enhances. To attain low value of taper angle, small value of tool feed rate is preferred. Figure 7 shows the micro hole drilled on quartz.

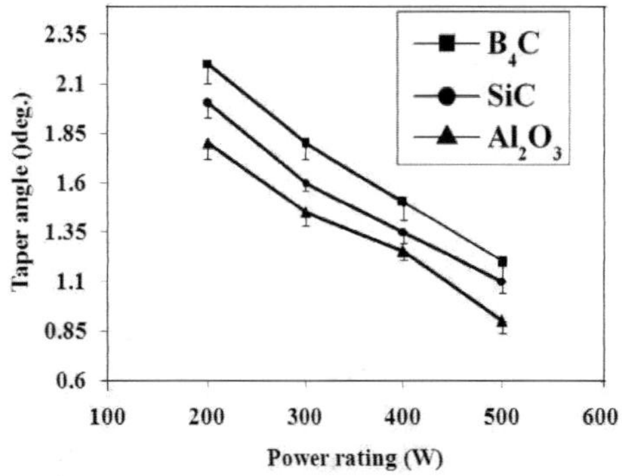

Figure 5. Effect of power rating on taper angle

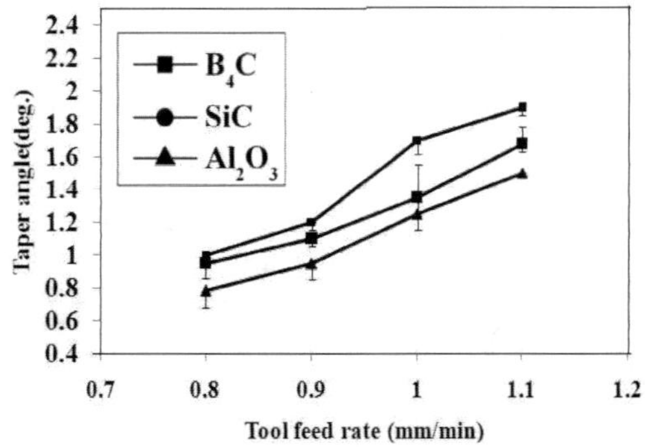

Figure 6. Effect of tool feed rate on taper angle

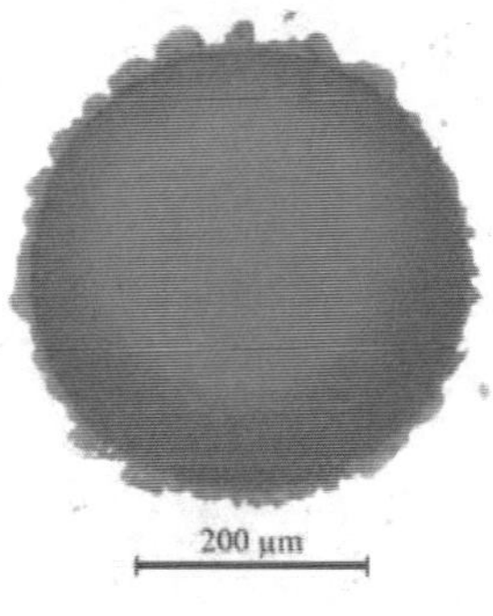

Figure7. Micro hole on quartz

4. General Conclusions

Micro ultrasonic machining (micro USM) process has been successfully useful to generate micro hole with the developed micro tool. The experimental investigation highlights that overcut and taper angle are influenced by machining parameters such as tool feed rate and power rating for generating through micro hole on quartz using different abrasives. Out of the observed experimental results lower overcut of 27 µm is observed as well as lower taper angle of 0.780 has been obtained. The micro hole diameter of 330 µm is observed during ultrasonic drilling of a quartz material. It is concluded that for achieving lower overcut the lower value of process parameters preferred. For attaining lower taper angle, larger value of power rating and smaller value of tool feed rate and abrasive of

Al_2O_3areselected.

Few research works are reported on generation of micro features on quartz by micro USM process. However more research has to be carried out in the area of generation of micro channel on quartz. Investigation of surface roughness during micro machining of quartz by micro USM is also the area of research interest for wider application in modern micro manufacturing industry.

References

[1] Masuzawa T 2000 State of the art of micromachining. *Annals of the CIRP* 49 (2): 473–488.

[2] Thoe T, D. Aspinwall D, Wise M 1998 Review on ultrasonic machining *International Journal of Machine Tools and Manufacture,* 38(4): p. 239-255

[3] Klopfstein M J, Ghisleni R, Lucca D A, Brinksmeier E 2008 Surface characteristics of micro-ultrasonically machined (100) silicon, *International Journal of Machine Tools & Manufacture* 48, 473–476

[4] Jain Vivek 2014 Application of Taguchi Approach in the Optimization of Cutting Parameters in Micro-Ultrasonic Drilling Process. *International Journal of Engineering Research and Applications (IJERA)* ISSN: 2248- 9622

[5] Jingsi Wang, Keita Shimada, Masayoshi Mizutani, Tsunemoto Kuriyagawa 2018 Effects of abrasive material and particle shape on machining performance in micro ultrasonic machining, *Precision Engineering* 51, 373–387

[6] Kumar S, Hansda B, Das S, Doloi B and Bhattacharyya B 2017 Micro hole fabrication on quartz using ultrasonic micromachining process. *Int. J. Precision Technology*, Vol. 7, pp 222-236

Wear behaviour of Al-SiC metal matrix composite under lubricated environment

Samar Mondal[1], Tapan Kumar Barman[1] and Prasanta Sahoo[1]

[1]Department of Mechanical Engineering, Jadavpur University, 188 Raja S C Mallick Road, Kolkata 700032, West Bengal, India

[1]samar.ghurni@gmail.com, [1]tkbarman@gmail.com, [1]psjume@gmail.com

Abstract. The present study investigates the wear behaviour of silicon carbide based aluminium metal matrix composite under lubricated environment. The composite reinforced with 2.5% SiC by weight was prepared through stir casting method. The experiments were conducted on a pin on disc type tribotester under lubricated condition using five levels of load and sliding velocity varying from 10 N to 90 N and 0.26 m/s to 0.78 m/s respectively. The experiments were carried out based on the combination available in Taguchi's L_{25} orthogonal array (OA). Then the effect of design factors on wear rate was investigated. Taguchi's method was implemented to find the optimal parameter combination. Statistical analysis was made to find out the relative influence of design factors on wear rate. Finally, field emission scanning electron microscopy (FESEM) was carried out to determine the wear mechanisms.

Keywords. AlSiC, Wear, Friction, Lubricated Environment, Taguchi

1. Introduction

Metal matrix composite (MMC) is one of the emerging class of materials for modern days engineering applications. Composite materials have two parts - matrix and reinforcement. When composites are subjected to the real time applications, matrix and reinforcement act as a single material system. The composites acquire many advantages compare to the monolithic material because of their unique properties like corrosion resistance, creep resistance, lower coefficient of thermal expansion, dimensional stability and thermal conductivity etc. In general, composite materials are used in various engineering applications, particularly in aerospace applications, automobile industry, automotive industry, military and sport application etc. [1-3]. In electronics industry, MMCs are used as heat sink or heat spreader [4]. In particular, metal matrix composites (MMCs) were developed because of their low density and efficient mechanical behaviours. The presence of hard reinforcement phase in the matrix in particulates, fibers or whiskers forms provide excellent friction and wear characteristics to the MMCs. Aluminium, magnesium, titanium and their alloys are generally used as matrix material. But for weight minimization, aluminium is extensively used among them because of its low density. Most commonly used reinforcement materials are SiC, aluminum oxide, TiB_2 and graphite [5-6]. Nowadays, brake drums of two wheelers are also being manufactured by aluminium silicon carbide (AlSiC) metal matrix composites. Incorporation of SiC provides elastic strengthening of the composite [7]. But in various industries, it is one of the major concern to develop aluminium based metal matrix

composites for high tribological performance. Researchers have broadly investigated the mechanical and tribological behaviour of AlSiC MMCs [8-30]. It was reported that, incorporation of SiC in aluminium improves the mechanical properties of AlSiC MMCs like tensile strength, hardness, specific modulus, fatigue strength, strength to weight ratio, corrosive resistance and dimensional stability [8-14]. Literature study shows that, friction and wear behaviours depend on various intrinsic and extrinsic factors like normal load, sliding distance, sliding velocity, reinforcement size, reinforcement orientation etc. [15-30]. Pramanik [8] reported that, AlSiC is more wear resistant compare to the pure aluminium. Lee et al. [11] reported that, tremendous improvement on wear rate of aluminium alloy can be expected with addition of SiC_P reinforcement. With increase in SiC volume percentage, hardness of AlSiC MMCs generally increases and tensile strength first increases and then decreases [12]. Under combined loading like tension and torsion, torsion and compression, AlSiC composites give better mechanical properties [13-14]. Rehman et al. [15] reported that, presence of SiC particles enhance the coefficient of friction. At lower load, friction mechanism generally adhesion and de-lamination controlled [16]. It was observed that, friction and wear mechanisms can also be dependent on temperature [17] and reinforcement orientation [25]. Pradhan et al. [27] reported the wear behavior of AlSiC under various corrosive environments. The physical separation can occur in three different ways i.e. particle detachment, third body behavior and particle ejection [19]. Vingsbo [19] described wear as a tribo-fracture phenomenon. Huching [20] investigated the effect of third body on friction and wear mechanisms under dry condition. The two body and three body abrasion mechanisms were represented by Al-Rubaie et al. [21] and Sasada et al. [22] respectively. The effect of normal load and sliding velocity was reported by several researchers. It was noticed that, with increase in load wear rate increases satisfying Archard theory [24]. When the normal load was increased monotonically, jump of wear rate from mild wear to severe wear was observed. The transformation of mild wear to severe wear was dependent on normal load and sliding velocity [25-26]. It was also reported that, with increase in reinforcement volume percentage, wear performance improves [28]. As minimization of wear rate is one of the major concerns in many engineering applications, optimization of wear rate by Taguchi approach [31], Genetic Algorithm and Fuzzy logic was performed by several researchers.
From the literature review it was observed that, most of the researchers have investigated the tribological behavior of AlSiC MMCs under dry condition. But, there are engineering applications like cylinder block and piston liner, brake drum and brake caliper etc. where AlSiC MMCs are subjected to wear under lubricated condition. Some researchers conducted their experiment using solid lubricant like graphite [32]. The improvement of wear rate of AlSiC was reported under solid lubrication [32]. But, tribological behavior of AlSiC MMCs under lubrication was unexplored. So, keeping these applications of AlSiC MMCs in mind, an attempt was made in the current study to investigate the wear behaviour of AlSiC MMC under fully lubricated condition at various levels of normal load and sliding velocity. Taguchi approach was implemented to find out the optimal parameter combination for wear rate minimization. Finally, field emission scanning electron microscopy (FESEM) was used to find the wear mechanisms.

2. Experimental procedure

For preparing the samples, stir casting method was selected. Aluminium alloy with 3.2% Cu, 6.0% Mg, 0.05% Fe, 0.4% Mn, 0.2% Ni, 0.15% Zn alloying elements were used as matrix material. During casting, silicon carbide (SiC) particles were added in a predefined amount so that, 2.5% SiC by volume is ensured in the composite. After the solidification of the casting, subsequent machining operations were performed to produce the cylindrical samples of $\Phi6\times30$ mm for tribological tests.

From the literature review, it was observed that normal load and sliding velocity have significant effect on wear rate. For present study, these two parameters were taken as design factors. Design factors along with their different levels is represented in table 1. The design of experiment was performed based on Taguchi's L_{25} orthogonal array (OA). The tribological tests were performed in a pin-on-disc type tribotester (Ducom, India). Commercially available heavy duty engine oil Servo Pride

– 40, (a product of Indian Oil with Kin. Vis. @100°C = 13-15 and viscosity Index = 90) was chosen to lubricate the contact zone. Weight of the samples were measured using a digital weight balance before and after the experiments to calculate the weight loss. Specific wear rate was calculated using equation (1).

$$\text{Wear rate} = \frac{\Delta m}{\rho*(\frac{\pi*D*N}{60})*t*F}\ \frac{mm^3}{Nm} \quad (1)$$

where, ρ = density of AlSiC; D = track diameter; t = experiment duration; F = normal load. FESEM was carried out using Supra 35VP (Carl Geiss, Germany) to determine the wear mechanisms.

Table 1. Tribotesting parameters with their different levels.

Level	Normal Load (N)	Sliding velocity (rpm)	Sliding distance (m)
1	10	100	
2	30	150	
3	50	200	706.58
4	70	250	
5	90	300	

3. Results and discussions

3.1 Optimal parameter combination based on Taguchi's S/N ratio analysis

Tagichi's S/N ratio analysis is an optimization technique which can be used to find out the optimal parameter combination on which the objective of an engineering problem depends on. Here, S represents the signal and N represents the noise. So, S/N ratio is basically signal to noise ratio which is also a statistical measure of performance for Taguchi's method. In present investigation, an attempt was made to find the optimal parameter combination to minimize the wear rate. So, Lower-the-Better type function was chosen which can be written as equation (2).

$$S/N = -10\log\left(\frac{1}{n}\sum_{i=1}^{n} y^2\right) \quad (2)$$

where, n represents number of observation and y is the corresponding observed data. Parameter combination based on Taguchi's L_{25} design of experiment along with the wear rate and S/N ratio is represented in table 2.

The relative influence of each design parameters on wear rate can be investigated by means of main effect plot and interaction diagram. If a line in the main effect plot, representing the effect of a design factor, has highest inclination with respect to the horizontal line, it will have highest influence in output. The highest point in the main effect plot represents the best operating condition. Main effect plot for S/N ratio based on present investigation is shown in figure 1. It is clear from figure 1 that, normal load and sliding velocity both of the design factors are significant for controlling the wear rate under lubricated condition. A combination of 10 N normal load and 300 rpm slididing velocity provides minimum wear rate. But to determine the most dominant design factor, response table was generated and result shows normal load to be the most dominant design factor. Interaction of design factors was also studied and the result shows that both normal load and sliding speed has strong interaction effect. Finally, analysis of variance (ANOVA) was performed to find the significance of design parameter and it was observed that, both normal load and sliding velocity has significant effect on wear rate under lubricated condition. For the brevity of the paper full response table, ANOVA table and interaction plot are omitted.

To study the variation of wear rate with normal load and sliding velocity, 3D surface plot was generated using MATLAB. This surface plot is represented in figure 2. It may also be noticed that, with increase in normal load wear rate increases and the same decreases with respect to the sliding velocity.

Table 2. S/N ratio table for wear rate (smaller is better).

No of Exp.	Parameter combination		Wear rate	S/N ratio
	Load	Speed	$(\frac{mm^3}{Nm})$	
1	10	100	3.3832	-10.5866
2	10	150	4.0525	-12.1545
3	10	200	2.9798	-9.4837
4	10	250	2.5506	-8.1328
5	10	300	2.1009	-6.4479
6	30	100	5.5042	-14.8139
7	30	150	3.9028	-11.8274
8	30	200	3.6735	-11.3017
9	30	250	2.4205	-7.6781
10	30	300	2.6069	-8.3226
11	50	100	5.3911	-14.6336
12	50	150	4.4761	-13.0180
13	50	200	4.8172	-13.6558
14	50	250	3.5501	-11.0047
15	50	300	3.1760	-10.0376
16	70	100	5.3623	-14.5870
17	70	150	4.8253	-13.6705
18	70	200	4.4415	-12.9506
19	70	250	4.0931	-12.2411
20	70	300	3.8534	-11.7169
21	90	100	5.8394	-15.3273
22	90	150	5.7069	-15.1280
23	90	200	5.7079	-15.1296
24	90	250	4.8879	-13.7824
25	90	300	4.4845	-13.0342

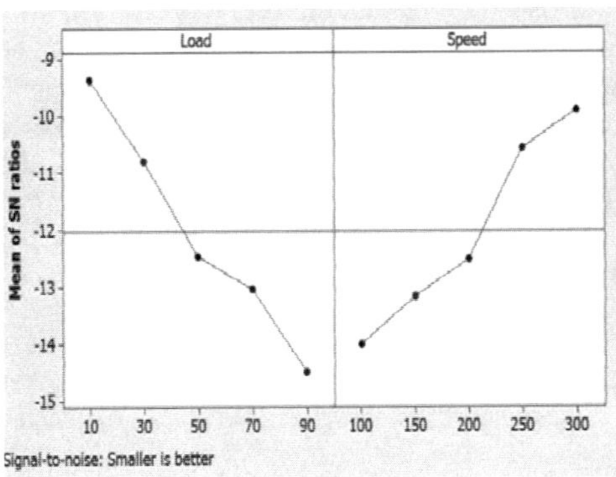

Figure 1. Main effect plot of S/N ratio for wear rate.

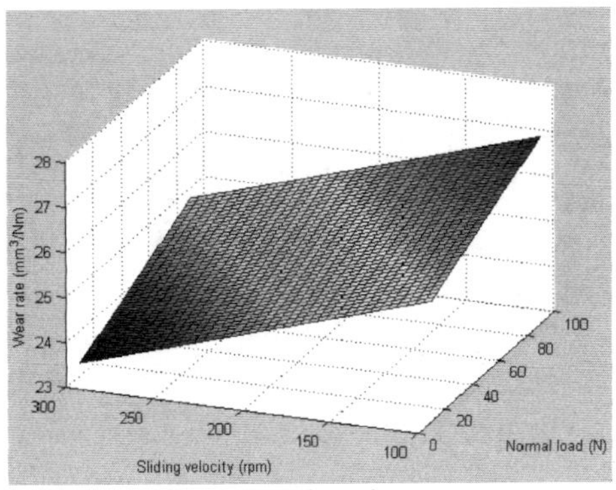

Figure 2. 3D surface plot of wear rate.

3.2 Microstructural characterization

FESEM of worn out AlSiC sample under lubricated condition is represented in figure 3. The FESEM micrograph was taken at optimal operating condition i.e. 10 N normal load and 300 rpm sliding velocity. From the FESEM micrograph it may be noticed that, large number of silicon carbide (SiC)

IOP Conf. Series: Materials Science and Engineering **653** (2019) 012028 doi:10.1088/1757-899X/653/1/012028

particles of different sizes are present on the worn out surface of AlSiC sample. So, it may be concluded that, optimal wear behaviour of AlSiC sample under lubrication is mainly abrasive in nature.

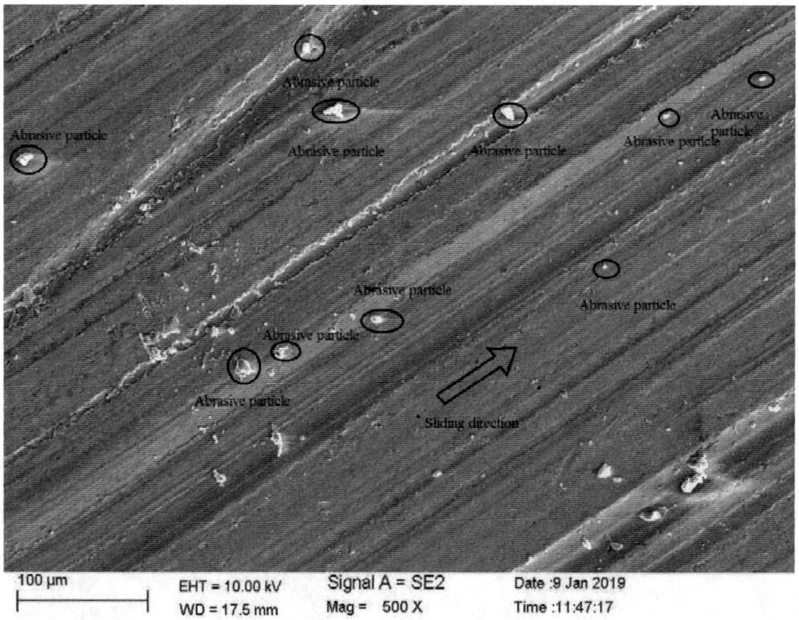

Figure 3. Microstructural characterization of wear rate at optimal operating point.

4. Conclusion

For present investigation wear behaviour of AlSiC metal matrix composite was studied under fully lubricated condition. AlSiC metal matrix composite was prepared by stir casting process with 2.5 weight percent of SiC. The tests were conducted based on Taguchi's L_{25} orthogonal array on a pin on disc type tribotester. Engine oil Servo Pride – 40 was used to lubricate the contact zone. Experimental results reveal that, with increase in normal load wear rate of AlSiC metal matrix composite increases but with increase in sliding speed wear rate of the same decreases. Optimal parameter combination is found to be 10 N normal load and 300 rpm sliding velocity i.e. at lower load and higher sliding velocity. Wear of AlSiC metal matrix composite under lubricated condition was strongly dependent on both normal load and sliding speed. Finally, FESEM micrograph was analysed to study the prevalent wear mechanisms associated with the wear rate of AlSiC MMC at optimal operating condition. It was observed that, wear rate at optimal point is mainly abrasive in nature.

References

[1] Pawar P.B and Utpat A 2014 Development of Al based silicon carbide particulate metal matrix composite for spur gear *Procedia Material Science* **6** pp 1150-56

[2] Urena A, Utrilla M.V, Rams J, Rodrigo P and Ferrer M 2007 Electroless multilayer coating on aluminium – silicon carbide composites for electronics packaging *Journal of the European Ceramic Society* 27 pp 3983-86

[3] Noguchi M and Fukizawa K 1993 Alternate material reduced weight in automobile *Adv. Mater. Process.* **143** pp 20-26

[4] Schobhel M, Michiel D.M and Hofmann M 2011 Internal stress and void in SiC reinforced alumina composite heat sink applications *Composite Science and Technology* **71** pp 724-733

[5] Surappa K.M 2003 Aluminium matrix composites: Challenges and opportunities *Sadhana* **28** pp 319-334

[6] Clyne T.W and Withers P.J 1995 An introduction to metal matrix composites *Cambridge university press*
[7] Nair S.V, Tien J.K and Bates R.C 1985 SiC-reinforced metal matrix composites *International Metals Reviews* **30** pp 275-290
[8] Pramanik A 2016 Effect of reinforcement on wear resistance of aluminium metal composite *Transactions of Nonferrous Matals Society of China* **26** pp 348-358
[9] Aberu C.M, Acuna R, Cabeza M, Cristobal M.J, Merino P and Verdera D 2017 Microstructure and mechanical properties of Al/SiC composite surface layer produced by friction star processing *Ciencia & Technology dos Materials* **29** pp 82-86
[10] Puhan D, Mahapatra S.S, Sahu J and Das L 2013 A hybride approach for multi-response optimization of non-conventional machining on $AlSiC_P$ *MMC Measurement* **64** pp 3581-92
[11] Lee H, Lu W.H and Chan S 1992 Abrasive wear of powder metallurgy Al alloy 6061-SiC particle compostes *Wear* **159** pp 223-231
[12] Balasubhramanian I and Maheswaran R 2015 Effect of inclusion of SiC particle on the mechanical behavior of stir cast Al6063 SiC composite *Materials and Design* **65** pp 511-520
[13] Rees D.W.A 1998 Deformation and fracture of matal matrix composites under combine loading *Composite Part A* **29** pp 171-182
[14] Padmanavan R, MacDonald B.J and Hashmi M.S.J 2004 Mechanical behavior of SiC reinforced Aluminium thin walled tube under combined axial and tensile loading *Journal of Material Processing Technology* **155-156** pp 1260-73
[15] Rehman A, Das S and Dixit G 2012 Analysis of stir die cast Al-SiC composite brake drum based on coefficient of friction *Tribology International* **51** pp 36-41
[16] Suh N.P and Sin H.C 1981 The genesis of friction *Wear* **69** pp 91-114
[17] Pan Y.M, Fine M.E and Gheng H.S 2008 Sliding wear of an alloy SiC whisker composite *Tribon Trans.***35** pp 482-490
[18] Godet M, Berthier Y, Lancaster J and Vincent L 1991 Wear modeling, how far we can get with first principle? In K.C. Ludema and R.g Bayer (eds) *Tribological Modeling for Mechanical Designer ASTM STP 1105 ASTM Philadelphia PM* pp 77-86
[19] Vingsbo O 1988 Fundamental of friction and wear, in FA Smith and P JBlow (eds) *Engineering material for advance friction and wear application ASM International MD* pp 1-19
[20] Huching I.M 1993 Mechanism of wear in powder technology: A Review *Powder Technology* **76** pp 3-13
[21] Al-Rubaie K.S, Yoshimura H.N and De Mello J.D.B 1999 Two body abrasive wear of Al-SiC composites *Wear* **33-35** pp 444-454
[22] Sasada T, Norose S and Mishina H 1981 The behavior of adhered fragment interposed between sliding surfaces and the formation process of wear particles *Journal Lubrication Technology* **103** pp 195-202
[23] Howell G.J and Ball A 1995 Dry sliding wear of particulate reinforced aluminium alloys against automobile friction material *Wear* **181–183** pp 379–390
[24] Archard J.F 1953 Contact and rubbing of flat surfaces *Journal of Applied Physics* **24** pp 981–988
[25] Wang A and Rack W.J 1991 Dry sliding wear in 2124 Al-SiC_W 17-4 PH Stainless steel systems *Wear* **148** pp 355-374
[26] Sato A and Mehrabian R 1976 Aluminum matrix composites: Fabrication and properties *Metallurgical Transaction* **7** 443-450
[27] Pradhan S, Ghosh S, Barman T.K and Sahoo P 2017 Tribological behavior of AlSiC composite under dry aqueous alkaline medium *Silicon* **9** pp 923-931
[28] Alpas A.T and Ghang J 1992 Wear rate transition in cast silicon alloys reinforced with Sic particles *Scripta Metallurgica* **26** pp 505-509
[29] Kowk J.K.M and Lim S.C 1999 High speed tribological properties of some $AlSiC_P$ composites: frictional and wear rate characteristics *Composites Science and Technology* **59** pp 55-63
[30] Poria S, Sutradhar G and Sahoo P 2016 Design of experiments analysis of friction behavior of Al-TiB_2 *Materials Today proceedings* **4** pp 2956-64
[31] Mukhopadhyay A, Duari S, Barman T.K and Sahoo P 2016 Wear behavior of electroless Ni-P-W coating under lubricated condition - a taguchi based approach *IOP Conference Series: Materials Science and Engineering 149012004 doi:10.1088/1757-899X/149/1/012004*
[32] Suresha S and Sridhara B.K 2010 Wear characteristics of hybrid aluminium matrix composites reinforced with graphite and silicon carbide particulates *Composite science and technology* **70** pp 1652-59

IOP Conf. Series: Materials Science and Engineering **653** (2019) 012029 doi:10.1088/1757-899X/653/1/012029

Electrochemical Discharge Machining Technology Applied for Turning Operation

M N Ali[1], B Doloi[2], B R Sarkar[3]

[1] Mechanical Engineering Department, Netaji Subhash Engineering College, Kolkata-152, India
[2-3] Production Engineering Department, Jadavpur University, Kolkata-32, India

Email: mnali.jupe@gmail.com

Abstract. In hybrid electrochemical discharge machining (ECDM) process machining of electrically conducting and non-conducting materials takes place in an alkaline electrolyte with the introduction of a tool electrode and an auxiliary electrode. In this paper an alternative of ECDM process, Electrochemical Discharge Turning (ECDT), is proposed to machine a cylindrical job which rotates continuously against a stationary or moving tool electrode. The material removal mechanism of ECDT process is similar to ECDM process, but ECDT employs ECDM components with rotary cylindrical workpiece immersed in an electrolytic solution. The present paper highlights the design and development of ECDT set-up which can be applied for turning operation. The proposed ECDT set up can also be utilized for making grooves of different shapes and sizes on cylindrical parts of electrically non-conducting materials.

1. Introduction

The limitations of conventional machining process in the area of complex shapes, miniature sizes, three dimensional (3D) complex parts and the need of machining of brittle harder and newer materials lead to the development and establishment of non-conventional machining processes in the industry as efficient and economic alternatives to conventional machining processes. Demand and need of research in these areas are also growing to develop precise understanding and knowledge of different aspects. Electrochemical Discharge Turning (ECDT), an alternative process of ECDM process, can be used to machine a cylindrical job in which job rotates continuously. The material removal mechanism of ECDT process is similar to ECDM process but ECDT employs ECDM components with rotary cylindrical workpiece immersed in an electrolytic solution. Sarkar *et al.* [4] studied the parametric analysis of ECDM process on silicon nitride ceramics. Furutani *et al.* [5] studied the performance of lathe type ECDM process for machining of glass rod revolving in an electrolyte as a working fluid. Singh *et al.* [9] developed ECDM based triplex hybrid methods and presented machined features with its complexity and accuracy level. Furutani *et al.* [8] constructed the control method of the electrode feeding for machining of a thin rod with the lathe-type ECDM process and showed the application to a fine part fabrication. Sabahi *et al.* [11] showed that the presence of surfactant in electrolyte helped in improvement of the material removal rate (MRR), accuracy, heat affected zone (HAZ) and better surface quality of micro-channels were obtained. Lie *et al.* [10] developed a new wire electrochemical discharge machining (WECDM) set up by using a rotating type helical tool-electrode and documented the influence of differents machining parameters on the side-gap during the machining of glass workpiece. Singh *et al.* [12] documented that a texture tools for machining of micro-channels on borosilicate glass material with ECDM process. Mehrabi *et al.* [13] demonstrated a hollow electrode with high-pressure electrolyte injection system to provide the electrolyte to the machining area in deep holes for machining of an optical glass using ECDM process. Arya *et al.* [14] developed a pressurized flow system and a fresh electrolyte was injected into the machining zone through a tubular electrode with appropriate controlled electrolyte flow rate. From the past literature survey, it is clear that very little work has been carried out on ECDT. Therefore the present paper deals with the application of ECDM technology for turning operation to machine cylindrical objects. The paper also focuses on the design and development of ECDT set-up.

2. Working Principle of ECDM Process

Electrochemical discharge machining (ECDM) is a nonconventional hybrid machining process used for generating various micro-features such as micro-holes, micro-channels, etc. in electrically conducting as well as non-conducting materials. During ECDM, the removal of material from the surface of workpiece takes place due to the simultaneous interaction of electric spark discharges as well as electrochemical dissolution. In this process two electrodes are used, one is anode and the other is cathode. Tool electrode is generally connected to the negative terminal, i.e. to cathode whereas the positive terminal i.e. anode, is connected to an auxiliary electrode. Both the tool and auxiliary electrodes are connected to a DC power source. The auxiliary electrode has a greater surface area than the tool electrode. When the voltage is applied, electrolysis takes place between the cathode and anode. The bubble generation starts at both the electrodes. The density of bubbles generated at cathode increases with the increased potential difference. At cathode the bubbles grow and combine together to form a thin gas film which encloses the tool electrode. The gas film formed doesn't allow the current to pass through the tool electrode to the electrolyte. As a consequence, high electric field is developed which results in the generation of an electric discharge (As shown in Figure 1). During the spark discharge a focused stream of electrons initiating from cathode strikes the workpiece surface with very high speed creating mechanical compression waves. As the workpiece is kept in close vicinity to the tool electrode, material removal from the surface of workpiece occurs due to melting and vaporisation. A stable discharge phenomenon is very much important for getting desired machining accuracy and high-reliability. In case of electrochemical discharge machining (ECDM) a large inter-electrode gap of about 20-50 mm is maintained between the auxiliary electrode and the tool electrode. Moreover, the mechanism of material removal in the ECDM process is very complex in nature and is governed by various process parameters that are still not very clear to researchers.

ICAMME 2019 IOP Publishing
IOP Conf. Series: Materials Science and Engineering **653** (2019) 012029 doi:10.1088/1757-899X/653/1/012029

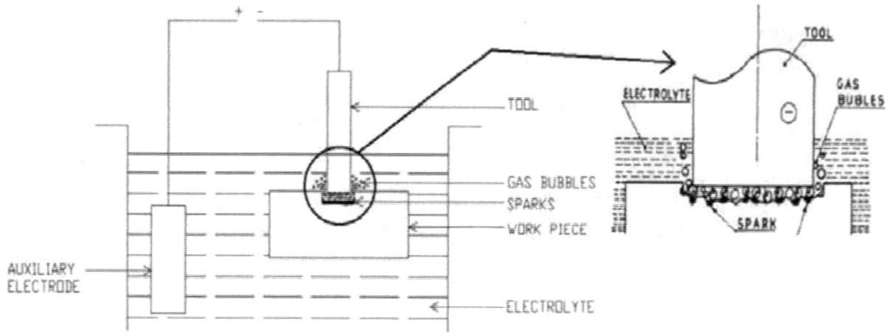

Figure 1. Bubble formation and sparking phenomena in ECDM

3. Applications of ECDM Technology

ECDM technology generally applied for machining various electrically conducting and non-conducting advanced engineering materials such as silicon nitride ceramics [3] and glass etc. The technology of ECDM process can be used for milling operation on Pyrex Glass [7]. This process has the emerging application to micro-machining and fabrication of nano-particles and several applications of interest in catalysis and biomedical domain [6]. Another important application of ECDM technology is its popular variant i.e. travelling wire electrochemical discharge machining (TW-ECDM) for cutting and slicing of high-strength non-conductive materials such as optical glass and quartz bars [3]. Electrochemical Discharge Technology is one of the best technology for trueing and dressing of metalic diamond tools that is used for the grinding of advanced materials [1]. Electrochemical Discharge Turning (ECDT) is another application of ECDM process. It is used to machine a cylindrical job which rotates continuously in an electrolytic bath [5]. During the turning operation by ECDT process the rotational motion of the workpiece feeds the fresh electrolyte across the narrow gap between the tool and the workpiece and the performance of the process is influenced by different process variables. Therefore, it is important to identify the relative importance of process parameters on the performances of ECDT process. Literature survey reveals the enough scope of research exists in the area of ECDT process. Hence, exhaustive experimentations are urgently needed to investigate thoroughly on various aspects of ECDT process. The rest of discussion includes the design and development of ECDT set up, which can be used for turning operation of electrically non-conducting materials.

4. Design and Development of ECDT Set-up

To fulfill the objectives of present research work an indigenously designed ECDT set-up has been developed. The proposed ECDT setup consists of following sub-system. a)DC power supply unit, b)Main Machining Chamber, c)Rotating Job Holding unit, d)Tool Feeding Control Unit and e)Electrolyte supply unit. The main machining chamber is made of Perspex as it is a transparent material and also possessing the shock resistance property. In this process, the auxiliary electrode is a flat rectangular graphite plate which is plunged into the electrolyte solution and it is connected to the positive terminal of the D.C. power supply. The workpiece is rotated with spindle speed driven by a stepper motor. For linear movement tool electrode is mounted on a XY table and feeding is given to the tool by spring feed mechanism. Fresh electrolyte solution is drawn into the Inter-electrode Gap (IEG) by the rotational movement of job materials. In so far as ECDT is concerned many factors come into the consideration and selection of the factor by trial and error method is very important. The influence of some important parameters like applied voltage, electrolyte concentration, pulse frequency, duty ratio, tool polarity, rpm of spindle speed, Inter-Electrode Gap(IEG) between tool and workpiece and some other parameters on various machining responses like MRR, accuracy, overcut and surface finish are identified. based on trial and error method. The rotational speed of workpiece material is one of the most predominant process parameter and control of this rotational speed is a challenging task. The Figure 2 shows the schematic diagram of ECDT set-up.

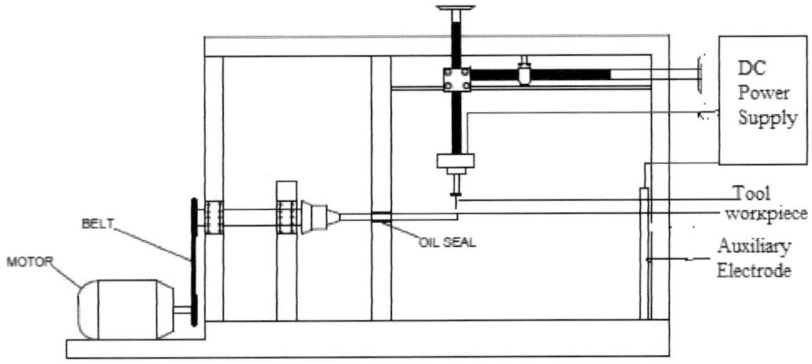

Figure 2. Schematic diagram of Electrochemical Discharge Turing Set up.

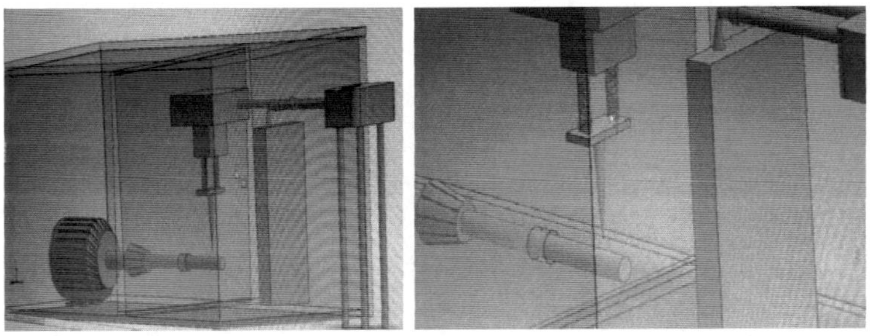

Figure 3. 3D Model of ECDT Set up

The most important part of ECDT set up cylindrical type workpiece holding unit. Tool holding unit is guided springs by which feed is given to the tool while it moving along the vertical direction for cross feed and along horizontal direction for longitudinal feed. The Figure 3 shows the feeding mechanism in ECDT set-up. The dimension of the tool may vary from mm to micron range. However, this is not well known established process so it is required to design of experiments and measure machining responses and analysis through different statistical and optimization methods.

5. Conclusions

The design and developed of Electrochemical Discharge Turning System setup can be used for turning of cylindrical workpieces of various engineering materials. The workpieces machined are of electrically non-conducting materials such as glass, ceramics and composites, etc. ECDT process can be utilized for making grooves of different shapes and sizes on cylindrical parts. Even the ECDT set-up can be utilized for cutting and slicing operation. An ex-haustive experimental investigation can be carried out to find out the influence of various process parameters on various machining criteria using the developed experimental ECDT set-up. The ECDM technology can be used for dressing of cylindrical grinding wheel .This process can also be used for cleaning and finishing of cylindrical workpiece surfaces.

References

[1] Schöpf M, Beltram I, Boccadoro M , Kramer D and Schumacher B 2001 ECDM (Electro Chemical Discharge Machining), a New Method for Trueing and Dressing of Metal Bonded Diamond Grinding Tools, *CIRP Annals Manuf.Technol.,* 50,125–128
[2] Kulkarni A, Sharan R and Lal G K 2002 An experimental study of discharge mechanism in electro-chemical discharge machining, *Int. Journal of Machine Tools & Manuf,* 42,1121–1127.
[3] Peng W Y and Liao Y S 2004 Study of electrochemical discharge machining technology for slicingnon-conductive brittle materials, *Journal of Materials Processing Technol,*. 149,363–369
[4] Sarkar B R, Doloi B and Bhattacharyya B 2006 Parametric analysis on electrochemical discharge machining of silicon nitride ceramic, *Int. Journal Adv. Manuf. Technol*, 28, 873-881
[5] Furutani K and Maeda H 2008 Machining a glass rod with a lathe-type electrochemical discharge machine. *J Miceomech Microeng*,18, 065006.
[6] Wüthrich R and Allagui A 2010 Building micro and nanosystems with electrochemical discharges, *Electrochimica Acta*, 55, 8189–8196
[7] Changjian L G, An G ,Meng L and Shengyi Y 2012 Themicro-milling Machining of pyrex glass using the electrochemical discharge machining process, *Adv.Mater.Res*. 403,738–742
[8] Furutani K and Kojima S 2016 Prototyping of Acceleration Sensor by Using Lathe-type Electro-chemical Discharge Machine, *Procedia CIRP, 42* ,772 – 777
[9] Singh T and Dvivedi A 2016 Developments in electrochemical discharge machining: A review on electrochemical discharge machining, process variants and their hybrid methods, *Int. Journal of Machine Tools & Manuf.,*105, 1-13
[10] Liu Y, Wei Z, Wang M and Zhang J 2017 Experimental investigation of micro wire electrochemical discharge machining by using a rotating helical tool, *Journal of Manuf. Processes,* 29, 265-271
[11] Sabahi N, Razfar M R and Hajian M 2017 Experimental investigation of surfactant-mixed electrolyte into electrochemical discharge machining (ECDM) process , *Journal of Materials Proc. Technol,* 250, 217-229
[12] Singh T and Dvivedi A 2018 on Performance evaluation of Textured tools during microchanneling with ECDM , 32 , 699-713
[13] Mehrabi F, Farahnakian M, Elhami S and Razfar M R 2018 Application of Electrolyte Injection to the ECDM on the Glass , *Journal of Materials Processing Technol.*,255, 665-672
[14] Arya R K and Dvivedi A 2019 Investigations on quantification and replenishment of vaporized electrolyte during deep micro-holes drilling using pressurized flow-ECDM process, *Journal of Materials Processing Technol.,* 266, 217–229

IOP Conf. Series: Materials Science and Engineering **653** (2019) 012030 doi:10.1088/1757-899X/653/1/012030

State of the Art on Laser Assisted Electrochemical Machining

Indranil Mandal*and B Doloi

Production Engineering Dept., Jadavpur University, Kolkata, India

*mandal.indranil2@gmail.com

Abstract. Laser Assisted Electrochemical Machining (LAECM) has proven its advantages and applications over almost all the range of conductive materials. It offers its adroitness to generate simple shapes to complex shapes from macro to micro machining. Laser assisted ECM is a hybrid approach of machining which has been manifested to obtain improved results in terms of surface integrity, thermal damage, geometrical tolerance and material removal rate. Laser assisted Electrochemical Machining (LAECM) is a hybrid machining process in which laser enhances the electrochemical dissolution process by increasing the temperature and hence the current density of the electrolyte. Laser helps to remove material from the particular machining zone and stray machining effect is reduced and thus improves the precision and efficiency of LAECM, also productivity. New route of laser assisted hybrid machining processes has been built in the past decade by few of the researchers to reduce the intrinsic problems of ECM. This paper reviews on the investigation into LAECM process. The article also highlights various possibilities and applications of LAECM process. There are various challenges in LAECM process which have been also discussed in this paper.

1. Introduction

Hybrid machining process has been developed by combining two or more processes to take the advantages and also to overcome the problems of individual processes. Hybrid process is more advantageous in every aspect than a single machining process. Imperative role is played by the advanced material in modern fabricating industries like automotive, aerospace, and tool and die manufacturing industries. HMP'S are accomplished to encounter the requirements of ultra precision machining. It also meets the high productivity requirements along with dimensional tolerance and surface quality for manufacturing of various advanced components [1].

Laser assisted machining is one of the heat-assisted paramount category of HMPs, in which the energy of laser beam is utilized to soften the work material during traditional machining, i.e. milling and turning. Beside that in ECM process the application of laser beam is a well-ordered way for increasing the temperature of the machining zone and thus the assistance of laser enhances the electrochemical dissolution rate.

Classifications of Laser-assisted HMPs are as follow:

a. **Laser Assisted Mechanical Machining Processes**: During conventional machining processes i.e. milling, grinding, turning etc. the job material is heated ahead the cutting tool with the help of laser. Laser beam softens the material before conventional machining.
b. **Laser Assisted Advanced Machining Processes**: The energy of laser beam assists the material removal in Advanced Machining Processes (AMPs) [2].

2. Laser Assisted Electrochemical Machining (LAECM) Process

In laser assisted electrochemical machining process (LAECM) laser beam is utilized to enhance the machining rate and accuracy of Electrochemical Machining process. The fundamentals of laser assisted ECM are discussed herein after.

2.1 Fundamentals of Electrochemical Machining (ECM)

In ECM process material removal takes place due to chemical dissolution of electrolyte. The main advantages of ECM include machining of almost all conductive materials with better surface finish, does not apply any stress in the workpiece and no tool wear. Due to stray machining effect the precision and accuracy of ECM are decreased. So this effect must be controlled by removal of material form specific machining zone to get better precision and accuracy. ECM is applied in various fields such as aerospace, medical industries, defence etc.

2.2. Lasers Used in LAECM Process

In LAECM laser improves localization of the dissolution process by increasing the temperature of electrolyte and increases MRR, precision and accuracy. Most widely used types of laser in LAECM are low power (375 mw-570 mw) Nd: YAG lasers and fiber lasers etc [4]. The laser is a non-contact thermal energy technology and it is also used in material processing. For the heating and preheating of various materials the laser heat source can be used.

2.3. Fundamentals of Laser Assisted ECM Process

Due to stray machining effect the precision and accuracy of Electrochemical Machining process are decreased. Another drawback in ECM includes complex tool design. LAECM eliminates this problem by focusing laser in the machining zone. machining zone. Laser assists the ECM process. Hence the accuracy and precision of LAECM process are increased [7].

Laser assisted Electrochemical Machining (LAECM) is a hybrid machining process in which laser beam assists the ECM process and increases its machining performance by increasing the temperature of electrolyte which further increases its current density. The laser by its thermal activation assists dissolution and laser doesn't remove any material.

The main tasks of the laser beam in LAECM are as follows: i. According to the Arrhenius law the rate of dissolution is increased by raising the temperature of the electrolyte in the inter electrode gap ii. The temperature of electrolyte is to be increased for getting higher current density and conductivity. iii. Surface passivation is eliminated, which increases the material removal rate intensification. Thus stray machining effect can be reduced by focusing the laser in particular machining zone [4]. In the machining zone the temperature is raised with the help of laser beam. This effect has a positive effect on treatment by increasing electrolyte conductivity hence current density, the ECM reactions are initiated easier, and transportation of the removed material and the diffusion of ions are intensified to enhance the machining precision.

3. Research Status on Laser Assisted Electrochemical Machining (LAECM) Process

Research work on laser assisted electrochemical machining (LAECM) has been performed by various researchers and some of the important work has been highlighted in this present paper.

Applications of individual advanced machining processes are limited in some areas such as machining of advanced and difficult to cut materials, requirements of ultra precision machining etc. Hybrid machining processes reduce these problems in which one process energy assists the main process energy for obtaining better material removal rate. Thus precision and accuracy of main process are increased. Machining of novel materials and various complex shapes can be obtained by hybrid processes. In LAECM process laser assists the electrochemical process and increases the temperature and current density of electrolyte and improves machining performance. For obtaining high level of localization in machining diode pumped solid state pulse laser can be applied [1, 2, 4, 8]

In laser assisted ECM laser energy and ion energy together are used for machining conductive materials from particular machining zone and stray machining effect is reduced. Various novel materials can be machined by LAECM process and laser helps in removal of passivating layer in ECM process [3]. LAECM process can be applied for machining of steels, titanium alloys and super alloys. Micro dimensional features can be fabricated by LAECM process [3].

In LAECM process laser increases the chemical dissolution and hence accuracy is increased. Green laser of wavelength 470-560 nm can be applied for low absorption coefficient electrolyte in ECM process. A mathematical model has been developed for the selection of laser radiation intensity. Laser enhanced electrochemical machining process can be applied in various engineering fields such as biomedical etc. LAECM process can be also applied for surface structuring of workpiece [5].

Point-by-point and scan methods are two main LAECM methods. Better etching surface quality can be achieved by scan method and various complex profiles can be made by this method. LAECM process has various advantages such as better machining efficiency, removal of passivating layer, lesser HAZ etc. Complex microstructures and integrated structures can be made by this method [6].

In LAECM laser enhances chemical dissolution process by increasing the current density of electrolyte. LAECM process can be applied for machining of difficult to cut conductive materials. LAECM improves machining performance and also accuracy of ECM process [7].

Silicon nitride ceramics and metallic alloys can be machined by laser assisted machining. Higher material removal rate can be achieved by hybrid machining process [9].

In LAECM laser reduces activation energy level and increases the temperature of electrolyte. The current density of electrolyte is increased and hence chemical dissolution process is increased. Thus higher material removal rate can be obtained [10].

Conductive materials can be machined by Jet Electrochemical Machining (JECM) process and effect of stray machining affects this machining by reducing the effectiveness. Hybrid Laser Assisted Jet Electrochemical Machining (LAJECM) process can eliminate this problem and material removal occurs from particular machining zone due to the metal dissolution by electrochemical action that follows the main principle of Faraday's laws of electrolysis. Laser beam assists the ECM process and increases its machining performance by increasing the temperature of electrolyte which further increases its current density. Laser assistance increases accuracy and precision of LAJECM process. If supply voltage, duty cycle and electrolyte concentration are increasing MRR and taper are also increasing. The main significant parameters are supply voltage and inter electrode gap for MRR and electrolyte concentration and supply voltage for taper [11].

4. Challenges in LAECM process

The wavelength of laser radiation should be properly selected so that the electrolyte layer should absorb minimal energy and for this reason the electrolyte layer should be as thin as possible. In LAECM, one of the challenges is to investigate the effect of laser radiation on the current efficiency of the anodic dissolution which is not still properly understood. In LAECM, shaping of small elements (5-500 μm) with high accuracy (1-10 μm) that are made from difficult-to-machine materials such as ceramics, composites and alloys is also a challenging area. Another challenging area in LAECM is to minimize the formation of high-absorptive layer of dissolution products.

5. Experimental Set-up of LAECM

The schematic diagram of Laser Assisted Electrochemical Machining set-p is shown in Figure 1. It describes LAECM set-up where laser is focused in the machining zone from the side of the electrode.

A low power laser beam is focused through reflector, focusing lens and polarizer to the specific machining zone during LAECM process. Laser strikes the workpiece and due to absorption of laser energy the temperature of material raises and heats up the adjacent electrolyte, which increases the rate of electrochemical reactions and enhances material removal rate. Electrolyte is supplied through pump to the machining zone. X and Y axis are controlled by servo motor and Z axis is controlled by stepper motor for tool feeding system.

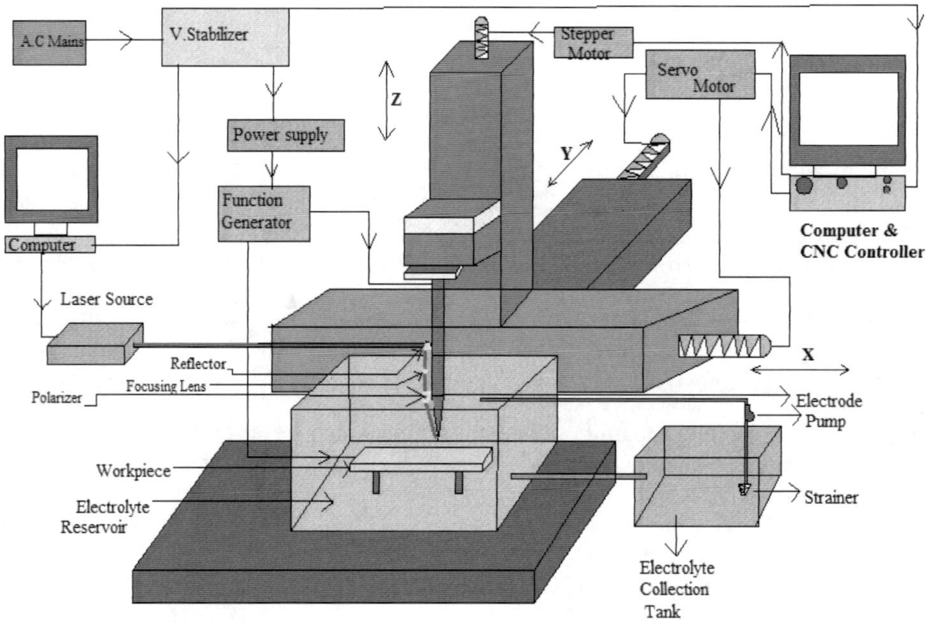

Figure1. Schematic diagram of the experimental set-up of LAECM.

6. Various possibilities and applications of LAECM process

The laser hybridized with ECM process improves surface finish, removes thermal defects and increases precision and accuracy of ECM process. Laser enhances electrochemical dissolution by increasing the temperature of electrolyte and hence current density. LAECM process can be applied for machining of aluminium , titanium alloy, stainless steel etc. LAECM process can be applied for surface structuring and various micro-dimensional features can be manufactured [3].

7. Conclusions

The effect of hybrid machining processes has greater influence than the sum of the advantages of a single machining process. Application of laser based hybrid machining processes is recently increasing in various industries.

Laser assisted Electrochemical Machining (LAECM) method combines two different sources of energy at the same time: ions energy and photons energy. Laser has been used to improve localization of the dissolution phenomena in electrochemical machining process. Surface layer as well as the machined surface of the job material undergoes through several chemical and physical phenomena due to the heat generated by laser. Faster chemical dissolution is achieved by the interaction of laser as it intensifies the kinetics of the chemical reactions.

LAECM process enhances precision, accuracy and surface finish of ECM process and also reduces thermal defects. LAECM process can be applied for machining of hastelloy, titanium alloy, stainless steel, and aluminum materials. LAECM has potential applications to fabricate micro-cavities and micro-structures cavities of different size and shapes needed for electronics, bio-medical, aviation and optical industries etc. Further investigations can be done on LAECM by studying the several process parameters that affect the performance of LAECM process, minimization of stray current of ECM by laser assistance and comparative study between LAECM and Jet Electrochemical Machining (JECM) process.

References

[1] Saxena KK, Bellotti M, Qian J, Reynaerts D, Lauwers B and Luo X 2018 Overview of Machining Processes chapter 2 pp 21-41

[2] Gupta K, Jain NK, Laubscher RF 2016 Hybrid Machining Processes perspective on Machining and Finishing ISBN 978-3-319-25920-8 PP 45-58

[3] Saxena KK, Bellotti M, Camp DC, Qian J and Reynaerts D 2018 Electrochemical Based Hybrid Machining Chapter 5 pp 111-129

[4] Wyszynski D, Skoczypiec S, Grabowski M, Ruszaj A and Lipiec P 2012 Electrochemical Microprocessing Assisted by Diod Pumped Solid State Nd:YAG Pulse Laser *Journal of Machine Engineering* **12** pp 131-142

[5] Skoczypiec S 2015 Application of laser and electrochemical interaction in sequential and hybrid micromachining processes *Bulletin of the polish Academy of Sciences* **63** pp 305-314

[6] Zhang Z, Cai M, Feng Q and Zeng Y 2014 Comparison of different laser-assisted electrochemical methods based on surface morphology characteristics *The International Journal of Advanced Manufacturing Technology* **71** pp 565-571

[7] Malik A and Manna A 2018 Multi-response optimization of laser-assisted jet electrochemical machining parameters based on gray relation analysis *Journal of the Brazilian Society of Mechanical Sciences and Engineering* **40** pp 1-21

[8] Lee CM, Woo WS, Kim DH, Oh WJ, and Oh NS 2016 Laser-Assisted Hybrid Processes: A Review *International Journal of precision Engineering and Manufacturing* **7** pp 257-267

[9] Chryssolouris G, Anifantis N and Karagiannis S 1997 Laser Assisted Machining An Overview *Journal of Manufacturing Science and Engineering* **119** pp 766-769

[10] Pajak P T, De silva AKM, Harrison DK and Mc geough JA 2006 Precision and efficiency of laser assisted jet electrchemical machining *Precision Engineering* **3** pp 288-298

[11] Malik A and Manna A 2018 Investigation on the laser-assisted jet electrochemical machining process for improvement in machining performance *The International Journal of Advanced Manufacturing Technology* **96** pp 3917-3932

Mechanical Behavior of Ultrasonic Spot Welded Aluminum to Cupro-Nickel Sheets for Battery Tabs

Soumyajit Das[1]*, Mantra Prasad Satpathy[1], Bharat Chandra Routara[1], Susanta Kumar Sahoo[2], Subhra Gantayat[3]

[1]School of Mechanical Engineering, KIIT Deemed University, Bhubaneswar-751024, Odisha, India

[2]Department of Mechanical Engineering, National Institute of Technology, Rourkela-769008, Odisha, India

[3]NIIS Institute of Business Administration, Bhubaneswar- 752054, Odisha, India

*E-mail: soumyajitdasofficial@gmail.com

Abstract. Ultrasonic metal welding process has received noticeable attention throughout the last few years because of their proper application in comparison to the conventional fusion welding process. Fusion welding of dissimilar metals like copper and nickel, copper and aluminum alloy is always a difficult task due to the different physical, chemical, thermal properties, poor control of grain size and it has a deleterious effect on mechanical properties. Automotive industries have been widely accepted the ultrasonic metal welding process for assembling lithium-ion battery tabs. The aim of this current work is to investigate the weldability of two dissimilar metals namely aluminum (AA1060) and cupronickel (UNS C71500) sheets at various levels of welding time, vibration amplitudes and welding pressures with a fixed ultrasonic frequency of 20 kHz. The highest ultimate tensile shear strength test of joint achieved at highest vibration amplitude with moderate welding pressure and weld time. Weld quality determining characterization is introduced for the weld region to establish a relationship between physical attributes with the microstructural phenomenon.

Keywords: Lithium-ion battery, ultrasonic metal welding, tensile shear strength, aluminum, cupronickel

1. Introduction

The energy crisis is a critical problem in recent years. The non-renewable energy sources are fast draining, but renewable energy resources are still underdeveloped to meet the ever-growing energy demands. Nowadays, electric vehicles (EVs) grab attention in many of the automobile makers to meet the challenges like unsteady fuel price and anthropogenic environmental pollution [1]. The functioning of these EVs depends on the energy and power capacities of their batteries. Lithium-ion (Li-ion) batteries have advantages over various battery technologies because of its high energy density capacity and compactness for the same power requirement. To produce these battery packs for EVs, there is a definite need for reliable and efficient welding technique which can weld hundreds of cells. Joining of these battery packs have faced many challenges like highly conductive materials such as copper, nickel, aluminum with different thickness, physical and mechanical properties [2]. Numerous joining technics are used in the battery pack industry like laser welding, ultrasonic welding and resistance welding [3]. Depending on reliability, cost and easy-to-use, ultrasonic metal welding (USMW) process is the most widely used technology for manufacturing battery packs. Dissimilar metals like aluminum-copper, aluminum-nickel can be joined by this method at a relatively low cost, low resistance and without any filler material or heat affected zone. The USMW was originated in the 1950s for thin metal foil joining, tube sealing and wire bonding [4]. It is a solid-state joining process, which applies the high-frequency ultrasonic vibration to produce joints under a static clamping force. USMW bonding is generally achieved approximately at 300°C temperature in a short welding cycle (typically <0.5 s) [5].

During the last few years, some researchers have been studied the welding quality in USMW for battery tab joining. Jahn et al. [6] examined for the effect of weld energy on the growth of weld strength and microstructure in USMW. Kong et al. [7] performed experiments on ultrasonic welding, the various effects of battery packs on the sonotrode force. Bakavos et al. [8] have studied that USMW with advantages of high efficiency and low energy consumption is an emerging welding process. Imai and Matsuoka [9] investigated welding time and optimal clamping pressure by using an aluminum alloy for high-quality ultrasonic welding. Lee et al. [2] observed the physical characteristics of the weld quality which affect the welding performance of USMW. Kim et al. [10] characterized the weld performance on test loads, and the weld lobe is defined for dissimilar and thin metals using full factorial experiments. As a result, maximum peel force and type of failure loads in the T-peel test were determined according to the range of robust performance. Wu et al. [11] investigated the microstructure of ultrasonic welded three layers of lithium-ion battery bars (Al or Cu). They revealed that the main mechanism was friction-induced physical bonding in the Al/Cu interfaces, and no intermetallic compound layer was found in the weld zone.

Currently, only few works have been done between cupronickel sheet with aluminum alloy. In this experiment vibration amplitudes have been varied by using full factorial design method. The objective of this study is to weld aluminum (AA1060) with cupronickel (C71500) coupons using ultrasonic welding process and to study the effect of different welding parameters on the tensile shear strength of the welded joint and a microstructural analysis for the weld interface is presented to determine the weld quality characterization.

2. Experimental Procedure

In this experiment, 0.3 mm thick cupronickel (UNS C71500) coupons lap welded with commercial 0.8 mm thick AA1060 aluminum alloy by using ultrasonic welding. A lateral drive ultrasonic spot welder (Telsonic® M4000) was used to perform experiments with the maximum output of 3 kW, and the frequency of 20 kHz (Figure 1(a)). The knurled patterns sonotrode tip of 9 mm x 11 mm was employed are also present in Figure 1(b). The dimension and shape of the specimen were rectangular and 20 mm wide × 75 mm length. The weld coupons are overlapped for 20 mm along its length and formed an overlap area of 20 × 20 mm^2 for the lap welding. The schematic diagram is shown in Figure 1 (d). The AA1060 coupons are softer than UNS C71500 coupons, so that aluminum sample was placed over the cupro-nickel sample. Table 1 presents the mechanical and thermal properties of the weld samples. Previous literature and experimental runs that the three process parameters such as weld pressure (WP), weld time (WT) and vibrational amplitude (A) have a significant effect on the weld strength. In this experiment, WP, WT, and A have been divided into three, six and four stages respectively. The domain of factors with their levels are shown in Table 2.

Table 1. Mechanical and thermal properties of weld materials

Materials	Ultimate tensile strength	Yield strength	Thermal Conductivity	Specific heat ($Jg^{-1}°C^{-1}$)

	(MPa)	(MPa)	(Wm^{-1}°C^{-1})	
AA1060	68.9	17	234.2	0.9
UNS C71500	380	125	29	0.38

The full factorial design is employed for systematic conducting of experiments. Total 72 number of tests were performed with five repetitions of each level. The AA1060 was placed on the sonotrode side, and cupronickel (UNS C71500) was positioned at the anvil side. Both sheets were polished with P120 grit polished emery papers followed by cleaning with acetone. In order to calculate the weld strength of the specimen, lap shear tensile test was performed for every welding conditions by using FIE® universal testing machine with a load cell of 40 kN. Aluminum and cupronickel sheets were clamped in the upper and lower jaws respectively (Figure 2).

Table 2. Factors and levels for Experimental design

Factors	Level 1	Level 2	Level 3	Level 4	Level 5	Level 6
Amplitude (μm)	47	54	60	68	...	...
Weld pressure (MPa)	0.26	0.3	0.36	...	...	...
Weld time (Sec)	0.4	0.45	0.5	0.6	0.65	0.7

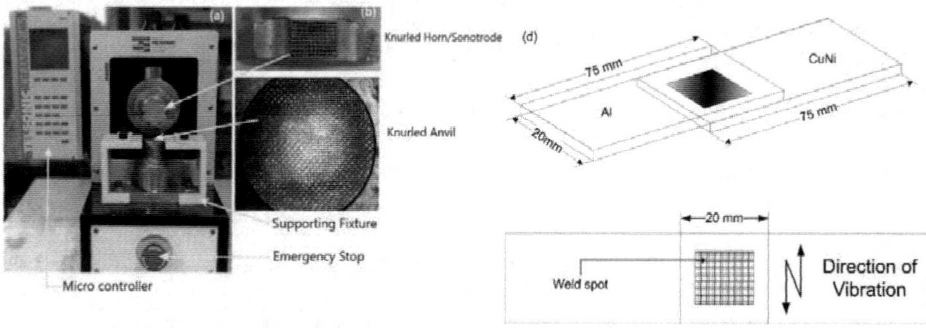

Figure 1 (a) Lateral drive USW machine, (b) Sonotrode tip with knurls, (c) Knurled anvil, (d) lap configuration of weld materials

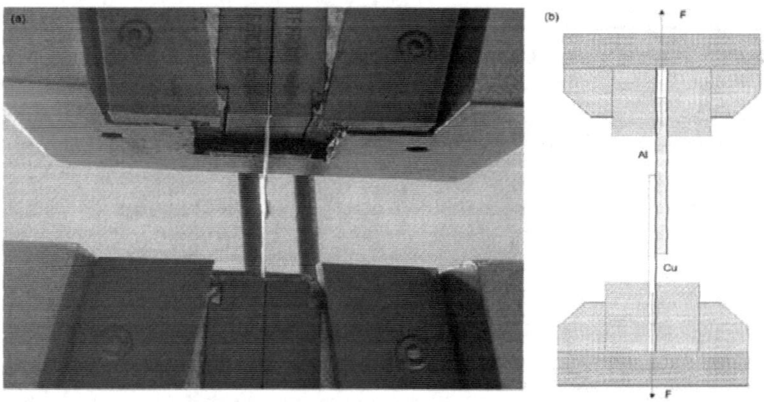

Figure 2 (a) Weld samples clamping image, (b) Schematic diagram of lap shear test

3. Experimental results

3.1 Tensile lap shear strength analysis:
The results show that peak tensile shear failure load of ultrasonic welded aluminum and cupronickel coupons joints at 47μm vibration amplitude with different welding times are shown in Figure 3(a). This tensile shear load increased with increasing welding time and pressure 0.60s and 0.3MPa and reached a peak value of maximum load 538N. Thereafter, the tensile shear failure load decreases with a further increase in weld time due to the crack formation occurs at weld spot. Similarly, the lap shear tensile load also same nature as described above at 54μm amplitude with different welding time are shown in Figure 3(b). The peak value of failure load slightly increased 570N at same weld pressure and time. Likewise, the tensile lap shear load increased rapidly at 60μm amplitude at welding time varies from 0.4s to 0.6s as shown in Figure 3(c). The maximum load of 710N is achieved at same input parameter values. The most notable observation in this experiment is at 68μm amplitude with 0.3MPa of weld pressure and 0.6s of weld time. The maximum tensile shear failure load as 809N are shown in Figure 3(d). The reason is that when vibration amplitude increases the relative motion between the sheets increases and it gives rise more amount of interface temperature. Due to this temperature, the more amount of plastic deformation occurred at the weld spot and the weld interface is filled with dense micro-bonds.

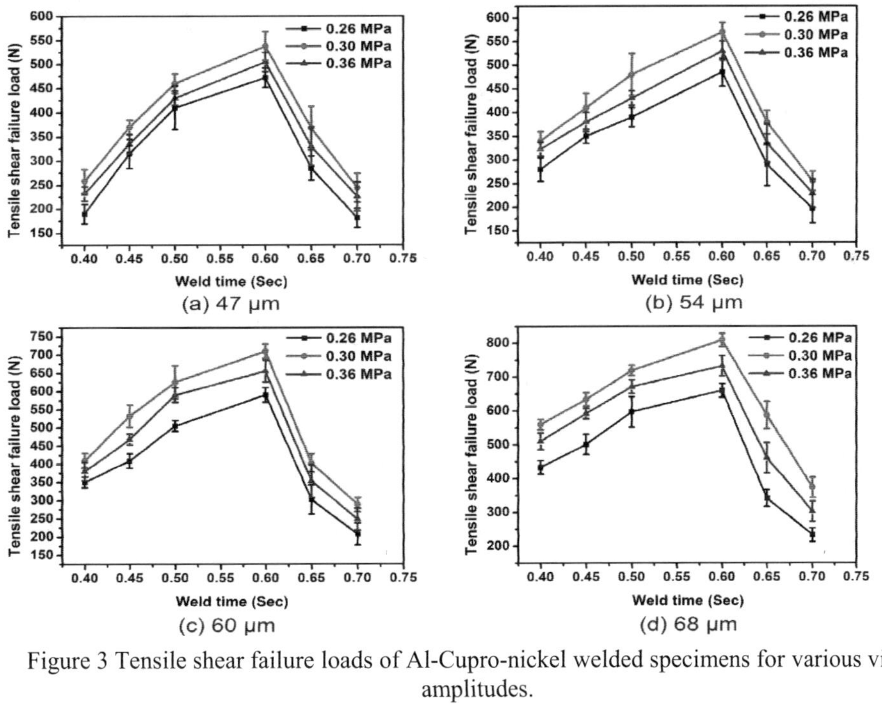

Figure 3 Tensile shear failure loads of Al-Cupro-nickel welded specimens for various vibration amplitudes.

3.2 Welding quality analysis:
Weld strength can be expressed by the physical attributes of the weldment. To identify the relationship between weld performance and weld attributes multiple weld coupons of varies weld quality were cross-sectioned and categorized by SEMs analysis. In SEM images of the bonding line for various weld time, 0.4s (under) 0.6s (good) 0.7s (over) are shown in Figure 4(d), (e) and (f) respectively. In

weld interface along to the bonding line, the gap is clearly visible in the under-weld sample as shown in Figure 4(a). If weld time increase the gap become less and fully covered by micro-bonds. In case of good welding condition, the coupons were perfectly intimate contact with each other as shown in Figure 4(b). At higher weld time 0.7s several cracks are observed in the weld interface due to the diffusion throughout the weld zone, mechanical interlocking and local melting at weld surface as shown in Figure 4(c).

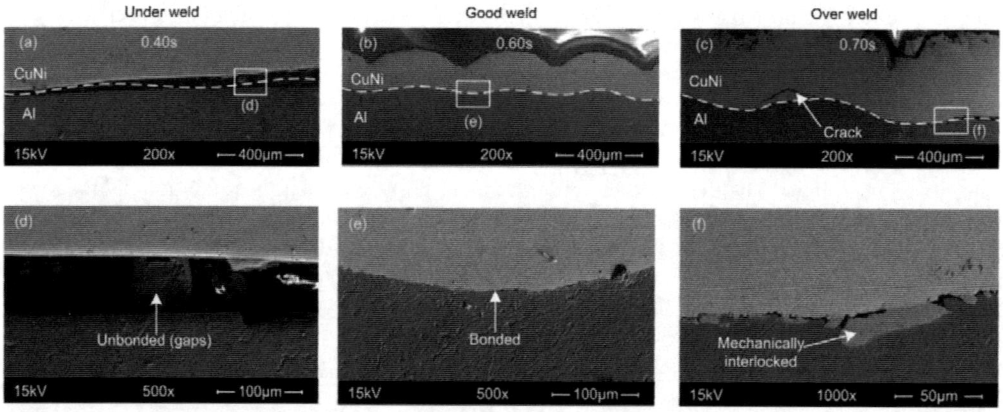

Figure 4 SEM images of ultrasonically welded joints made in various weld qualities (under, good, over weld).

4. Conclusions:

The USW of Aluminum (AA1060) with Cupronickel (UNS C71500) has been explored, and the weld qualities of the joints were studied using weld attributes and performance. The major conclusions of this paper extorted from the current experiment:

I. Tensile lap shear strength was increased with increasing weld time and pressure up to their minimal value of 0.6s and 0.3MPa. The maximum tensile shear failure load of 809N has been noted at vibration amplitude 68µm then decrease with further increase the weld time.

II. The SEM analysis discloses various levels of weld quality (under, good, over weld), depends upon different weld time, pressure and amplitude. At the highest weld time, it shows that the joints of two sheets occur with various laminar cracks at the periphery of the welding surface.

References

[1] Graff, K.F., 1974, November. Process applications of power ultrasonics-a review. In *1974 Ultrasonics Symposium* (pp. **628**-641). IEEE.

[2] Li, S., Wang, H., Lin, Y.T., Abell, J. and Hu, S.J., 2010, January. Benchmarking of high capacity battery module/pack design for automatic assembly system. In *ASME 2010 International Manufacturing Science and Engineering Conference* (pp. **505**-517). American Society of Mechanical Engineers

[3] Lee, S.S., Kim, T.H., Hu, S.J., Cai, W.W. and Abell, J.A., 2010, January. Joining technologies for automotive lithium-ion battery manufacturing: A review. In *ASME 2010 international manufacturing science and engineering conference* (pp. **541**-549). American Society of Mechanical Engineers.

[4] Hetrick, E.T., Baer, J.R., Zhu, W., Reatherford, L.V., Grima, A.J., Scholl, D.J., Wilkosz, D.E., Fatima, S. and Ward, S.M., 2009. Ultrasonic metal welding process robustness in aluminum automotive body construction applications. *Weld. J*, ***88*(7),** pp.149-158.
[5] Macwan, A. and Chen, D.L., 2015. Microstructure and mechanical properties of ultrasonic spot welded copper-to-magnesium alloy joints. *Materials & Design*, ***84*,** pp.261-269.
[6] Jahn, R., Cooper, R. and Wilkosz, D., 2007. The effect of anvil geometry and welding energy on microstructures in ultrasonic spot welds of AA6111-T4. *Metallurgical and Materials Transactions A*, ***38*(3),** pp.570-583.
[7] Kang, B., Cai, W. and Tan, C.A., 2014. Vibrational energy loss analysis in battery tab ultrasonic welding. *Journal of Manufacturing Processes*, ***16*(2),** pp.218-232.
[8] Bakavos, D. and Prangnell, P.B., 2010. Mechanisms of joint and microstructure formation in high power ultrasonic spot welding 6111 aluminium automotive sheet. *Materials Science and Engineering: A*, ***527*(23),** pp.6320-6334.
[9] Matsuoka, S.I. and Imai, H., 2009. Direct welding of different metals used ultrasonic vibration. *Journal of materials processing technology*, ***209*(2),** pp.954-960.
[10] Kim, T.H., Yum, J., Hu, S.J., Spicer, J.P. and Abell, J.A., 2011. Process robustness of single lap ultrasonic welding of thin, dissimilar materials. *CIRP annals*, ***60*(1),** pp.17-20.
[11] Wu, X., Liu, T. and Cai, W., 2015. Microstructure, welding mechanism, and failure of Al/Cu ultrasonic welds. *Journal of Manufacturing Processes*, ***20*,** pp.321-331.

IOP Conf. Series: Materials Science and Engineering **653** (2019) 012032 doi:10.1088/1757-899X/653/1/012032

An Evaluation on Green Manufacturing: It's Technique, Significance and Rationality

Bhubaneswari Bisoyi[1*], Divyajit Das[2], P Srinivas Subbarao[3], Biswajit Das[4]

[1]Sri Sri University, Cuttack, India
[2]KIIT University, School of Civil Engineering, Bhubaneswar, India
[3]Sri Sri University, Cuttack, India
[4]KIIT University, School of Management, Bhubaneswar, India

*Corresponding author: bhubaneswari.b@srisriuniversity.edu.in

Abstract
This research paper is focused on designing the environment through green manufacturing method, the use of eco-friendly products, conservation of energy, and the manufacturing of products with a reduction in the waste material. The research paper also draws attention towards the utilization of green manufacturing technique for modulating a product that is sustainable, reusable, and diminutive existence. The key objective of implementing green manufacturing is to protect the eco-system and to ease the expenditure. With unprecedented competition globally results in the improvisation of the manufacturing operation for providing better products and services. Emission of greenhouse gases is a major cause of climate change, global warming, and environmental degradation. It also results in unnecessary dissemination of energy that would be used for the betterment of society. One of the promising approaches to creating a balance in the eco-system is through the implementation of green manufacturing. The techniques of green manufacturing can be applied to reducing pollution level, minimizing waste, facilitates financial development and marmalade reserves. The present research paper focuses on accomplishing green manufacturing by means of a variety of procedures that has a tremendous influence on diminution in pollution and non-biodegradable waste products.

1.0 Introduction

With globalization, the environment condition is degrading gradually due to maximizing the use or natural resources and another factor is population. The imbalance of the planet earth is due to change in the climatic condition of our environment. The new process that has become the need of the current time is green manufacturing. The best strategy for sustainable development by adopting green manufacturing. The main objective of this research paper is to convey the manufactures that are following the principle of mass production [1, 2]. It has been observed that the power consumption by these industries is a high and a huge amount of waste material is released into our environment. These waste products are hazardous and the ultimate impact would be the termination of a human being. Poisonous hazards are in actual fact critical for a human being ruled on Waste Electrical and Electronic Equipment (WEEE) [3, 4]. Few companies looking to decrease recuperation and manufacturing expenses have started adapting designs of products and integrate end-of-life (EOL) product reprocess conception and design of the component. The impact on our environment occurs at every stage of the product life cycle starting from bringing out resources to produce, utilize, reprocess, recover and discarding. It has been proposed by researchers for including green practices such as green procurement, green built-up, resources management, green delivery/promotion and reverse logistics in green supply chain management (GSCM) [5,6]. Therefore, GSCM has come into sight as an imperative move towards a decrease of risk and burden on eco-system in manufacturing system dumping, in addition, to an increase in profit and competitive advantages. This study would help the growth of the automobile industry by adapting GSCM to add as a competitive advantage over other industries in developing countries.

2.0 Aim and Scope of Green Manufacturing

In the present scenario, one of the major issues is e-waste; therefore green technology is implemented to diminish the harmful blow of human association [7]. The application of green technology is a combination of ecological knowledge, green chemistry and building up a framework that monitors and conserves natural resources. The term green manufacturing is also used to illustrate technologies used for the generation of renewable energy, solar cells, wind turbines, bioremediation [8, 9]. There are several tools used for achieving the aim of green manufacturing for attaining a sustainable environment.

2.1. Tools for Management of Environment

The tool used for creating stability in the environment is done by mass balance. Mass balance considers the input and output of a process by creating equilibrium between effectiveness and wastage [10, 11]. In accounting full cost is considered to be the sum of material cost, energy, workers, disposal of waste another cost including sundry items. The other vital component of this tool is the product life cycle. The shorter product life cycle generates a less environmental loss. In case of a product manufactured by systematic engineering process consist of three stages: (1) theoretical, introductory and detail plan, (2) fabrication, creation and (3) functioning and system support [12]. The critical leverage point is achieved between business benefit and the environment by imposing comprehensive accountability on producers and manufactures. The manufacturers have been able to smooth the progress of product recuperation and reproducing through the unique ability to design the products for effortless dismantle and reprocess of a constituent.

2.2. Sustainable Technique of Manufacturing

The conception of sustainability appeared after a chain of congregation and information during the 1970s and 1980s was majorly stimulated by means of disasters and incidents related to environment and catastrophes in addition to trepidation in relation to chemical defect and reserve exhaustion. The term sustainability entails a great deal that consists of analyzing and transforming the ecological performance of the manufacturing industry in terms of their process and system [13]. A system is thought to be unsustainable when the outcome results in degradation of societal value in terms of consuming resources and producing waste products that are hazardous in nature. Firmly interpolation of sustainability can be examined in the framework of a closed system, as explained by figure 1[13]. The subsystem s that coexists together are a human, environmental system, and manufacturing is a part of it. As a result, the philosophy of sustainable manufacturing cannot be assumed to be self-sufficient of ecological and socioeconomic structure. With the calculation of the environmental and social performance exhibits an added exigent task for business and engineering. The sustainability associated impacts outcome commencing procedures and actions that manufacturing processes and systems utilize to translate input materials and energy into saleable products. The necessary input to the manufacturing process and system consists of material and energy; output includes processed products, wastes and carbon emission, are the inputs to the environment, society and economic condition.

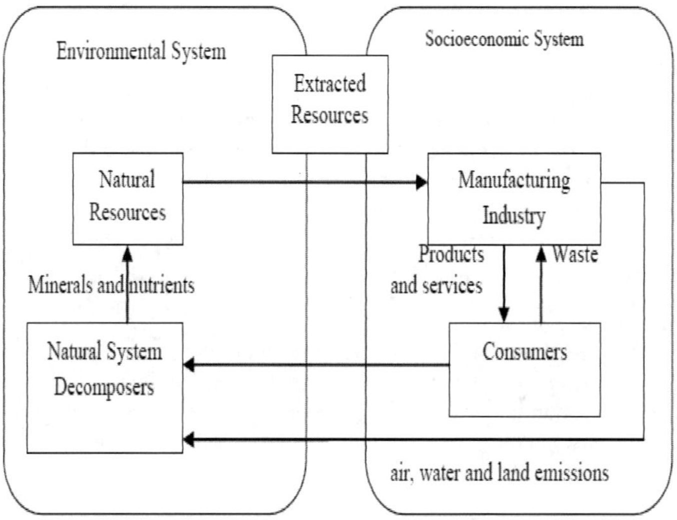

Figure 1. Sustainable system in manufacturing industry

3.0 Methods to Attain Green Manufacturing

3.1. Lean Manufacturing through Just-in-Time

The concept of a 'pull' model is considered to be the core of lean manufacturing. The implementation of this model helps to minimize stock and resources. It takes into account one party to purchase raw materials and manufacture and dispense products as soon as required. This process reduces the batch size, reduces the error and helps the production process to execute smoothly [1, 6].

3.2. Kanban

This process involves the people in the lean production process. The Just in Time model is supported by this system by developing nods that give the signal when it is needed to produce, replace and order [9].

IOP Conf. Series: Materials Science and Engineering **653** (2019) 012032 doi:10.1088/1757-899X/653/1/012032

3.3. Single Minute Exchange of Die (SMED)
This process facilitates you to construct suppleness in the production system. For instance with the automobile sector, it takes a number of days for changing a line to produce a variety of car models. But, through SMED the shifting of the line can be designed and the process also faster than another process [6].

3.4. 5S thinking
In the case of lean manufacturing, the process is said to be standardized. The whole arrangement in the workplace, instruments, and procedures is very simple and benchmarked. This improvises the system by reducing the inventory level. Therefore, the 5S system is implemented [11].

3.5. Zero Emission
This system gives emphasis on developing machinery, and progression with an aim to maximise the productivity of resources and to practically give no waste. The concept of zero-emission aims to contribute to design effectual technology, innovative manufacture method, preservation and reprocess of natural resources (reuse of waste for energy). This objective can be accomplished in a realistic manner by built-up natural balance, greenhouse gasses deterrence, industrial modernization, derivative synergy, spotless manufacturing [8, 9].

4.0 Sustainable Green Manoeuvre
This is an innovative approach to environmental management. The purpose of the green operation is to ensure the quality and compliance of the inputs given for processing into electronic components and products in the form of finished products, carbon emission, and waste. The green operation highlights the product and process to familiarize with practices that create a balance between environments, financial performance and also reduces pollution [13, 14].

5.0 Implementation of Green Supply Chain Management
The addition of green issues in the supply chain is termed as GSCM. The closed-loop of green supply chain management system includes supplier, manufactures, intermediaries, customers and reverse logistic [7]. The various activities carried out in the closed-loop of GSCM involves green designing of product considered for reuse, remanufactured, recycled, The other activities consist of practicing green procurement, total quality environmental management, environment-friendly covering, shipping, and different practices for each step in the product life cycle [6].

6.0 Green Applications
The foremost matter of concern in the world currently is the fuel used daily for smooth function of various activities. The fuel is extracted for non-renewable resources that shall finish one day. The solitary trust is on renewable energy that can be generated from solar, wind, tidal, biodiesel etc. that are considered to be eco-friendly products. Hence renewable energy can be used as a stimulus for attaining environmental sustainability. The other prime issue in human life is to get purified water for drinking. The water purification process can be carried out by using solar purification system [3]. A next issue is to reduce the carbon emission level by growing plants indoor and outdoor to provide pure air to breathe. The concept that is similar to water purification is sewage treatment. The

government has also introduced various policies for solid waste management in cities. Solid waste management assists in decontamination, reuse, recycle, and treatment for the solid wastes [9]. Energy conservation can be utilized to reduce the uses of non-renewable resources and conserve for the future generation.

7.0 Conclusion

This research paper discussed green manufacturing and tries to communicate about the motive of the researcher to implement green technology for the development of the environment. The energy consumption used for product manufacturing should also be generated from renewable resources for understanding the concept of sustainability. The paper also discussed the green operation and the tools used for the environmental management system. The tool used for improving the green image and competitive strategy is through a green supply chain. The Future research work shall centre on a broader assortment of green strategies that can be quantifiable within the industrial unit, e.g., illumination, HVAC, or under pressure air utilization, and classifying a means of linking fixed costs in the decision-making process. The key benefit of utilizing green manufacturing is to reduce cost because the company needs to eliminate the waste. Therefore, the waste should be eliminated in the first step by following the tool Zero Emission Strategy. The tools and techniques illustrated in this paper help in elevating the performance of the industry through the green manufacturing process.

References

[1] Agi M A and Nishant R 2017 Understanding influential factors on implementing green supply chain management practices: An interpretive structural modelling analysis *Journal of Environmental Management* **188** pp 351-363

[2] Alayón C, Safsten K and Johansson G 2017 Conceptual sustainable production principles in practice: Do they reflect what companies do? *Journal of Cleaner Production* **141** pp 693-701

[3] Bisoyi B and Das B 2018 Green technology for attaining environmental safety and sustainable development *International Journal of Mechanical Engineering and Technology* **9** Issue 3 pp 1087 – 1094

[4] Callahan J S, Dunne R A and Stanaback S 1997 New Jersey country dismantles recycling and emissions problems *World Wastes* **40** 3 pp 7–8

[5] Christmann P 2000 Effects of ''best practices'' of environmental management on cost advantage: the role of complementary assets *Academy of Management Journal* **43** 4 pp 663–680

[6] Chun-Jen Chung and Hui-Ming Wee 2008 Green-component life-cycle value on design and reverse manufacturing in semi-closed supply chain *Int. J. Production Economics* **113** pp 528-545

[7] David A Dornfeld, Jawahir I S and Andres F Clarens 1994 Environmental leadership: from compliance to competitive advantage *Academy of Management Executive* **8** (2) pp 7–20

[8] Ferguson M E and Toktay L B 2006 The effect of competition on recovery strategies *Journal of Production and Operations Management* **15** 3 pp 351–368

[9] Handfield R B, Walton S V, Seegers L K and Melnyk S A 1997 Green value chain practices in the furniture industry *Journal of Operations Management* **15** 3 pp 293–315

[10] Lee S Y and Klassen R D 2008 Drivers and enablers that foster environmental management capability in small- and medium-sized suppliers in supply chains *Journal Production and Operations Management* **17** 6 pp 573–586

[11] Pagell M, Krumwiede D W and Sheu C 2007 Efficacy of environment and supplier relationship investments—moderating effects of external environment *International Journal of Production Economics* **45** 9 pp 2005–2028

[12] Preuss L 2001 Is dirty chains? Purchasing and greener manufacturing *Journal of Business Ethics* **34** 34 pp 345–359

[13] Rogers D and Tibben-Lembke R 2001 An examination of reverse logistics practices *Journal of Business Logistics* **22** 2 pp 129–148

[14] Russo M V and Fouts PA 1997 A resource-based perspective on corporate environ- mental performance and profitability *Academy of Management Journal* **40** 3 pp 534–559

[15] Vachon S and Klassen R D 2007 Supply chain management and environmental technologies: the role of integration *International Journal of Production Research* **45** 2 pp 401–423

Experimental analysis of mechanical and structural properties of hybrid aluminium (7075) matrix composite using stir casting method

Anand Singh Rathaur[1], Jitendra Kumar Katiyar[2,*] and Vinay Kumar Patel[1, *]

[1] Department of Mechanical Engineering, Govind Ballabh Pant Institute of Engineering and Technology, Ghurdauri, Pauri Garhwal, Uttarakhand 246194, India [2] Tribology and Surface Interaction Laboratory, Department of Mechanical Engineering, SRM Institute of Science and Technology, Kattankulathur, Tamil Nadu- 603203, India

*Corresponding author: jitendrakumar.v@ktr.srmuniv.ac.in, vinaykrpatel@gmail.com

Abstract. Al-7075/Si_3N_2/ZrO_2 hybrid metal matrix composite is developed using stir casting method. The mechanical behaviour of the developed composite is carried out using impact test, tensile test and hardness test as per ASTM standards. The micro structural analysis shows that the grain boundaries is having inter-dendritic network of silicon nitride and zirconium oxide particles and it is distributed uniformly in an aluminium matrix. Furthermore, a considerable increment in the tensile strength, hardness and wear properties of the hybrid composite was observed with increasing the percentage reinforcements of zirconia (2, 4 and 6wt%) in Al-7075/Si_3N_4 composite (keeping a constant percentage of Si_3N_4 of 8wt%).

Keywords: Stir Casting, Metal Matrix Composite, Zirconium Oxide, Silicon Nitride, Aluminium

1. Introduction

In new era, the metal matrix composites have shown greater potential in fabrication of engineering components. In this series, hybrid aluminium composites are widely used in different engineering applications because of its enhanced mechanical properties, flexibility to conventional processing techniques and lower production cost. The act of these type of materials are generally dependent upon the selection of optimized concentration of reinforcing materials because the few output parameters depends upon the reinforcing particulates.

Since the last decades, the utilization of aluminium alloy in automobile industry has been rapidly increased because of its lower weight and density, high strength and wear resistance properties, and lower coefficient of thermal expansion. Aluminium matrix composites (AMCs) have shown excellent tribological performance among all materials, and due to this, it is extensively being used in practical as well as fundamental sectors. The strengthening is achieved in AMCs by the formation of continuous/discontinuous fibres, whisker or particulates in optimized concentration. According to the industrial application, the properties of AMCs can be tailored by optimized combinations of matrix, support and treating route. Presently, several grades of AMCs are manufactured using different routes. From the past research, it is observed that the physical, mechanical, thermo-mechanical and tribological properties of AMCs are very important for engineering application because it is widely used in aerospace, defence, automotive, and thermal management areas, as well as in sports and regeneration. The researchers blended different hard ceramic particles namely boron carbide, silicon carbide, alumina, silicon nitride, etc. in aluminium matrix and they achieved higher strength, results higher wear resistance property and further, they observed the decrement in weight to strength ratio as compared to conformist Al-alloy. Along with above advantages, it also has few disadvantages such as higher fabrication cost and higher brittleness behaviour from conformist Al-alloy, easy damaging property [1-7]. Moreover, the aluminium/silicon nitride composite exhibited higher specific strength from unreinforced alloy [8]. They used powder metallurgy method and pressure infiltration technique to fabricate the alumina composite. They observed high hardness from conventional Al-matrix and the hardness is improved by increasing the concentration of silicon nitride particles [8]. Further, AA6061 aluminium matrix is used to reinforcement of nickel coated silicon nitride particles which was manufactured by liquid metallurgy method. The reinforced aluminium composite has shown higher hardness and tensile strength from pure AA6061 [9]. Moreover, researcher [10] fabricated Al 356/ZrO_2 AMCs at various casting temperatures (750 °C, 850 °C and 950 °C) using stir cast method. It showed that when A356 aluminium alloy matrix composites reinforced with ZrO_2 by 5, 10 and 15 vol %, the mechanical properties such as hardness and tensile strength gain its maximum values of 70 BHN and 232 MPa, respectively [10]. The investigation showed that the addition of zirconia in alumina matrix, the composite achieved its higher density and higher mechanical properties such as flexural strength and fracture toughness. Further, the inverse dependency of K_{IC} on the hardness is observed. The pure zirconia and its composites have shown Palmqvist crack system at applied indentation of 98 N. The higher concentration of zirconia in alumina matrix showed higher mechanical properties such as flexural strength and fracture toughness from pure alumina. The aluminium hybrid metal matrix composites with various reinforcements find many advanced applications [11]. Form above literature, it was observed that methods used for fabrication of alumina metal matrix composites are very time consuming and sometimes costly. Few authors only proposed stir casting but the comprehensive study is not available in literature.

Therefore, in this present study Al7075- Si_3N_4-ZrO_2 hybrid metal matrix composite was fabricated using stir casting method. The zirconia weight percentage was varied by 2, 4 and 6 wt% with fixed 8 wt% of silicon nitride. In this study the composites abbreviation are used according to their percentage such as composite (1)-0%, composite (2) - 2%, composite (3)-4% and composite (4)-6%. The mechanical behaviour was determined by performing the following tests: Tensile test, Charpy Impact test, Rockwell hardness test at common parameters. Micro structural analysis was carried out for the study of structures.

2. Experimental Methods

2.1 Fabrication of Al-Si_3N_4-ZrO_2

Stir casting was used to fabricate the composite in this present work. For fabrication, the pure aluminium (7075) (Jindal Aluminium, Delhi), silicon nitride (size: 2 µm; J.D Powder Metallurgy, Andra Pradesh) and zirconia (size: 30 µm; M.K Ceramics, Delhi) has been taken, respectively. The powder was fed at constant feed rate. The stirring speed was kept in the range between 300-600 rpm and temperature was kept constant of 630 °C. Further, the matrix was preheated at a specified temperature of 500 °C for 30 min. Preheating of matrix removes the moisture or gases present inside.

2.2 Mechanical Properties

The universal tensile testing (UTM: HEICO-HL590 with a crosshead speed of 10 mm/min) was used to determine the ultimate tensile strength (UTS), maximum elongation and reduction in area. Moreover, The Rockwell hardness testing machine was used to measure the hardness of fabricated composite. The conversion table was used to convert HRB into MPa. Furthermore, the impact test was also performed on Izod/Charpy impact testing machine to determine the absorbed energy by the composite during the fracture.

2.3 Microstructure Analysis

The microstructural analysis was performed on Olympus optical microscope. The magnification was kept 20X. The main objective of this analysis was to investigate the shape and size of the grains present inside the alumina matrix composite.

IOP Conf. Series: Materials Science and Engineering **653** (2019) 012033 doi:10.1088/1757-899X/653/1/012033

2.4 Wear Test

The wear test was performed on pin-on-disc tribometer (Ducom Instruments Bangalore India). The pin was fabricated by using of composite and slide against a heat treated steel disc of 60 mm diameter. The disc was rotated at 480 rpm. The test was carried out for 30 min. The normal load of 2.5 kg and 3 Kg was applied over pin to calculate the wear resistant property. The wear rate is calculated from weight loss measurement.

3. Result and Discussion

3.1 Mechanical Properties

The influence of ZrO_2 concentration in alumina/silicon nitride matrix on ultimate tensile strength (UTS) were investigated using UTM as per the ASTM standard. From **Figure 1**, improvement of tensile strength was observed up to 4 wt% ZrO_2 after that small decrement in tensile strength was noticed. The dislocation density and uniform distribution of ZrO_2 particles in higher concentration might be responsible for decrement in tensile strength. But higher concentration of ZrO_2 increases the ultimate tensile strength. In Al-7075/Si_3N_4/ZrO_2 hybrid metal matrix composite, the direct reinforcement evolved by growth of dislocation density because of the difference in thermal expansion coefficient of zirconium oxide and aluminum alloy. The thermally induced dislocation density was also increased with increasing the concentration of ZrO_2. **Figure 2** shows the increment in displacement with respect to increasing concentration of ZrO_2.

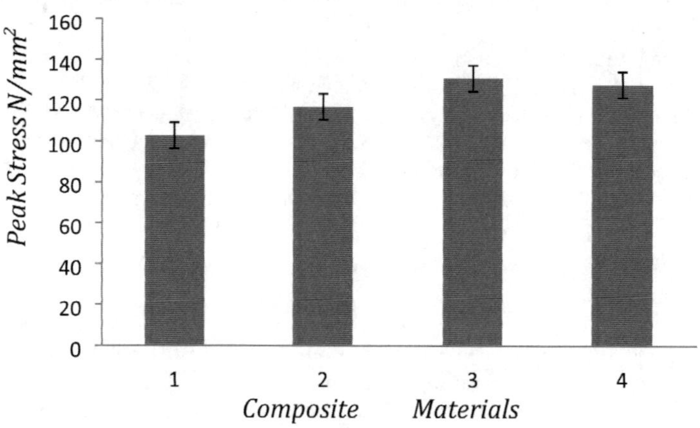

Figure 1 Average stress on different composites

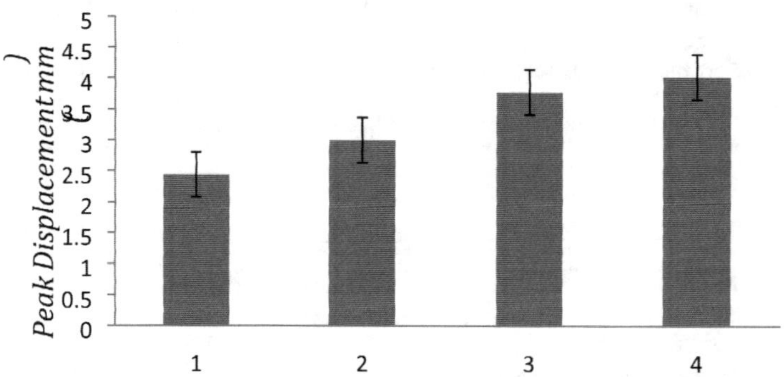

IOP Conf. Series: Materials Science and Engineering **653** (2019) 012033 doi:10.1088/1757-899X/653/1/012033

Composite Materials

Figure 2 Elongation of composite before fracture

Figure 3 shows the variation of hardness by changing the volume concentration of ZrO_2. It was observed that the mechanical properties of Al7075 was getting improved with the addition of Si_3N_4 and ZrO_2 assimilation. The hardness of fabricated composites increased linearly with respect to increasing concentration of zirconium oxide particles. This can be recognized by higher hardness of zirconia particles from that of aluminum alloy.

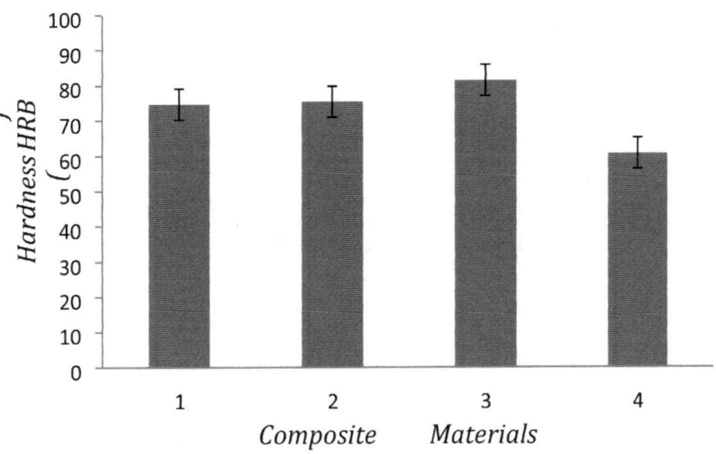

Figure 3 Hardness of various composite

Impact strength was used to measure the energy absorbing capacity of fabricated composite by applying sudden loading. **Figure 4** shows the influence of zirconia particle contend on the impact strength of various weight percent reinforcement (0, 2, 4, and 6wt%). It was observed that due to the reinforcement the impact strength gradually decreased with the increasing content of zirconia.

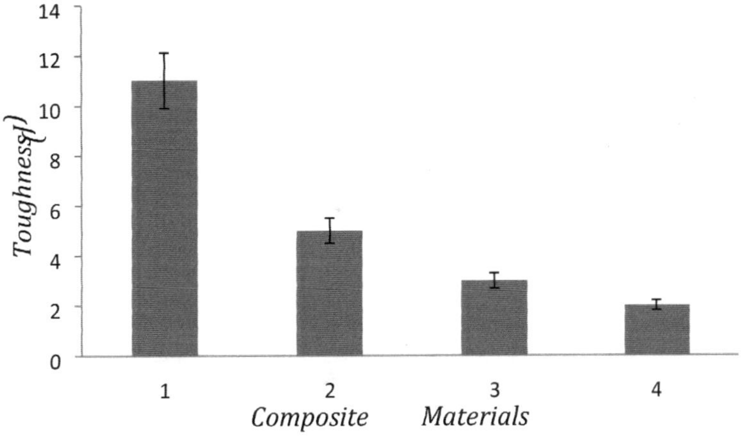

Figure 4 Graph showing impact test results

3.2 Sliding Wear Test

The **Table 1** shows the weight loss and wear rate of different composites tested at 480 rpm for 30 min. From the table 1, it is observed that the weight loss as well as wear rate decreases by increasing the percentage content (0, 2, 4, 6 wt%) of zirconia. Therefore, the wear resistance of the hybrid composite is found significantly higher than the matrix composite of Al7075/Si_3N_4.

Table 1. The weight loss and wear rate of fabricated composites at 480 rpm sliding speed for 30 min.

Samples	**Weight Loss (gm)**		**Wear Rate ($\times 10^{-5}$ g/m)**	
	2.5 Kg Normal Load	3 Kg Normal Load	2.5 Kg	3 Kg
1	0.0712	0.108	2.65	3.98
2	0.0771	0.107	2.84	3.94
3	0.0695	0.102	2.56	3.75
4	0.0654	0.073	2.41	2.70

3.3 Microstructure Analysis

The mounting of composite samples on Bakelite mould was carried out by applied pressure so that composite samples can be easily polished in buffing machine. After polishing of samples, the microstructure was studied by using optical microscope as shown in **Figure 5**. The uniform distribution of ZrO_2 and silicon nitride was observed in aluminium matrix. The percentage of white portion increased with increasing the concentration of ZrO_2. The distribution of grain boundary as well as interdendritic network of zirconia and silicon nitride particles in a matrix of aluminium was observed throughout.

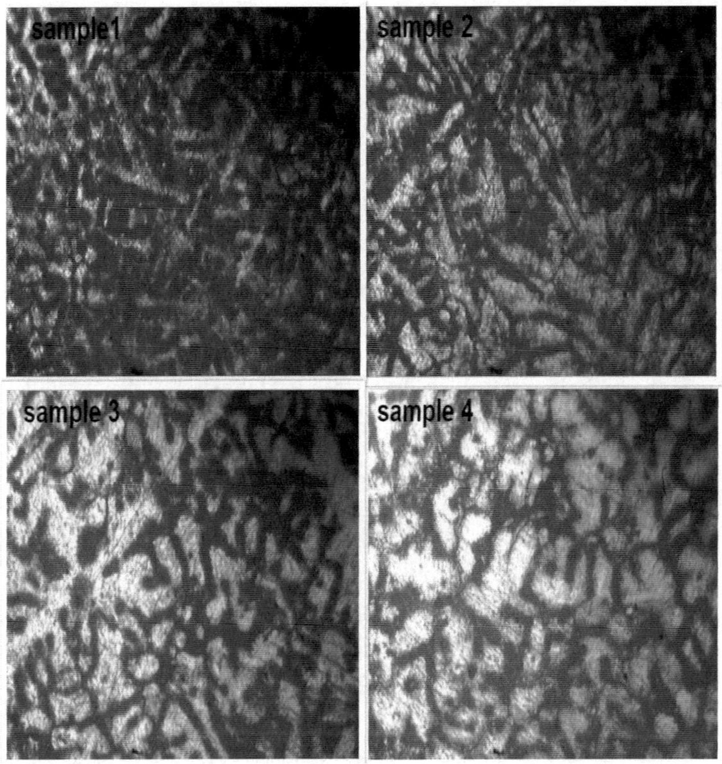

Figure 5. Optical Microstructures (Sample 1) (7075) + 8% Si_3N_2; (Sample 2) Al (7075) + 8% Si3N2 + 2% ZrO_2; (Sample 3) Al (7075) + 8% Si_3N_2 + 4% ZrO2; (Sample 4) Al (7075) + 8% Si_3N_2 + 6% ZrO_2)

4. Conclusion

The Al7075/Si_3N_4/ZrO_2 composites were fabricated using stir casting process with different concentration of zirconium oxide particles (0, 2, 4, and 6 wt%) keeping constant 8wt% Si_3N_4 particles. From this investigation, the following conclusions are drawn.

- The addition of ZrO_2 particles in Al7075/ Si_3N_4 composite improved the tensile strength from 102.74 MPa to 130.71 MPa. Further, the ductility was decreased with addition of reinforcement of zirconia particles. Moreover, the hardness of the composites increased from 74.66 HRB to 81.4 HRB with the addition of ZrO_2 particles.
- The microstructure shows the throughout distribution of grain boundary as well as interdendritic network of zirconia and silicon nitride particles in a matrix of aluminum.
- There is a considerable increase in the wear properties as compared to unreinforced aluminum alloy.
- As per the results obtained and compared, it is observed that toughness goes on decreasing with the increase in zirconia content.

References

1. Yigezu, B.S., Jha, P.K. and Mahapatra, M.M. 2013 The key attributes of synthesizing ceramic particulate reinforced Al-based matrix composites through stir casting process: A review. *Materials and Manufacturing Processes*, 28(9), 969-979.
2. Benedict, G. F. 1987 Non-traditional Manufacturing Process. Marcel Dekker Inc. 207-230, New York.
3. Abdizadeh, H. Mohd Baghchesara, A. 2012 Investigation of mechanical properties and fracture behavior of Al356 aluminium alloy based ZrO_2 particle reinforced metal matrix composite. *Ceramics International,* 2045-2050.
4. Brahim, I. A., Mohamed F. A. and Lavernia E. J. 1991 Particulate Reinforced Metal Composites-a-review", *Journal of Materials Science,* 26, 1137-1156.
5. Aggarwal, J. P 1990 Composite materials. DESIDOC, Delhi.
6. Patel, V.K. and Rani, K. 2017 Mechanical and wear properties of friction stir welded 0–6Wt% nAl_2O_3 reinforced Al-13Wt% Si composites. *Strojnícky casopis–Journal of Mechanical Engineering,* 67(1), 77-86.
7. Patel, V.K., Kumar, P. and Bhattacharya, S. 2018 Mechanical, Microstructural and Sliding Wear Properties of Friction Stir Welded AA6063-T6 and AA5052-H32 Aluminium Alloys. *Materials Focus*, *7*(1), 50-58.
8. Gajalakshmi, K. Kathiresan, S. Sreenivasan, K. S. and Ravindran, S. 2014 Investigation of microstructure and mechanical properties of silicon nitride reinforced AMMC using stir casting method. *IOSR Journal of Mechanical and Civil Engineering (IOSR-JMCE)*, 61-65.
9. Taya, M. and Arsenault, R. J. 1989 Metal Matrix Composite –Thermo-mechanical Behaviour, pregamon press.
10. Sharma, P., Sharma, S. and Khanduja, D. 2012 Production and some properties of Si_3N_4 reinforced aluminium alloy composite. *Journal of Asian Ceramic Societies*, 13 (3), 352-359.
11. Singh, J. and Chauhan A. 2016 Characterization of hybrid aluminum matrix composites for advanced applications–A review. *Journal of Materials Research and Technology*. 5(2):159-169.

IOP Conf. Series: Materials Science and Engineering **653** (2019) 012034 doi:10.1088/1757-899X/653/1/012034

Modeling and optimization of cutting parameters during machining of ZrO_2 filled MMC

T Mahanta[1, *], M P Satpathy[1] and A Swain[1]

[1]School of Mechanical Engineering, KIIT Deemed to be University, 751024, Odisha, India

[*]Correspondence Email: tpmahanta18@gmail.com

Abstract. In the era of composites and ceramics, reinforced metal matrix composites have paved its way to many engineering fields due to its superior mechanical and thermal properties. Zirconia (ZrO_2) is one such most researched ceramics, which leads to excellent bond strength, grain refinement, and improved tribological aspects of the composite. To simplify the post mold machining of such new composites, the present work emphasizes on the prediction of cutting force for different composition of ZrO_2 in an Aluminum metal matrix composite reinforced with SiCp and Graphite. The purpose is to streamline the complex analysis of induced cutting force during turning with quantifiable machining parameters for the developed composite. The influence of two machining parameters such as depth of cut and feed rate are analyzed with an increase in ZrO_2%. For the better prediction of the cutting force, the artificial intelligence techniques like adaptive neuro-fuzzy interface system (ANFIS) backed by a heuristic genetic algorithm (GA) is employed. The comparison between the predicted model values with the experiment results is exhibited ±1% error having R-square value 0.9927. The interaction plot and soft computing model stood in equal agreement that with the rise in both depth of cut and feed rate, the cutting force value decreases. Again for an increase in ZrO_2% and feed rate, a definite increase in cutting force is observed. The effect of an increase in ZrO_2% in the composite over the machining parameter is suggested through the main effect plot, which stands in equal agreement with the GA model.

1. Introduction

In the current automated manufacturing environment, there is an urgent need for advanced strategies to control and monitor the manufacturing process. Moreover, the optimal use of resources is one of the major concern for maximizing the production rate and quality with a minimal cost of the component. In metal cutting processes modeling and prediction of cutting forces stands crucial as it is governed by numerous interrelated parameters such as cutting speed, feed, depth of cut, cutting edge angles, nose radius, etc. Meanwhile, the knowledge on cutting force is also essential as it is directly related to the tool wear and breakage, power requirements, quality of the machined surface, induced vibration, etc. Proper selection of operating parameters, fixtures, and machining tools leads to an efficient manufacturing process. The accurate model of the cutting force is indeed a complex task due to highly interlinked variables associated with the composition of a material, i.e., a certain amount of hard and abrasive ceramics [1].

In this study, the developed metal matrix composite (MMC) is made up of Al, reinforced with SiC, graphite, and with various concentration of Zirconia (Zr). The primary reason for considering Al is that it provides the better support and distributes the load homogeneously on particulates apart from its unique properties like non-magnetic nature, high thermal conductivity, low density, excellent corrosion resistance and good damping capability. Meanwhile, ZrO_2 helps in grain refinement, bond strength and corrosion resistance. As the manufacturer does not provide the data for these recently developed materials; hence, the parameters for new material needs to be optimized. This MMC can find its application in aerospace, defense, recreation, atomic reactor and automotive sectors.

For the prediction of machining performance, numerous methods such as analysis of variance, regression analysis, adaptive neuro-fuzzy interface (ANFIS), artificial neural network (ANN), genetic algorithm (GA), etc. have emerged. Strategic implementation and selection of parameters are accomplished through different applications of artificial intelligence based modeling. A statistical tool such as analysis of variance preceded by orthogonal array distribution and signal to noise ratio was modeled to judge the significance of parameters and the_performance characteristics. To obtain stabilized optimal control of performances, Savadamuthu et al. [2] developed an ANFIS model to analyze its effectiveness with the control scheme. Pande et al. [3] utilized the ANFIS model for prediction of responses during machining of Inconel 718 considering feed and speed as the controlling parameters. Muthukrishnan and Davim [4] employed analysis of variance (ANOVA) and back propagation ANN technique to analyze surface roughness during turning experiment. For optimizing the multiple performance characteristics results in the drilling of hybrid metal matrix Al356/SiC-mica composite. Rajmohan et al. [5] employed the Taguchi method with grey rational analysis. The machinability characteristics modeled by Taguchi method with principal component analysis (PCA) and fuzzy interference system are investigated by Puhan et al. [6]. Such predictive models are applied in various fields to analyze complex responses. Samantaray et al. [7] predicted model equation to correlate drag coefficient and Reynold's number through regression analysis, artificial intelligence models and GA. The study revealed that ANFIS had predicted the desired output with minimal error. Similarly, Swain and Das [8,9] used artificial intelligence techniques for the prediction of flow boiling heat transfer coefficient on tube taking 5 variable input parameters from the different refrigerant. The results suggest that the developed ANFIS model stand prominently even with complex and multivariable inputs. Prediction of parameters for other manufacturing processes such as grinding, milling, and drilling are also analyzed by regression, neural network and ANFIS. In case of drilling of glass fiber, 5 input factors are optimized for 2 output factors where the ANFIS model performed better than a GP model [10]. In grinding, the material removal rate (MRR) and surface roughness were predicted by Varma et al. [11] using regression analysis, neural network and ANFIS. The comparison among the various correlation of process parameters was analyzed, and ANFIS was turn out to be a better optimizing tool. During the turning process, MRR and surface roughness and cutting zone temperature for dry turning are also modeled by using ANFIS with particle swarm optimization (PSO) by [12]. In case of end milling, Maher et al. [13] predicted the surface roughness by ANFIS modeling with process parameters such as spindle speed, feed and depth of cut. When the predicted values were compared with the measured data, it resulted in 6.25% of error. For modeling of cutting force parameters in turning operation, Cica et al. [14] used ANN and ANFIS in order to compare their predictive capability. The results demonstrated a good agreement among the two models with an average error of 2.39% for ANN and 1.97% for ANFIS model. Sparham et al. [15] researched the modeling of frictional force and ANFIS depicted a very low error percentage of 1.38%, proving the efficacy of ANFIS modeling.

From the literature survey, it is clear that artificial intelligence techniques have been widely used for better prediction of output parameters nowadays. These techniques may not give the best model, but they provide well-predicted results. This paper deals with the comparison of cutting force results during machining of a newly developed composite of an Aluminum metal matrix reinforced with Zirconia, Silicon carbide, and Graphite. Speed, feed, and depth of cut act as the significant influential parameters in determining the cutting force results during the conventional turning operation. The objective of the study is to study the effect of an increase in zirconia concentration on the cutting force keeping the speed as constant. Very less work depicts the impact of change in material concentration on the cutting force. The obtained experimental data are analyzed to predict the induced cutting force through different models, namely; ANOVA followed by regression analysis and ANFIS model. GA is applied to get the set of input parameters for a minimum value of cutting force.

2. Methodology

2.1. Adaptive neuro-fuzzy interface system (ANFIS)

Adaptive Neuro-fuzzy interface system (ANFIS) is a methodology to model non-linear system having features of neural network and fuzzy logic. The general ANFIS structure has five different layers; each of the layer contains nodes which represent some function to be carried out. The function of different layers is interlinked to construct the interference systems, obtaining inputs from the nodes in the previous layers. A basic model of ANFIS for a system of one output (Y) with m no of inputs (X_1, X_2......X_M) each with n membership functions based on a fuzzy rule base of R rule (see figure1). The functions of different layers are as follows.

Layer 1 (Fuzzification layer): The first layer changes the raw inputs X_i into levels of the linguistic tag such as medium, large, small, μ_{ij}, etc. The membership functions are applied to the output of node is expressed as follows:

$$O_{ij} = \mu_{ij}(X_{ij}), i = 1,2......, m, j = 1,2,.......n \tag{1}$$

Where π_{ij} is the Jth membership function for the input X_i

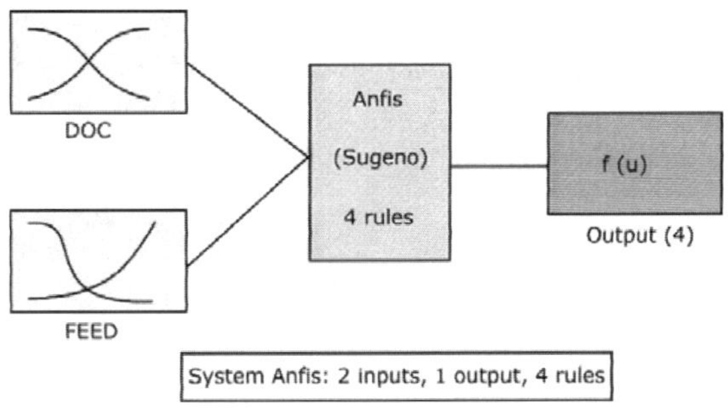

Figure 1. ANFIS modeling of membership functions

In this present model, two Gaussian membership functions are used for each input: depth of cut and feed. The Gaussian membership function is given as follows;

$$\mu(X) = e^{\frac{-(x-c)^2}{2\sigma^2}} \tag{2}$$

The parameters associated with the Gaussian function σ and c are adjusted while training the network or raw model.

Layer 2 (Product layer): The second layer of the model gets multiplied for each node in the layer, the output represents the weighting factor (cutting force) of the rule k. The output W_k is the product of all inputs as follows:

$$\sigma_k^2 = W_k = \mu_1 e_1(X_1)\mu_2 e_2(X_2)........\mu_m e_m(X_m) \tag{3}$$

Layer 3 (Normalized layer): The output of each node k in the normalized layer represents normalizing weighing factor W_k of the kth rule as follows:

$$\sigma_k^3 = \overline{W_k} = \frac{W_R}{W_1 + W_2 + \ldots\ldots + W_R} \tag{4}$$

Layer 4 (De-fuzzification layer): Each node of this layer gives a weighted output of the first order TSK-type fuzzy f then rule as shown below:

$$\sigma_k^3 = \overline{W_k} f_k \tag{5}$$

Where f_k represents the output of kth TSK-type fuzzy rules as follows:
If (X_1 is $A_1 e_1$), (X2 is $A_2 e_2$) and ….. (X_m is $A_m e_m$) then

$$f_k = \sum_{i=1}^{m} P_{1e_i} X_i + r_k \tag{6}$$

Where P_{1e_i} and r_k are called consequent parameters and e_1, e_2….e_m = 1...n, K=1….R

Layer 5 (Output layer): This single-node layer represents the overall output (Y) of the network as the sum of all weighed outputs of the rules (figure 2):

$$\sigma^5 = Y = \sum_{i=1}^{m} \overline{W_k} f_k \tag{7}$$

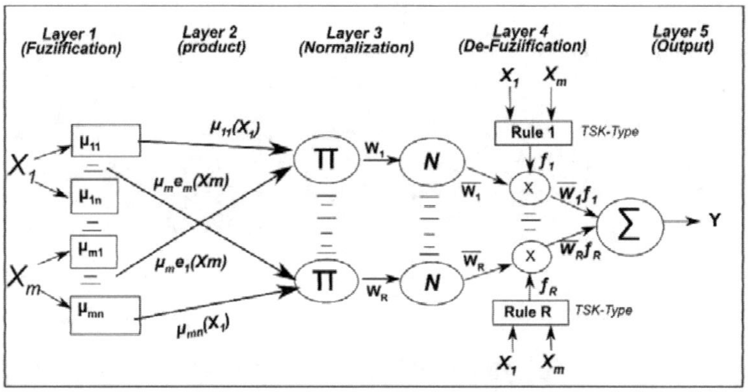

Figure 2. Schematic model of ANFIS

Linear-Back-propagation Learning Algorithm in ANFIS: When the premise parameters are fixed, all the overall output is a linear combination of subsequent parameters. In symbols, the output 'f' can be written as

$$f = \frac{W_1}{W_1 + W_2} f_1 + \frac{W_2}{W_1 + W_2} f_2 = \overline{W_1} f_1 + \overline{W_2} f_2 = (\overline{W_1} x) p_1 + (\overline{W_1} y) q_1 + (\overline{W_1}) r_1 + (\overline{W_2} x) p_2 + (\overline{W_2} y) q_2 + (\overline{W_2}) r_2 \tag{8}$$

Which is linear consequent parameter (p_i,q_i,r_i). A back-propagation algorithm adjusts the consequent parameters in a forward pass and premise parameters (a_i,b_i,c_i) in a reverse pass. In the forward pass, the network inputs propagate forward until layer 4, where the consequent parameters are identified by the least-squares method. In the reverse/backward pass, the error signals propagate backward, and the premise parameters are updated by gradient descent. Because the update rules for the premise and consequent parameters are decoupled in the back propagation learning rue, a computational speed up may be possible by using variants of the gradient methods or other optimization techniques on the premise parameters. Since ANFIS and radial basis function networks are functionally equivalent under some minor conditions, a variety of learning methods can be used for both of them [16].

2.2. Genetic algorithm methodology

The genetic algorithm is a stochastic optimization algorithm to find the global minimum value of a function and the corresponding value of inputs at which the minimum value is obtained. This process is very much established, and the steps involved are given as a flow chart in figure 3. The details of the methodology can be found in the book Pratihar [17].

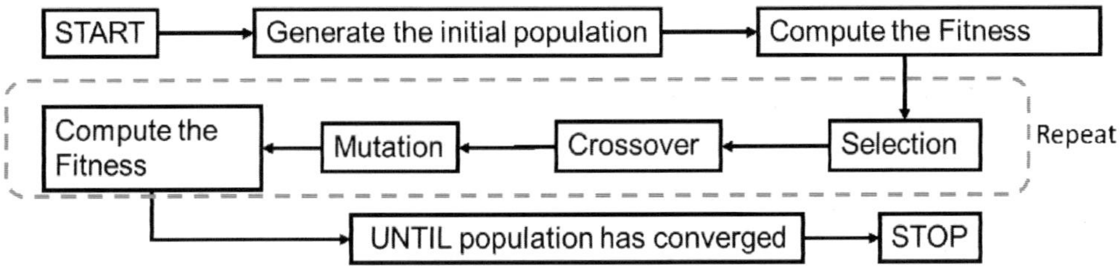

Figure 3. Genetic algorithm flowchart

3. Experimentation

3.1. Experimental design and setup

The experiments are carried out to measure the cutting force during turning process on a conventional lathe machine with the variation of zirconia percentage, depth of cut and feed. Entire experiments were carried out at a constant speed of 360 RPM. An HSS tool with positive rack angle was used for the study. This composite workpiece was prepared by stir casting method with the particle sizes of 40 μm SiC, 40 μm graphite and 80 μm ZrO_2. Furthermore, the experiment was carried out with a varied percentage of ZrO_2 (1.75%, 4.5% and 6%) with a fixed 7% SiC and 1.75% graphite content by volume. Samples prepared were in the form of cylindrical bars of diameter 20 mm and length 75mm. The positive rake angled cutting tool was used for getting high surface finish and low cutting force. Experimental trials were carried out in dry condition 1min duration. The cutting force was measured by Kistler lathe tool dynamometer (model no-9257B) connected with the strain gauge dial indicator. figure 4 shows the schematic representation of the experimental setup. Certain mechanical characteristics such as microhardness and tensile strength with respect to the specimens are measured as shown in Table 1.

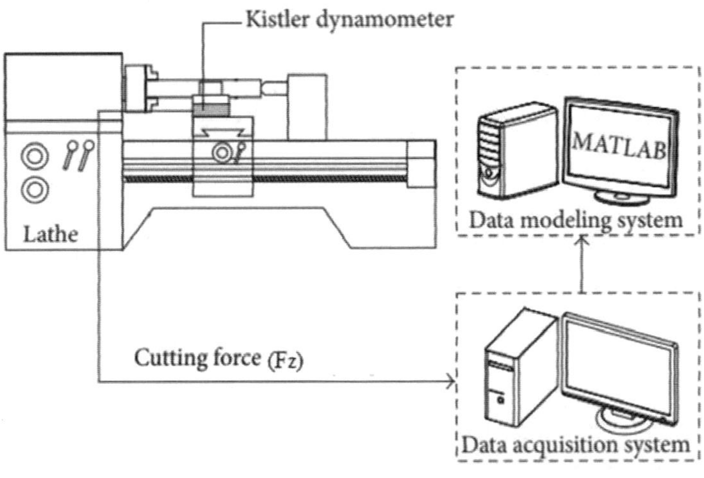

Figure 4. Schematic diagram of experimental setup

Table 1. Machining parameters and mechanical properties of specimens

Levels	Machining parameters				
	% of ZrO_2 (vol %)	Depth of Cut (mm)	Feed rate (mm/rev)	Microhardness (VHN)	Tensile strength (MPa)
Level 1	1.75	0.5	0.0142	80.25	91
Level 2	4.5	0.75	0.0313	103.22	121
Level 3	6	1	0.05592	103.77	132

The experimental design was arranged to conduct 27 no of experiment according to the design of experiments of Taguchi L27 orthogonal array. Machining parameters were considered in 3 levels, with 3 factors as depicted in Table 1. The results were implemented to predict a general model for determining the impact of the cutting force.

4. Results and Discussions

4.1. Effect of process parameters

The data generated from the experiments are used to train and test the ANFIS model. Three numbers of ANFIS models are developed from the experimental data for three compositions of metal matrix composites having different percentages of ZrO_2. The depth of cut and the feed rate are taken as input to the models predicting the cutting force as output. Two Gaussian membership functions are employed for each input in the model. The training of the raw model is carried out with the "ANFIS" routine in MATLAB which uses a combined algorithm of least mean square and back-propagation gradient descent method. Figure 5 shows the comparison between the cutting forces predicted from the ANFIS model with those obtained from the experiment, and it can be observed that they are within ±1 % for the above mentioned three models. The obtained R-square value is 0.9927.

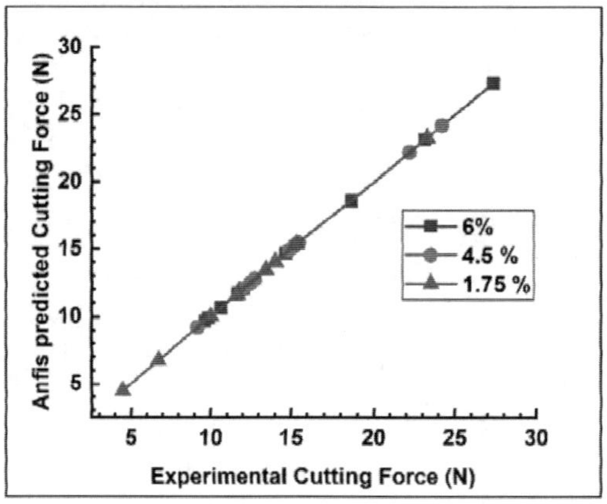

Figure 5. Comparison of the cutting force values from the ANFIS model with the experimental data.

After the cutting force is modeled with the ANFIS methodology with a good accuracy level, the objective is now to minimize the cutting force with respect to the depth of cut and the feed rate. The Genetic Algorithm (GA) is adopted for getting the minimum value of the cutting force for each percentage of the ZrO_2 material in the metal matrix composites. The GA algorithm tries to find the minimum value of the cutting force over the iterations. The change in the value of the output of the algorithm is plotted with the iteration number. The program converges when there is no significant change in the output. Figure 6 shows the convergence of the genetic algorithm for different values of the ZrO_2 percentage. The lower bound and upper bound for the inputs during the GA program are set 0.0142 to 0.05592 mm/rev for feed and 0.5 to 1 mm for depth of cut. The maximum number of iterations is set to be 1000. With the increase in order of percentage of ZrO_2, the minimum value for cutting force was depicted for 3 different sets of data in Table 2.

It is observed from Table 2 and Figure6 that GA has predicted the optimum parameter by reassigning values multiple times through the ANFIS model and has depicted the output parameters at an intermediate state which suggests the interaction between the parameters of cutting force with an increase in the percentage of ZrO_2. Keen observation from Table 2 also depicts that increase in material concentration requires more feed even to generate minimum cutting force whereas medium feed rate with moderate depth of cut suggested a minimum value for cutting force with an increase in % of ZrO_2.

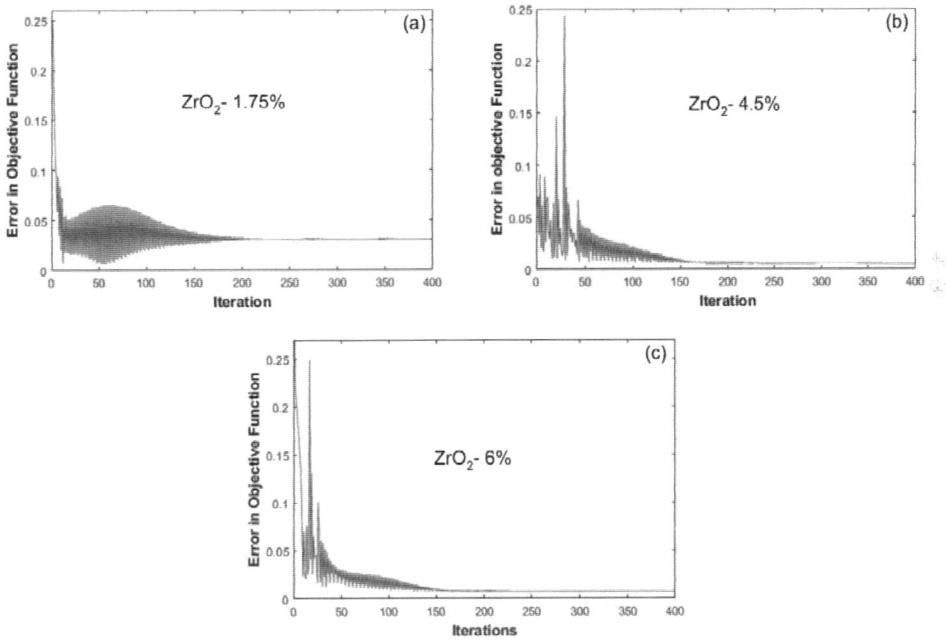

Figure 6. Error in predicted value of cutting force with iterations for (a) 1.75% ZrO_2 (b) 4.5% ZrO_2 (c) 6% ZrO_2

Table 2. Optimum parameters for minimum cutting force developed from GA

Levels	Machining parameters			
	% of ZrO_2 (vol %)	Depth of Cut (mm)	Feed rate (mm/rev)	Minimum cutting Force (N)
Level 1	1.75	0.5	0.0159	4.4598
Level 2	4.5	0.5	0.0142	9.1913
Level 3	6	0.833	0.0221	9.332

4.2. Main effects plot of parameters

The interaction of cutting force parameters with the change in material concentration has been carried out through the main effect plot to get a better understanding of the influential parameters. From Figure 7, it can be observed that there is a definite increase in cutting force with the rise in ZrO_2% and feed rate. However, interaction plot suggests that with the rise in both depth of cut and feed rate, the cutting force value decreases. This has also been predicted by GA in Table 2 for optimum value of the minimum cutting force with respect to change in material concentration. Hence, the soft computing model and interaction plot stood in equal agreement.

IOP Conf. Series: Materials Science and Engineering **653** (2019) 012034 doi:10.1088/1757-899X/653/1/012034

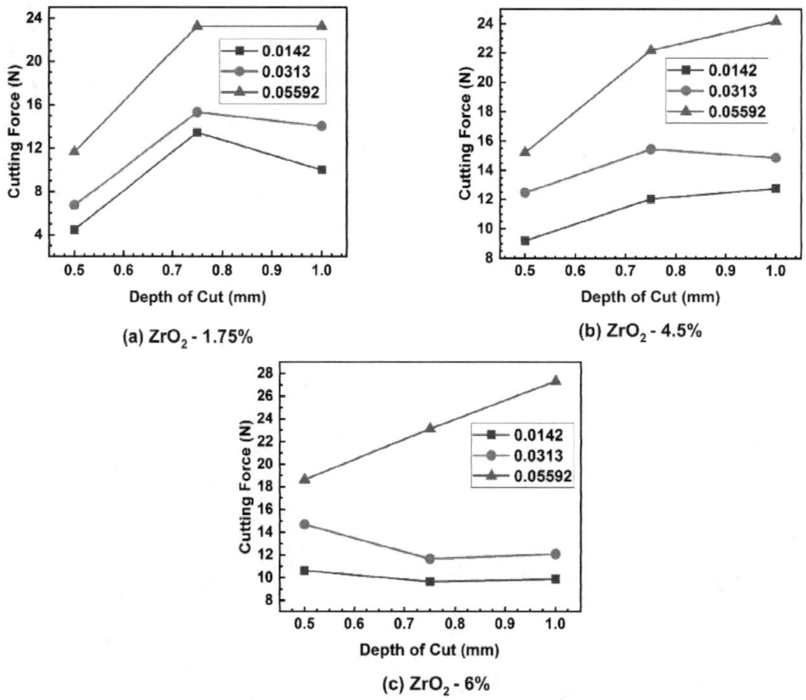

Figure 7. Main effects plot of cutting force for three different ZrO_2%

5. Conclusions

- In the present investigation, experiments are carried out to determine cutting force with respect to different concentrations of ZrO_2. Machining parameters such as depth of cut and feed rate were varied with a change in ZrO_2%. It is observed that for improved machining by minimum cutting force with respect to increasing in ZrO_2; feed rate should also increase. However, when both the feed and depth of cut are increased cutting force decreases for 4% ZrO_2. Furthermore, cutting force again increases with an increase in material concentration and the increase in values of machining parameter. Such critical interaction was properly addressed through optimization developed by GA and analyzed by ANFIS modeling.
- Three ANFIS models are developed for three different compositions of ZrO_2 in the MMC. The effective cutting parameter was predicted by Genetic algorithm through ANFIS modeling. From a different set of parameters with varying zirconia concentration, the optimum value to obtain minimum cutting force.
- The effect of an increase in ZrO_2% in the composite over the machining parameter was suggested through the main effect plot, which stands in equal agreement with the GA model.

References

[1] Mahanta T and Mishra A 2017 Optimization of cutting parameters of Al-ZrO2- SiCp and graphite hybrid metal matrix composites *J. Prod. Eng.* **20** 45–53

[2] Savadamuthu L, Muthu S and Vivekanandan P 2012 Optimization of cutting parameters for turning process using genetic algorithm *Eur. J. Sci. Res.* **69** 64–71

[3] Pande P P and Sambhe R K 2017 Artificial Intelligence Machining Predictions by Fuzzy GUI For Inconel 718 with PVD Coated Carbide Cutting Tool *Int. J. Mech. Prod. Eng. Res. Dev.* **7** 313–20

[4] Muthukrishnan N and Davim J P 2009 Optimization of machining parameters of Al/SiC-MMC with ANOVA and ANN analysis *J. Mater. Process. Technol.* **209** 225–32
[5] Rajmohan T, Palanikumar K and Kathirvel M 2012 Optimization of machining parameters in drilling hybrid aluminium metal matrix composites *Trans. Nonferrous Met. Soc. China* **22** 1286–97
[6] Puhan D, Mahapatra S S, Sahu J and Das L 2013 A hybrid approach for multi-response optimization of non-conventional machining on AlSiCp MMC *Measurement* **46** 3581–92
[7] Samantaray S K, Sahoo S S, Mohapatra S S and others 2017 Prediction of Drag Coefficient of Spherical Particle Using ANN, ANFIS, Regression and GA Optimization *Int. J. Chem. Sci.* **15** 146
[8] Swain A and Kumar das M 2016 Prediction of heat transfer coefficient in flow boiling over tube bundles using ANFIS *Heat Transf. Eng.* **37** 443–55
[9] Swain A and Das M K 2014 Development of Generalized ANFIS Model for Flow Boiling of Refrigerants on Plain Tube Bundles *International Conference on Production and Mechanical Engineering (ICPME'2014) Dec* pp 30–1
[10] Abhishek K, Panda B N, Datta S and Mahapatra S S 2014 Comparing predictability of genetic programming and ANFIS on drilling performance modeling for GFRP composites *Procedia Mater. Sci.* **6** 544–50
[11] Varma N S K, Varma I, Rajesh S, Raju K S R and Raju V V M K 2018 Prediction of surface roughness and MRR in grinding process on Inconel 800 alloy using neural networks and ANFIS *Mater. Today Proc.* **5** 5445–51
[12] Ayd\in M, Karakuzu C, Uçar M, Cengiz A and Çavu\cslu M A 2013 Prediction of surface roughness and cutting zone temperature in dry turning processes of AISI304 stainless steel using ANFIS with PSO learning *Int. J. Adv. Manuf. Technol.* **67** 957–67
[13] Maher I, Eltaib M E H, Sarhan A A D and El-Zahry R M 2014 Investigation of the effect of machining parameters on the surface quality of machined brass (60/40) in CNC end milling—ANFIS modeling *Int. J. Adv. Manuf. Technol.* **74** 531–7
[14] Cica D, Sredanovic B, Lakic-Globocki G and Kramar D 2013 Modeling of the cutting forces in turning process using various methods of cooling and lubricating: an artificial intelligence approach *Adv. Mech. Eng.* **5** 798597
[15] Sparham M, Sarhan A A D, Mardi N A, Hamdi M and Dahari M 2017 ANFIS modeling to predict the friction forces in CNC guideways and servomotor currents in the feed drive system to be employed in lubrication control system *J. Manuf. Process.* **28** 168–85
[16] Natarajan U, Palani S and Anandampilai B 2012 Prediction of surface roughness in milling by machine vision using ANFIS *Comput. Aided. Des. Appl.* **9** 269–88
[17] Pratihar D K 2013 *Soft computing: fundamentals and applications* (Alpha Science International, Ltd)

IOP Conf. Series: Materials Science and Engineering **653** (2019) 012035 doi:10.1088/1757-899X/653/1/012035

Evaluation of Performance and Emissions of Ceramic Coated Turbocharged Direct Injection Diesel Engine operated with Diesel and Biodiesel Blends

B. Rajendra Prasath[1] and P. K. Devan[2]

[1]Jeppiaar Engineering College, Anna University, Chennai 600 119, India.
[2]RMK College of Engineering & Technology, Anna University, Chennai 601 206, India
br_prasath@rediffmail.com

Abstract. It is evident that the main derivatives of biodiesel obtained from the transesterification of vegetable oils are identified as a best suitable alternate fuel for conventional diesel fuel. However higher viscosity and poor volatility of biodiesel causes engine problems for long-term use in diesel engines. Preheating of biodiesel is one of the technique reducing the problems associated with the higher viscosity of biodiesel in diesel engines. The positive effect of increased operating temperature of ceramic-coated engine concept is incorporated to avoid the preheating and maximize the effective heat release of biodiesel. The combination of increased operating temperature associated with LHR engine and oxygenated nature of biodiesel ensure nearly complete combustion. The experiment was conducted on a four cylinder four stroke direct injection turbocharged diesel engine and the combustion chamber was coated with partially stabilized zirconia (PSZ) of 0.35 mm thickness over 0.15 mm thickness of NiCrAl bond coat. The performance and emissions of conventional turbocharged engine and ceramic coated turbocharged engine were analyzed using diesel and biodiesel blends.

1. Introduction

Due to the oil crisis in the past few decades, research interests on piston engines have amplified towards alternate fuels with expectation like renewable in nature, locally available and cheaper than petrodiesel. The great deal of research yields, vegetable oils holds a place to replace the diesel fuel in diesel engines. However, many research outcomes are explicit that usage of straight vegetable oils (SVO) in diesel engines causes severe engine deposits for long run, injector coking, gum formation and piston ring sticking [1-8]. However, transesterification of vegetable oil called biodiesel decreases the viscosity and makes it suitable to use in diesel engines becoming popular and increasing the attention of researchers and automotive manufacturers because of biodegradability, no sulfur content, non-toxic, ease of handling, oxygenated and renewable in nature. The fuel oxygen content about 10 to 11% by weight mandatorily reduces the harmful emissions like hydrocarbon (HC), carbon monoxide (CO) and particulate matter (PM).

Thermally insulated combustion chamber using ceramic coating named as low heat rejection (LHR) engine that crack down the heat transfer to the engine components and increase the engine thermal efficiency. The insulated combustion chamber includes piston top, cylinder head, engine valves and outside of cylinder liner that retain the considerable part of energy produced and utilized, thus increases the engine thermal efficiency. The present investigation commits the evaluation of performance and emissions of diesel and biodiesel blends using turbocharged direct injection diesel engine using diesel and biodiesel derived from Jatropha seed oil. The test engine converted to LHR engine owing to the following advantages [9-13] like better atomization and fuel economy due to increased wall temperature, oxygenated nature reducing emission levels except oxides of nitrogen and multi-fuel capability especially low cetane number and high viscous fuel.

2. Ceramic Coatings

Ceramic coatings are providing heat insulation for many applications especially engine components. The desired properties of ceramic coating material includes the lower value of thermal conductivity, specific heat, friction, wear resistance and higher values of strength, fracture toughness, thermal shock resistance, temperature capability and expansion coefficient. Chemical inertness and high resistance to erosion and corrosion also to be considered for high temperature applications.

Silicon Nitride (Si_3N_4) and Silicon Carbide (SiC) was used as materials for engine cylinder applications in the first phase of development of LHR engine. Granular form of Silicon carbide possess good characteristics especially half the density of steel. More brittleness and shrinking nature during sintering not suitable for all the applications. Though it was successfully used in few applications and advanced research brings out new materials for engines. The other materials like Aluminum titanate (Al_20_3 TiO_2) and Aluminum Magnesium Silicate (AMS) was also used in automotive engines. Table 1 shows few ceramic materials with characteristics [10]. Out of many materials, zirconia having high flexure strength similar to Si_3N_4, low thermal conductivity closely with AMS and Al_20_3 TiO_2 and high thermal expansion coefficient to with stand high temperature. However the phase change nature at higher temperature would limit the engine applications. Elimination of phase change nature of zirconia at high temperature would make it possible to use in engine [11].

The newly developed Partially Stabilized Zirconia (PSZ) that drastically decrease the magnitude of phase nature and desirable for low heat rejection applications. By the addition of Magnesium (Mg) and Nickel (Ni) with PSZ, ductility nature and ultimate strength of PSZ increases. Magnesium (Mg) added PSZ suitable for cylinder liners, piston caps, valve guides and tappet inserts of engine components because of its closely equivalent values of elastic modulus and expansion coefficient like steel and iron [12].

Table 1. Ceramic materials with characteristics

Material	Ultimate Flexure Strength, MPa	Thermal Expansion Coefficient at 1260 K, 10^{-6}/K	Thermal Conductivity, W/m K	Density, g/cc	Young's Modulus at 1260°C, GPa
Silicon Nitride (Si3N4)	300	3.2	12.0	3.10	300
Silicon Carbide (SiC)	450	4.5	40.0	3.15	400
Aluminum Magnesium Silicate (AMS)	20	0.6	1.0	2.20	12
Zirconia Oxide (ZrO2)	300	9.8	2.5	5.70	200
Aluminum titanate (Al2O3 TiO2)	20	3.0	2.0	3.20	23
Partially Stabilized Zirconia (PSZ)	545	10.0	2.5	5.6	350

3. Transesterification process

Based on the local availability, Jatropha seed oil was preferred to produce biodiesel [14-18].Sufficient quantity of Jatropha seeds were collected from the formers, they were manually separated from the shell. Jatropha oil seeds were cracked and removed the shells. The kernels were used for Jatropha oil preparation. Approximately 2 kg of Jatropha oil seed yields 1 kg of clean kernel. Raw Jatropha oil was prepared by grinding Jatropha seed kernels.

A sample of 100 g of the straight Jatropha oil was considered for analysis. For transesterification process methanol (CH_3OH) was chosen as alcohol and NaOH as a catalyst. Figure 1 shows the basic transesterification chemistry. After weighing the Jatropha seed oil, methanol and NaOH in a separate flask, NaOH and methanol mixture stirred in order to obtain the complete homogeneity. The final homogeneous mixture was gently added into the Jatropha seed oil and the mixture was stirred about an hour in the warm water bath. The final mixture was stirred vigorously using mechanical stirrer, after an hour, the mixture is allowed for settling 24 hours. On the top golden yellow liquid called biodiesel and glycerol at the bottom was separated. The pH value of biodiesel reduced by washing with distilled water. After washing, biodiesel is heated to 100°C in order to remove the water particles.

CH_2OOCR	CH_3OH		CH_3OOCR		CH_2OH
$CH\ OOCR$	$+ CH_3OH$	$\longrightarrow$	CH_3OOCR	$+$	$CH\,OH$
CH_2OOCR	CH_3OH		CH_3OOCR		CH_2OH
Triglycerides	Methanol		Methyl ester		Glycerol

Figure 1. Basic Transesterification process with methanol

The various fuel and chemical properties are presented in Table 2.

Table 2. Properties of diesel and biodiesel

1) Properties	ASTM method	**DF**	**B100**	**B20**
Density at 15°C (kg/m^3)	--	830	880	840
Viscosity at 40°C (cSt)	D 445	2.8	4.6	3.16
Flash point (°C)	D 93	55	170	78
Cetane number	D 613	45	50	46
Lower Heating Value (MJ/kg)	--	42	36	41

4. Experimental Setup and Test Engine Development

Four cylinder, four stroke turbocharged DI diesel engine was chosen for the present study. The other details of the engine are: bore 111 mm, stroke 127 mm, compression ration 16, Cylinder volume 4.9134 litre, Power 75 HP @ 1500 rpm and fuel injection pressure 200 bar. Figure 2 shows the test engine and test setup.

At full load condition, the tests were conducted for the engine speed varying from 1000 to 1500 rpm with increment of 100 rpm. K type thermocouple was used to measure exhaust gas temperature, inlet and exit temperatures of cooling water that connected to a multi-channel temperature indicator for digital display. Constant volume burette method was used to measure the total fuel consumption and mass of air flow was measured using air box and U-tube water manometer. The turbocharged engine referred as standard (STD) engine and ceramic coated engine referred as low heat rejection (LHR) engine. Firstly, diesel fuel used in STD engine and repeated the experiment using biodiesel (20% biodiesel blend with diesel). After the completion of test on STD engine, the various parts of engine components especially inlet and exhaust valves, engine cylinder head, piston top and outside of cylinder liner was coated with thermal plasma sprayed PSZ of 0.35 mm thickness over 0.15 mm thickness of NiCrAl bond coat [19-21]. A similar test procedure as like base engine test procedure was followed on LHR engine using diesel and biodiesel blends and the results were analyzed.

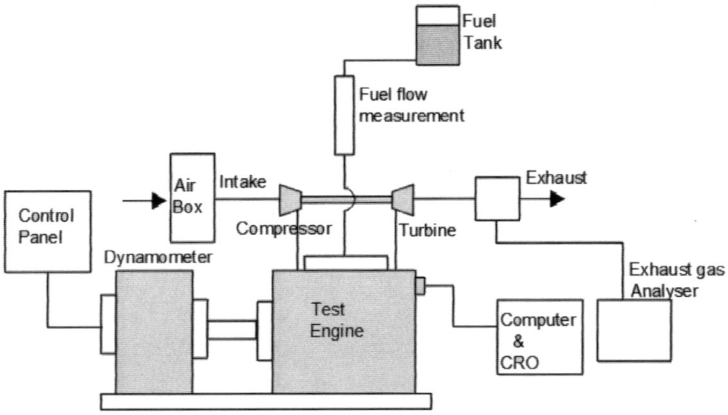

Figure 2. Schematic of test engine system

5. Results and Discussion

The experiments were conducted at four different engine and fuel combinations under identical operating conditions.

- Conventional turbocharged engine with diesel (STD diesel)
- Conventional turbocharged engine with biodiesel (STD biodiesel)
- LHR turbocharged engine with diesel (LHR diesel)
- LHR turbocharged engine with biodiesel (LHR biodiesel)

All results are compared with LHR biodiesel operation. The discussion includes the cylinder peak pressure (P_{max}), engine power (P), brake thermal efficiency (BTE), specific fuel consumption (SFC), exhaust gas temperature (EGT), oxides of nitrogen NO_x), hydrocarbon (HC) and smoke intensity (BSN).

The variation of cylinder peak pressure with engine speed is shown in Figure 3 for STD engine operated with diesel (STD Diesel), STD engine operated with biodiesel (STD biodiesel), LHR engine operated with diesel (LHR diesel) and LHR engine operated with biodiesel (LHR biodiesel). The peak cylinder pressure increases as increase of fuel injection rate with respect to speed. The peak cylinder pressure increases marginally as increasing the speed from 1000 rpm to 1500 speed. The measured percent increases are 8.4 %, 8.6%, 8%, and 8.3% for STD diesel, STD biodiesel, LHR diesel and LHR biodiesel respectively. The variation of engine power with engine speed is shown in Figure 4. The engine power is lower for LHR biodiesel than LHR diesel and higher than STD diesel and STD biodiesel. The LHR biodiesel engine power is decreases an average of 1% than LHR diesel and increases about 2% and 1% than STD diesel and STD biodiesel. As the speed increases from lower level to higher level the percent variation of engine power output gradually decreases. Increasing the turbulence at higher speeds narrow down the engine power output. Higher viscosity of biodiesel lowers the fuel atomization and lower calorific value brings down the effective heat release rate.

The variation of specific fuel consumption with engine speed is shown in Figure 5. Specific fuel consumption is calculated as the ratio of total fuel consumption in kg/h to brake power in kW at the particular condition. Specific fuel consumption for LHR biodiesel is lower than STD biodiesel and higher than STD diesel and LHR diesel. Lower calorific value of the biodiesel which is about 0.93 times of diesel fuel is responsible for the higher fuel consumption for biodiesel operation in both engines. The higher cycle temperature of LHR turbocharged engine reduces the specific fuel consumption than conventional turbocharged engine.

The heat input corresponding to the fuel is the product of fuel mass (m_f) and lower heating value (C.V), the corresponding brake thermal efficiency with engine speed is shown in Figure 6. Higher the power out for LHR biodiesel leads to increase the thermal efficiency than the STD engine operation. The engine brake thermal efficiency of LHR biodiesel is higher about 1% - 2.5% than STD diesel and 0.5% -1 % than STD biodiesel. Brake thermal efficiency for LHR biodiesel decreases about 1.3%-1.7% than LHR diesel. The increase in brake thermal efficiency for LHR biodiesel and LHR diesel conditions due to the reduction of specific fuel consumption.

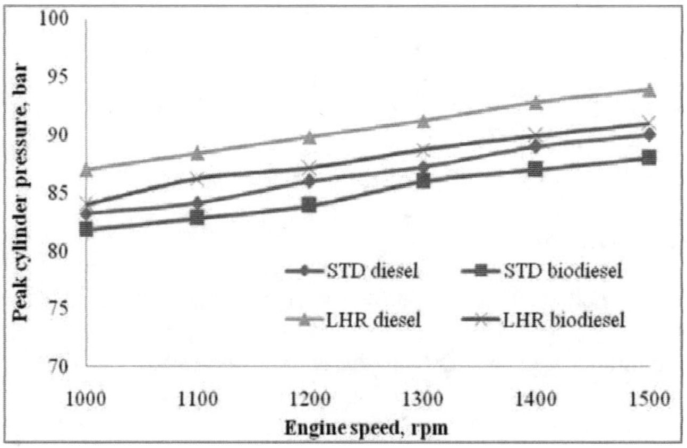

Figure 3. Peak cylinder pressure with engine speed

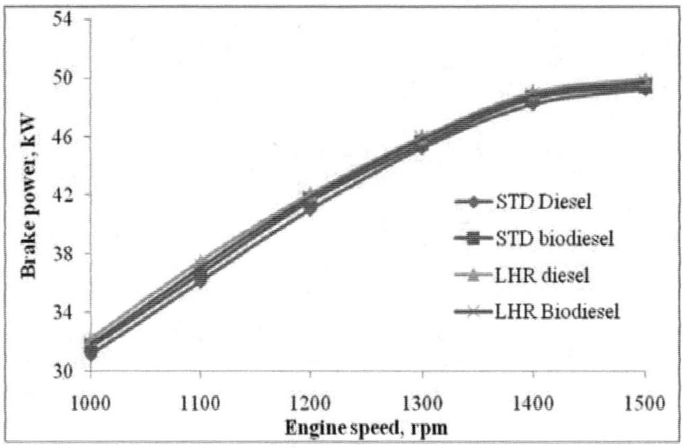

Figure 4. Brake power with engine speed

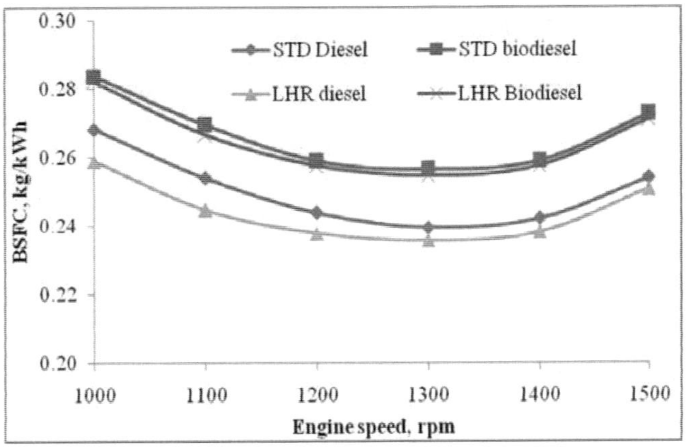

Figure 5. Specific fuel consumption with engine speed

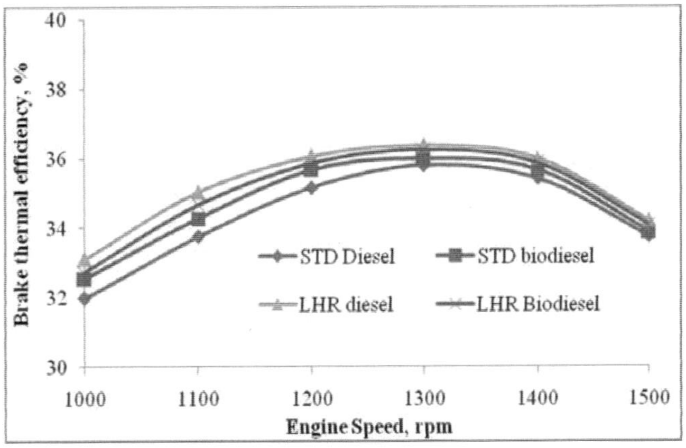

Figure 6. Brake thermal efficiency with engine speed

The variation of exhaust gas temperature with engine speed is shown in Figure 7. Due to ceramic coating exhaust temperature for LHR engines are higher than STD engine. Lower calorific value of biodiesel decrease the exhaust gas temperature when compared to diesel fuel irrespective of the engine. Considerable rise in exhaust gas temperature was observed throughout the speed for LHR biodiesel and LHR diesel which is about 9.1% and 7.7%. Figure 8 shows the variation of NO_x emissions with engine speed for LHR diesel and LHR biodiesel. It shows that, NO_x emission increases as increasing engine speed for four different operating conditions. The NO_x emission increases by 23% and 19% for LHR diesel and LHR biodiesel when compared to STD diesel at 1500 rpm. Higher oxygen mole concentration during combustion and higher heat retainment of heat are the reasons for high NOx. It is evident that the formation of NOx in the function of oxygen availability and high temperature. The high exhaust gas temperature of LHR engine and oxygenated nature of biodiesel increases the NOx. A method of retarding the injection timing is the viable technique to reduce the effect of NOx [28].

Figure 9 shows the variation of HC emissions with engine speed for STD and LHR diesel and biodiesel. It shows that, hydrocarbon (HC) emission decreases as increasing engine speed for four different operating conditions. The HC emission decreases by 24.5% and 15.1% for LHR biodiesel and LHR diesel respectively when compared to STD diesel at 1500 rpm. The decrease in HC emission due to the oxygenated nature of biodiesel. Figure 10 shows the variation of smoke intensity with engine speed. It shows that, the smoke intensity decreases by 28.5% and 31% for LHR diesel and LHR biodiesel when compared with STD diesel at 1500 rpm. At full load condition, for lower speeds the LHR engines provides relatively low exhaust gas temperature which causes lower turbocharger performance. On the other hand at full load condition and higher speeds with the support of fuel oxygen an average of 30% reduction was noticed due to relatively higher exhaust gas temperature.

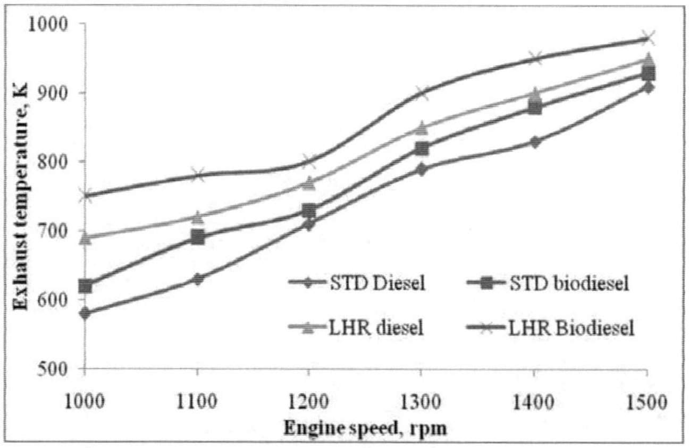

Figure 7. Exhaust gas temperature with engine speed

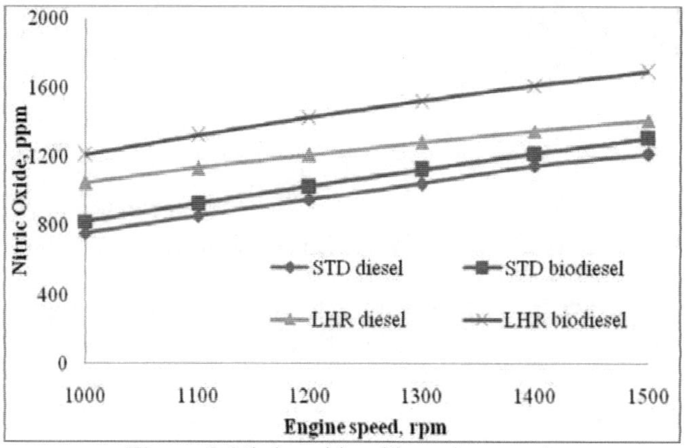

Figure 8. Nitric oxide with engine speed

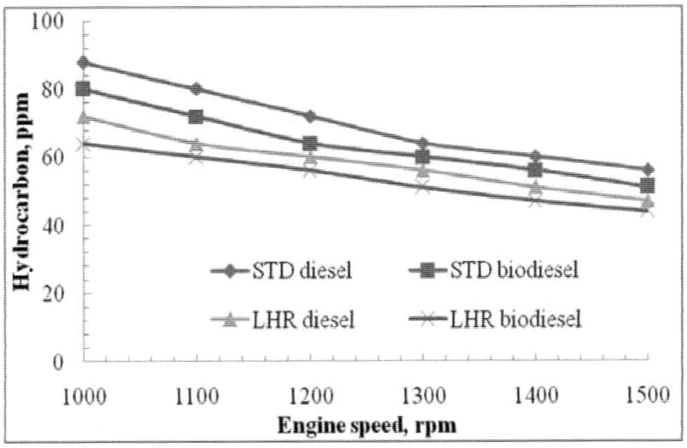

Figure 9. Hydrocarbon with engine speed

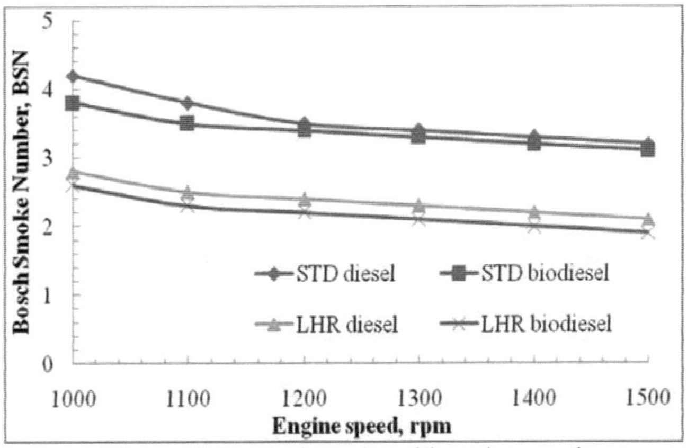

Figure 10. Smoke intensity with engine speed

6. Conclusion

The engine power increases for LHR engine than STD engine due to ceramic coating with increment in exhaust gas temperatures. The specific fuel consumption of biodiesel is higher than diesel fuel owing to lower heating value having the direct impact on brake thermal efficiency. The volumetric efficiency of biodiesel operation increases irrespective of engine due to less heat transfer to engine parts. The combined effect of increasing cylinder temperature and oxygenated nature of biodiesel increases oxides of nitrogen than any other case. The HC and smoke intensity decreases drastically for LHR operation when compared with diesel.

References

[1] Ziejewski M, Kaufman KR 1983 Vegetable oils as a potential alternate fuel in direct injection diesel engines *SAE* paper no. 831357

[2] Hemmerlein N, Korte V, Richter H, Schroder G 1991 Performance, exhaust emissions and durability of modern diesel engines running on rapeseed oil *SAE* paper no. 910811

[3] Ryan T.W, Bagby M.O 1993 Identification of chemical changes occurring during the transient injection of selected vegetable oils *SAE* paper no. 930933

[4] Schlick M L, Hanna MA, Schinstock JL 1988 Soybean and sunflower oil performance in a diesel engine *Trans ASAE* vol. 31 no. 5 pp 1345–1349
[5] Dorado MP, Ballesteros E, Arnal JM, Gomez J, Lofez FJ 2003 Exhaust emissions from a diesel engine fueled with transesterified waste olive oil *Fuel* vol. 82 pp 1311–1315
[6] Lang X, Dalai AK, Bakhshi NN, Reaney MJ, Hertz PB 2001 Preperation and characterization of bio-diesels from various bio-oils *Biosource Technology* vol. 80 pp 53 –62
[7] Al-Widyan M, Tashtoush G, Abu-Qudais M 2002 Utilization of ethyl ester of waste vegetable oils as fuel in diesel engines *Fuel Process Technology* vol. 76 pp 91–103
[8] Monyem A, Van Gerpen JH 2001 The effect of biodiesel oxidation on engine performance and emissions *Biomass Bioenergy* vol. 20 pp 317–325
[9] R.A.Churchill, J.E.Smith, N.N.Clark, and R.A.Turton 1998 Low-Heat Rejection Engines- A Concept Review *SAE* Paper No.880014
[10] R.Kamo and W.Bryzik 1979 Uncooled, Unlubricated Diesel *Automotive Engineering* vol. 87 no. 6 pp 59-61
[11] M.Marmach, D.Server, R.H.J.Hannink, M.J.Murrray, and M.V.Swain 1983 Toughened PSZ Ceramics-Their Role as Advanced Engine Components *SAE* Paper No.830316
[12] Moorhouse, Peter and Johnson, Michael P 1987 Development of Tribological surfaces and Insulating coatings for Diesel Engines *SAE* Paper No.870161
[13] Carr, Jeffrey and Jones, Jack 1984 Post Densified Cr2O3 Coatings for Adiabatic Engines *SAE* Paper No.840432
[14] Dennis Y.C. Leung, Xuan Wu, M. K.H. Leung 2010 A review on biodiesel production using catalyzed Transesterification *Applied Energy* vo. 87 no. 4 pp 1083-1095
[15] Syed Ameer Basha, Raja Gopal. K, Jebaraj. S 2009 A review on biodiesel production, combustion, emissions and performance *Renewable and Sustainable Energy Reviews* vol. 13 pp 1628–1634
[16] Jon Van Gerpen 2005 Biodiesel processing and production *Fuel Processing Technology* vol. 86 pp 1097– 1107
[17] A.A. Apostolakou, I.K. Kookos, C. Marazioti, K.C. Angelopoulos A.A. Apostolakou, I.K. Kookos, C. Marazioti, K.C. Angelopoulos 2009 Techno-economic analysis of a biodiesel production process from vegetable oils *Fuel Processing Technology* vol. 90 pp 1023–1031
[18] HoufangLu, YingyingLiu, HuiZhou, YingYang, MingyanChen, Bin Liang *2009* Production of biodiesel from Jatropha curcas Oil *Computers & Chemical Engineering vol.* 33 no. 5 *pp 1091-1096*
[19] Rajendra Prasath, B. Tamilporai, P. Mohd.F. Shabir 2010 Analysis of Combustion, Performance and Emission Characteristics of Low Heat Rejection Engine Using Biodiesel International Journal of Thermal Sciences vol. 49 pp 2483-2490
[20] Rajendra Prasath, B. Tamil Pori, P. Mohd. F. Shabir, Devan, P. K, Vigneshvaran. S 2012 Combustion and Performance Analysis of Single Cylinder DI Diesel Engine Using Jatropha Biodiesel and its Blends *Applied Mechanics and Materials* vol. 110 pp 3-7
[21] Tamilporai, P 1998 Simulation and analysis of combustion and heat transfer in low heat rejection diesel engine using two zone combustion model and different heat transfer models, PhD thesis, Anna University, India
[22] Vinay Kumar Domakonda, Ravi Kumar Puli 2012 Application of Thermal Barrier Coatings in Diesel Engines: a Review *Energy and Power* vol. 2 no. 1 pp 9-17
[23] Abedin, M. J, Masjuki, H.H. Kalam, M. A, Sanjid, A. Ashraful, M 2014 Combustion, performance, and emission characteristics of low heat rejection engine operating on various biodiesels and vegetable oils *Energy Conversion and Management* vol. 85 pp 173-189
[24] Pratap S. Kulkarni G. Sharanappa M. R. Ramesh N. R. Banapurmath S. V. Khandal 2017 Experimental investigations of a low heat rejection (LHR) engine powered with Mahua oil methyl ester (MOME) with exhaust gas recirculation (EGR). doi.org/10.1080/17597269.2017.1345356.

[25] Modi, A. and Patel, D 2015 Experimental Study on LHR Diesel Engine Performance with Blends of Diesel and Neem Biodiesel *SAE* Technical Paper 2015-26-0052
[26] M.V.S. Murali Krishna, T. Ohm Prakash, P. Ushasri, N. Janardhan and P.V.K. Murthy 2016 Experimental investigation on direct injection diesel engine with ceramic coated combustion chamber with carbureted alcohol and crude jatropha oil *Renewable and sustainable energy reviews* vol. 53 pp 606-628
[27] Debasish Das, K.R. Sharma and Gautam Majumdar 2013 Review of Emission Characteristics of Low Heat Rejection Internal Combustion Engines *International Journal of Environmental Engineering and Management* vol. 4 no. 4 pp 309-314
[28] Ekrem Buyukkaya, Muhammet Cerit 2008 Experimental study of NOx emissions and injection timing of a low heat rejection diesel engine *International Journal of Thermal Sciences* vol. 47 pp 1096–1106

IOP Conf. Series: Materials Science and Engineering **653** (2019) 012036 doi:10.1088/1757-899X/653/1/012036

Passive Constrained Layer Damping: A State of the Art Review

Ambesh Kumar[1,*] and Ranjan K. Behera[2,*]

[1,2]School of Mechanical Engineering, Kalinga Institute of Industrial Technology, Deemed to be University, Bhubaneswar, Odisha-751024, India

*Email: mbeshh@gmail.com; ranjan.beherafme@kiit.ac.in

Abstract. In this paper authors discuss the advancements made over years in the passive constrained layer damping (PCLD) technique used for structural vibration control. To start with, the paper presents a brief literature survey on passive damping techniques using viscoelastic materials (VEM) emphasizing unconstrained layer damping (UCLD) followed by PCLD. The discussion on PCLD technique includes different mathematical models and experimental investigations related to it. Finally, the various design modifications proposed by the different researchers in order to improve the performance of conventional PCLD configuration in presented.

1. Introduction

Viscoelastic materials possess the inherent property of dissipating energy under a transient deformation. Based on this property, viscoelastic materials have long been utilized for suppression of structural vibrations. Generally, these materials are used in the form of a layer that is either freely attached to the surface of a host structure or constrained between a constraining layer and the host structure-surface. These two kinds of arrangements of a viscoelastic layer for passive damping of structural vibration are commonly known as UCLD and constrained layer damping (CLD). In some references, a CLD is also called as PCLD. The energy dissipation in a UCLD treatment occurs mainly due to alternate extension and compression of the viscoelastic layer, while the same in the CLD/PCLD treatment appears due to the transverse shear strains of the constrained viscoelastic layer. [1].

In this paper, at the outset, a brief introduction on passive damping technique using VEM is presented. Discussion starts with UCLD arrangement followed by PCLD arrangement. The research works related to PCLD arrangement is presented exhaustively mainly emphasizing its mathematical models, experimental investigations and design modifications in the conventional PCLD arrangement.

2. Passive Damping of Structural Vibration

In many common structural vibration applications where cost is a prime factor along with reliability, passive damping serves as the vibration control mechanism, and it is usually achieved by the use of viscoelastic materials. Several techniques in the use of viscoelastic materials for suppression of structural vibration have been proposed in the literature such as vibration absorbers, UCLD treatment or free layer damping, CLD treatment, tuned damper, edge damping etc. [2,3]. Among these various techniques, the UCLD and CLD treatments are the popular ones for control for bending mode of vibration of thin-walled flexible structures. In the following section, the damping mechanism, as well as a literature review on UCLD techniques, is presented.

2.1. Unconstrained Layer Damping (UCLD) Treatment

In the UCLD treatment, a viscoelastic layer is freely attached to the surface of a flexible structure (substrate layer) as shown in Figure 1. As the overall structure undergoes bending deformation, the viscoelastic layer experiences extensional/compressional strain (Figure 1) that leads to the energy-dissipation from the overall structure. This damping treatment was proposed in 1952 [4], and has been utilized substantially for passive damping of vibration of different engineering structures.

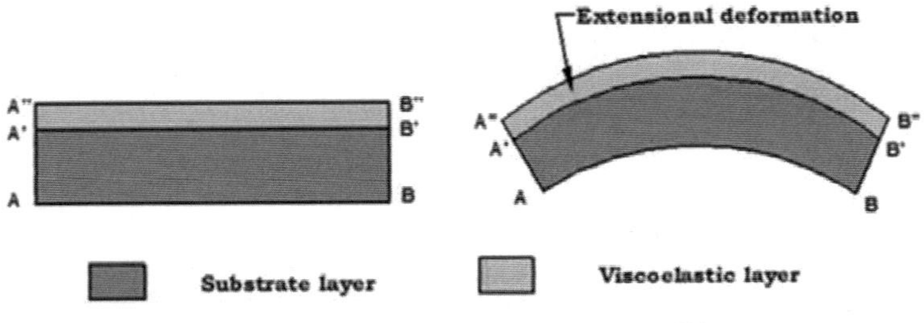

Figure 1. Schematic diagrams of (a) un-deformed and (b) deformed substrate layer integrated with a UCLD layer.

3. Passive Constrained Layer Damping (PCLD) Treatment

Although the UCLD treatment is easy to implement and also economically cheap passive damping arrangement, it possesses very low damping capacity that limits its applications in the suppression of structural vibration. In comparison to the UCLD treatment, a better passive damping treatment is the PCLD treatment [5, 6]. This damping treatment is found to be very useful in automobile and aircraft engineering applications especially for door panels, floor panels, dash panels, brake covers, skin, stringers, frames of fuselage section in commercial aircraft, flutter suppression of aeronautical panels. The damping (PCLD) treatment is also applied substantially in railway wheels, outlet guide van, computer hardware industry, squeal noise in trams, machine tool, spinning disks compression blades [7-14] etc.

A typical geometric configuration of the PCLD treatment is shown in Figure 2 where a viscoelastic layer is constrained between the surface of the host structure and a stiff constraining layer. As the overall structure undergoes bending deformation (Figure 2), the viscoelastic layer experiences transverse shear deformation/strain that leads to the energy-dissipation from the overall structure during its vibration.

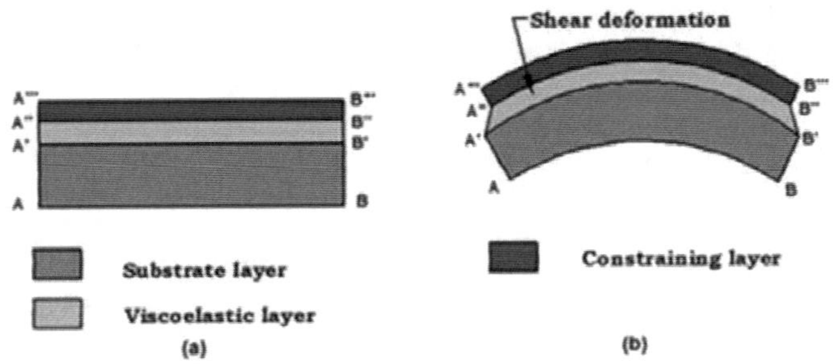

Figure 2. Schematic diagrams of (a) undeformed and (b) deformed substrate layer integrated with a PCLD layer.

The PCLD/CLD treatment was first introduced by Swallow [5] and its first mathematical formulation was addressed by Kerwin [6] considering travelling sinusoidal waves of transverse displacements in beams. Afterward, various theoretical and experimental studies on the PCLD treatment of structures appeared in the literature particularly for investigating its damping mechanisms and damping capacity for different kinds of structures. The various mathematical models (FE and analytical) and their experimental verification related to PCLD are reported in the literature [15-35].

In most of the analyses of the PCLD/CLD treatment, extensional/compressional strains of the viscoelastic layer are not accounted for characterizing the passive damping since the damping appears mostly due to the shear strain of the damping layer [36]. However, it has also been shown in various studies that the compressional damping cannot be neglected especially for the damping treatment with a thick viscoelastic layer [37-41].

3.1. CLD/PCLD Treatment in Different Configurations

In the quest of improved damping capacity of the CLD/PCLD treatment, it (CLD/PCLD) appeared with different geometric and/or material configurations. One such effort to increase the damping is the introduction of segmented PCLD arrangement which was achieved by cutting the constraining layer to the appropriate lengths followed by its optimization. [42-46]. Apart from the segmentation of constraining layer, multilayer CLD configurations were also proposed [47-52]. Whittier [53] introduced a standoff/spacer layer between the damping layer and the surface of host structure for improved damping in the CLD treatment. The standoff layer enhances the shear strain in the constrained viscoelastic layer, and that results in the improved damping capacity of the CLD treatment. It was further modified and verified in many of the works. [54-56].

In some of the works viscoelastic composite material (VCM) layers in place of VEM have been proposed in order to increase the extensional as well as shear damping efficiency of the conventional PCLD treatment [57-59].

In many engineering designs of vibrating structural elements, the constraints of mass and economic cost of the overall structure are imposed. In this circumstance, partial PCLD treatment may be an obvious choice, and hence several works have been carried out on this aspect [60-72].

4. Conclusion

A literature review is presented with an objective of understanding the state-of-art research in the broad area of structural vibration control using PCLD arrangement. This literature survey reveals a substantial number of research reports on the passive damping of structural vibration using the viscoelastic materials in layer-form through CLD/PCLD arrangement. In these studies on the CLD treatment, the mathematical modeling and experimental verification of damping in the treatment have been ad-dressed for different kinds of structures. Along with these theoretical and experimental studies, the research in the quest of improved damping capacity of the CLD treatment has also been reported using standoff/spacer layer, segmented constraining layer, segmented damping layer, multiple damping layers, viscoelastic composite layers, partial PCLD closely spaced axial beams through VE layer, etc.

References

[1] Jones D I 2001 Handbook of viscoelastic vibration damping *John Wiley & Sons*

[2] Grootenhuis P 1970 The control of vibrations with viscoelastic materials *Journal of Sound and Vibration* **11(4)** 421-33

[3] Nakra B C 2000 Structural dynamic modification using additive damping *Sadhana* **25(3)** 277-89

[4] Oberst H and Frankenfeld K 1952 Damping of the bending vibrations of thin laminated metal beams connected through adherent layer *Acustica* **2** 181-94

[5] Swallow W 1939 An improved method of damping panel vibrations *British Patent Specification* GB513171A

[6] Kerwin Jr E M 1959 Damping of flexural waves by a constrained viscoelastic layer *The Journal of the Acoustical society of America* **31(7)** 952-62
[7] Tomlinson G R 1990 The use of constrained layer damping in vibration control *International Journal of Mechanical Sciences* **32(3)** 233-42
[8] Marsh E R and Hale L C 1998 Damping of flexural waves with imbedded viscoelastic materials *Journal of Vibration and Acoustics Transaction ASME* **120(1)** 188-93
[9] Seubert S L, Anderson T J and Smelser R E 2000 Passive damping of spinning disks *Journal of Vibration and Control* **6(5)** 715-25
[10] Jones C J and Thompson D J 2000 Rolling noise generated by railway wheels with viscoelastic layers *Journal of sound and vibration* **231(3)** 779-90
[11] Rao M D Recent applications of viscoelastic damping for noise control in automobiles and commercial airplanes *Journal of Sound and Vibration 2003* **262(3)** 457-474
[12] Sun J and Kari L 2010 Coating Methods to Increase Material Damping of Compressor Blades—Measurements and Modeling *ASME Paper No.* GT2010-44014
[13] Cunha-Filho A G, de Lima A M, Donadon M V and Leão L S 2016 Flutter suppression of plates using passive constrained viscoelastic layers *Mechanical Systems and Signal Processing* **79** 99-111
[14] Shi J, Song Q, Liu Z and Wan Y 2017 Formulating a numerically low-cost method of a constrained layer damper for vibration suppression in thin-walled component milling and experimental validation *International Journal of Mechanical Sciences* **128** 294-311
[15] Yu Y Y 1963 Viscoelastic damping of vibrations of sandwich plates and shells *Polytechnic Institute of Brooklyn NY*
[16] Yin T P, Kelly T J and Barry J E 1967 A quantitative evaluation of constrained-layer damping *Journal of Engineering for Industry* **89(4)** 773-82
[17] Pan H H 1969 Axisymmetrical vibrations of a circular sandwich shell with a viscoelastic core layer *Journal of Sound and Vibration* **9(2)** 338-48
[18] Mead D J and Markus S 1969 The forced vibration of a three-layer, damped sandwich beam with arbitrary boundary conditions. *Journal of sound and vibration* **10(2)** 163-75
[19] Lu Y P, Killian J W and Everstine G C 1979 Vibrations of three layered damped sandwich plate composites *Journal of Sound and Vibration* **64(1)** 63-71
[20] Johnson C D and Kienholz D A 1982 Finite element prediction of damping in structures with con-strained viscoelastic layers *AIAA journal* **20(9)**1284-90
[21] Sun C T, Sankar B V and Rao V S 1990 Damping and vibration control of unidirectional composite laminates using add-on viscoelastic materials *Journal of Sound and Vibration* **139(2)** 277-87
[22] Shin Y S and Maurer G J 1991 Vibration response of constrained viscoelastically damped plates: Analysis and experiments. *Finite elements in analysis and design* **7(4)** 291-97
[23] Imaino W and Harrison J C 1991 A comment on constrained layer damping structures with low viscoelastic modulus *Journal of sound and vibration* **149(2)** 354-59
[24] Wilson C J, Carnevali P, Morris R B and Tsuji Y 1992 Viscoelastic damping calculations using a p-type finite element code *Trans. ASME, J. Appl. Mechanics* **59** 696-99
[25] Ramesh T C and Ganesan N 1992 A finite element based on a discrete layer theory for the free vibration analysis of cylindrical shells *Computers & structures* **43(1)** 137-43.
[26] Yi S and Hilton H H 1994 Dynamic finite element analysis of viscoelastic composite plates in the time domain *International Journal for Numerical Methods in Engineering* **37(23)** 4081-96
[27] Ramesh T C and Ganesan N 1995 Vibration and Damping Analysis of Cylindrical Shells With Constrained Damping Treatment--A Comparison of Three Theories. *Transactions-ASME Journal Of Vibration And Acoustics* **117** 213
[28] Korjakin A, Rikards R, Altenbach H and Chate A 2001 Free damped vibrations of sandwich shells of revolution *Journal of Sandwich Structures & Materials* **3(3)**:171-96

[29] Cai C, Zheng H and Liu G R 2004 Vibration analysis of a beam with PCLD patch. *Applied Acoustics* **65(11)** 1057-76
[30] Moreira R A, Rodrigues J D and Ferreira A J 2006 A generalized layerwise finite element for multilayer damping treatments *Computational Mechanics* **37(5)** 426
[31] Alvelid M and Enelund M 2007 Modelling of constrained thin rubber layer with emphasis on damping *Journal of Sound and Vibration* **300(3)** 662-75
[32] Shin W H, Lee S J, Oh I K and Lee I 2009 Thermal post-buckled behaviors of cylindrical composite shells with viscoelastic damping treatments *Journal of Sound and Vibration* **323(1)** 93-111.
[33] Cao X T, Zhang Z Y and Hua H X 2011 Free vibration of circular cylindrical shell with constrained layer damping *Applied Mathematics & Mechanics* **32(4)**.
[34] Ferreira A J, Araújo A L, Neves A M, Rodrigues J D, Carrera E, Cinefra M and Soares C M 2013 A finite element model using a unified formulation for the analysis of viscoelastic sandwich laminates *Composites Part B: Engineering* **45(1)** 1258-64
[35] Filippi M, Carrera E and Regalli A M 2016 Layerwise Analyses of Compact and Thin-Walled Beams Made of Viscoelastic Materials *Journal of Vibration and Acoustics* **138(6)** 064501
[36] Xie Z and Shepard W S 2009 An enhanced beam model for constrained layer damping and a parameter study of damping contribution *Journal of Sound and Vibration* **319(3)** 1271-84.
[37] Douglas B E 1986 Compressional damping in three-layer beams incorporating nearly incompressible viscoelastic cores *Journal of Sound and Vibration* **104** 343–47
[38] Sylwan O 1987 Shear and compressional damping effects of constrained layered beams *Journal of sound and vibration* **118(1)** 35-45.
[39] Lee B C and Kim K J 1995 Consideration of both extensional and shear strain of core material in modal property estimation of sandwich plates. *In Proceedings of the American Society of Mechanical Engineers Design Technical Conferences Boston MA* 701-8
[40] Huang P Y, Reinhall P G, Shen I Y and Yellin J M 2001 Thickness deformation of constrained layer damping: an experimental and theoretical evaluation *Journal of Vibration and Acoustics* **123(2)** 213-21
[41] Sisemore C L and Darvennes C M 2002 Transverse vibration of elastic-viscoelastic-elastic sandwich beams: compression-experimental and analytical study *Journal of Sound and Vibration* **252(1)** 155-67
[42] Plunkett R and Lee C T 1970 Length optimization for constrained viscoelastic layer damping *The Journal of the Acoustical Society of America* **48(1B)** 150-61
[43] Trompette P and Fatemi J 1997 Damping of beams- Optimal distribution of cuts in the viscoelastic constrained layer *Structural and Multidisciplinary Optimization* **13(2)** 167-71
[44] Al-Ajmi M A and Bourisli R I 2008 Optimum design of segmented passive-constrained layer damping treatment through genetic algorithms *Mechanics of Advanced Materials and Structures* **15(3-4)** 250-57
[45] Lepoittevin G and Kress G 2010 Optimization of segmented constrained layer damping with mathematical programming using strain energy analysis and modal data *Materials & Design* **31(1)** 14-24
[46] Tian S, Xu Z, Wu Q and Qin C 2016 Dimensionless Analysis of Segmented Constrained Layer Damping Treatments with Modal Strain Energy Method *Shock and Vibration*
[47] Torvik P J and Strickland D Z 1972 Damping additions for plates using constrained viscoelastic layers *The Journal of the Acoustical Society of America* **51(3B)** 985-91
[48] Alam N and Asnani N T 1984 Vibration and damping analysis of a multilayered cylindrical shell, Part I: Theoretical analysis *AIAA journal* **22(6)** 803-10
[49] Rao M D, Echempati R and Nadella S 1997 Dynamic analysis and damping of composite structures embedded with viscoelastic layers *Composites Part B: Engineering* **28(5-6)** 547-54
[50] Teng T L and Hu N K 2001 Analysis of damping characteristics for viscoelastic laminated beams *Computer methods in applied mechanics and engineering* **190(29)** 3881-92

[51] Le Maout N, Verron E and Bègue J 2011 Simultaneous geometrical and material optimal design of hybrid elastomer/composite sandwich plates *Composite Structures* **93(3)** 1153-57
[52] Zheng L, Qiu Q, Wan H and Zhang D 2014 Damping analysis of multilayer passive constrained layer damping on cylindrical shell using transfer function method *Journal of Vibration and Acoustics* **136(3)** 031001
[53] Whittier J S 1959 The effect of configurational additions using Viscoelastic interfaces on the damping of a cantilever beam *Wright Air Development Command, US Air Force*
[54] Garrison M R, Miles R N, Sun J Q and Bao W 1994 Random response of a plate partially covered by a constrained layer damper *Journal of Sound and Vibration* **172(2)** 231-45
[55] Rogers L and Parin M 1995 Experimental Results for Stand-off Passive Vibration Damping Treatment. Passive Damping and Isolation *Proceedings SPIE Smart Structures and Materials* **2445** 374-83
[56] Masti R S and Sainsbury M G 2005 Vibration damping of cylindrical shells partially coated with a constrained viscoelastic treatment having a standoff layer *Thin-walled structures* **43(9)** 1355-79
[57] Kumar A and Panda S 2016 Design of a 1-3 viscoelastic composite layer for improved free/constrained layer passive damping treatment of structural vibration *Composites Part B: Engineering* **96** 204-14
[58] Kumar A and Panda S 2017 Optimal Damping in Circular Cylindrical Sandwich Shells With a Three-Layered Viscoelastic Composite Core *Journal of Vibration and Acoustics* **139(6)** 061003
[59] Kumar A, Panda S, Narsaria V and Kumar As 2018 Augmented constrained layer damping in plates through the optimal design of a 0-3 viscoelastic composite layer *Journal of Vibration and Control*. 2018 **24(23)** 5514-24
[60] Nokes D S and Nelson F C 1968 Constrained layer damping with partial coverage *Shock and Vibration Bulletin* **38** 5-10
[61] Markus S 1974 Damping mechanism of beams partially covered by constrained viscoelastic layer *Acta Technica CSAV* **19(2)** 179-94
[62] Lu Y P 1977 Forced vibrations of damped cylindrical shells filled with pressurized liquid. *AIAA Journal* **15(9)** 1242-49
[63] Lall A K, Asnani N T and Nakra B C 1987 Vibration and damping analysis of rectangular plate with partially covered constrained viscoelastic layer *Journal of vibration, acoustics, stress, and reliability in design* **109(3)** 241-47
[64] Lu Y P, Roscoe A J and Douglas B E 1991 Analysis of the response of damped cylindrical shells carrying discontinuously constrained beam elements *Journal of sound and vibration* **50(3)** 395-403
[65] Marcelin J L, Trompette P and Smati A 1992 Optimal constrained layer damping with partial coverage *Finite elements in Analysis and Design* **12(3-4)** 273-80
[66] Kung S W and Singh R 1998 Vibration analysis of beams with multiple constrained layer damping patches *Journal of Sound and Vibration* **212(5)** 781-805
[67] Chen L H and Huang S C 1999 Vibrations of a cylindrical shell with partially constrained layer damping (CLD) treatment *International Journal of Mechanical Sciences* **41(12)** 1485-98
[68] Saravanan C, Ganesan N and Ramamurti V 2000 Study on energy dissipation pattern in vibrating fluid filled cylindrical shells with a constrained viscoelastic layer *Computers & Structures* **75(6)** 575-91
[69] Chen L H and Huang S C 2001 Vibration attenuation of a cylindrical shell with constrained layer damping strips treatment *Computers & Structures* **79(14)** 1355-62
[70] Wang H J and Chen L W 2004 Finite element dynamic analysis of orthotropic cylindrical shells with a constrained damping layer *Finite Elements in Analysis and Design* **40(7)** 737-55
[71] Cao X, Hua H and Zhang Z 2013 Acoustic Radiation From Stiffened Cylindrical Shells With Constrained Layer Damping *Journal of Vibration and Acoustics* **135(1)** 011005
[72] Khalfi B and Ross A 2016 Transient and harmonic response of a sandwich with partial constrained layer damping: A parametric study *Composites Part B: Engineering* **91** 44-55

Recent Trends and Future Perspectives on Vibration Assisted Turning: A Brief Review

Shuvam Mohapatra, Amlana Panda*, Ramanuj Kumar, Ashok Kumar Sahoo

School of Mechanical Engineering,
Kalinga Institute of Industrial Technology (KIIT),
Deemed to be University, Bhubaneshwar-751024, Odisha, India

*amlana.pandafme@kiit.ac.in

Abstract. Over the years, machining Industry has been facing significant challenges concerning enhanced machining characteristics. The main focal objective in machining technology is to augment the measured performance characteristics and also to diminish manufacturing costs, which is important for successful production planning. Vibration-assisted machining is generally used to improve the process by applying a less amplitude and more frequency vibration to the cutting tool or work piece motion. This process requires many advantages, such as superior mechanical properties, lower machining forces and stress levels over the traditional methods that lead to the notable and universal use of ultrasonic vibrations in machining. This brief overview has indicated out research gap and technological confronts for the betterment of vibration assisted techniques. This present article is proposed to demarcate the vibration assisted turning principle, factors and its benefits that can be further used to assess for the machining distinctiveness. Furthermore, based on the brief overview, future research possibilities have been identified to reinforce the capabilities of the vibration assisted turning process to make it efficient for industrial and academic metal cutting research community.

Keywords: High-Temperature Super Alloy, Mathematical Model, Optimisation

1. Introduction

Generally, traditional machining practice of hard alloys produces high cutting temperatures and large cutting forces consequence to the rapid growth of tooling wear and residual stresses at lower cutting speeds [1, 2]. Precision machining of new hard alloy is a challenge to research communities and industrial firm. Poor surface finish and dimensional inaccuracy are the foremost problems in machining of hard to cut materials due to their lower thermal conductivity, higher toughness [3].

Vibration Assisted Machining has been potentially applied to complex purposes such as diamond turning of brittle materials, optical parts and or cost-effective production of precision macro scale components in the machining of Inconel or titanium hard alloys [1]. Moreover, the growing demand for precision components by brittle materials cannot meet up the conditions by traditional cutting processes. However, to enhance precision cutting practice and to achieve enhanced surface accuracy at a lower cost, the vibration is incorporated in several types of work material, cutting tool and on operational medium termed as Vibration Assisted Machining [4].

Continuous endeavours have been performed to develop the machining performance of this innovative machining that has revealed for favourable outcomes in augmenting the part quality and machining effectiveness [1]. The main purpose of this machining is that the tool-workpiece contact happens to intermittent because of applied vibration assistance, reduces in temperature, valuable stresses and cutting forces, tooling wear, and roughness [4]. Tool vibration depicts the relative displacement periodic variation among cutting tool and workpiece while its natural source is not specifically clear [5]. Ultrasonic vibration turning can be applied proficiently to machining biodegradable magnesium-based materials [6]. In Vibration-assisted machining, the cutting tool

IOP Conf. Series: Materials Science and Engineering **653** (2019) 012037 doi:10.1088/1757-899X/653/1/012037

travels with a little amplitude and providing improved machining performance for hard-to- cut materials [7]. In this efficient cutting process, the tooltip is driven in either reciprocating or elliptical direction. The tool loses regular contact with the chip for the proper combination of the speed, amplitude, and frequency [1, 8].

In this machining, external energy improves the material removal process by implementing to smaller amplitude and more frequency vibration to the motion of workpiece or tool [9]. Vibration assisted turning increases the tool lifespan, surface qualities and reduces cutting forces. Herein, vibration is primarily applied with ultrasonic frequency. Most ultrasonic vibrations are generated by a piezoelectric transducer [10]. Contrasted with traditional turning process, when the ultrasonic vibration is appended to the machining direction, this demonstrates the gains in the cutting of the hard-to-cut material. Further, the machining systems are classified into one- direction and two-direction. Moreover, the advantages of this technique are as follows [11]:

- Machining forces are significantly reduced by up to 50% in a few cases.
- The surface roughness is decreased perceptibly.
- Cutting Tool wear is reduced which in turn extend the tool life.

In the vibration assisted efficient cutting process, a cutting tool travels at a higher frequency in the lower amplitude path. If the motion of the path has linear, it is termed as one-directional machining and if the path movement is elliptical then it is known as two-directional machining [12]. The needed principle is to either simulation of the tool or the workpiece [13]. This paper deals with a brief review on emphasizing on Vibration Assisted Turning on its principle, benefits, and industrial applications.

2. Insights of Vibration Assisted Turning

Vibration assisted turning is an innovative efficient practice to entail external controlled vibration on machining motion of turning process to improve surface quality, machined part accuracy, and tooling life [2]. Vibration assisted turning is an advanced technique to implement controlled vibration on original cutting motion so as to increase tooling life and surface accuracy [5]. Herein, low amplitude and large frequency are led to focus on the tool to achieve the intermittent machining during the cutting of hard to cut material [14, 15].

In metal machining processes increasing the temperature interface in the tool–work grounds the thermal softening of the cutting tool and then generate tooling wear [16]. In the ultrasonic vibration technique, the cutting quality mainly depends on tool–workpiece contact ratio and tool–workpiece relative speed. The mechanism demonstrates that the tool–workpiece contact ratio depends on cutting speed, the tool vibration frequency, and the tool vibration amplitude. The workpiece revolves at the constant cutting speed when the cutting tool oscillates at an ultrasonic frequency with small vibration amplitude a where the work material revolves at a constant cutting speed [17]

3. Related Literature Review

The following section illustrates the review of literature executed by past researchers on the machining performance of hard-to-machine materials. The summary of Vibration Assisted Turning incorporating various vibration conditions is presented in table 1.

Table 1. Summary of Vibration Assisted Turning

Material and Cutting Tool	Process Parameters	Vibration Condition	Remarks	Reference

Ti-6Al-4V alloy, cemented carbide	Depth of cut 0.75 mm, Rotational speed17.6, 35.2,52.8, 70.4 m/min, Feed rate 0.15 mm/rev	longitudinal Direction, Amplitude : 10 μm, Frequency : 28 kHz	Minimum Quantity Lubrication based ultrasonic vibration demonstrates less cutting force range and tool wear rate be- cause of the joint effects of vibration and lubrication	[29]
Ti-6A-l4V alloy, Tungsten carbide	Cutting speed: 30,45,60 m/min Feed rate : 0.05, 0.1, 0.15 mm/rev	Frequency (Hz):200, 400,600 Amplitude (μm):50, 100, 150,	Cutting temperature force, equivalent stress and compressive stress developed are lower when contrasted to the traditional turning.	[31]

Fuchs et al. signified that vibration assisted machining has an important influence on chip breaking mechanism leads to complete prevention on continuous chips [13]. In cutting, tool wear is a common factor and ultimately causing tool failure. Cutting temperature is an additional parameter that influences the tooling wear [18]. Over the years, vibration-assisted machining process has been significantly ascertained to capable of restraining the diamond tooling wear [19].

Sivareddy et al. [20] observed that amplitude and frequency play a significant role than speed and feed rate in Vibration Assisted Turning of Ti6Al4V alloy. Adnan and Subbiah [21] viewed that the surface finish reduces as the vibration amplitude is excelled while turning Al-2024T3 aerospace alloy using uncoated tungsten carbide triangular insert.

Yassin et al. [22] summerized that providing vibrations from the tool side in machining practice can be beneficial in turning and grinding processes. Muhammad et al. [23] revealed better extended tool life for tungsten carbide fine-grained (KC5510) is achieved when compared to conventional CP-500 tools using Ti15333 alloy in ultrasonically assisted turning. Yan et al. [24] observed that cutting forces are appreciably lessened using ultrasonically vibration assisted machining, though the lower range of cutting force is achieved in Minimum quantity lubrication system with ultrasonic vibration at different cutting speed levels due to advanced lubrication effect. During ultrasonic-assisted turning Amini and Kazemiyoun [25] suggested that, the cutting speed value should be at least one-third value of vibration velocity.

From the experimental analysis by Zhang et al. [30] it was revealed that cutting speed has more effect on the output characteristics. Moreover, the surface roughness advances significantly with a decrease in the cutting speed. For an improved understanding of the diamond tooling wear method Zhang et al. suggested for the development of novel effective techniques that can contain the chemical wear rate [26]. During the vibration assisted turning, Amini et al. [27] found that, the cutting forces are significantly minimized and the adept surface quality was achieved in contrast with the traditional turning.

Further, the machining performance of the alumina using a PCD cutting tool is examined. For vibration-assisted turning process, Rao and Krishna [12] implemented a useful strategy for real-time tool monitoring systems. According to Pervaiz et al. [28] the elevated magnitude of machining forces with cutting tools with modified geometry under vibration aided machining is obtained. Turning along with ultrasonic vibration process will have its momentous practical uses in precision machining. Moreover, in Ultrasonically-assisted turning, ultra- sonic vibration is superimposed on a cutting tool motion transforming essentially the traditional continuous cutting route in a multiple-impact process [32].

4. Discussion

Vibrations can physically happen in an engineering system and may be envoy of its free and natural dynamic performance. The frictional vibration (self-excited) at the tool flank caused by high contact stresses is the prime source of vibration [33]. Vibration assisted turning process has occupied an imperative role in modern metal cutting practice. With the emergence of Vibration assisted turning technologies, turning practice have matured enough to tackle critical issues precisely. The turning principle helps to achieve an in-depth analysis of the turning process, address its challenging problems, and suggest an outline to devise highly performance equipment. In a nut shell, the outcomes of the vibration-assisted machining have evidently demonstrated that the freshly machined specimen quality can be enhanced by implementing more vibration frequency to the workpiece or tool. To evaluate the significance of favorable outcomes, the design of experiments technique should be explored further. However, this novel turning technology is becoming more pervasive within machining sector and there are still some concerns need to be scientifically addressed for proper machine tool systems, cutting process parameters and mechanisms for cost-effective and stable manufacturing solution. VAM is based on the principle of applied frequency values is classified into non-resonant and resonant types. Non-resonant utilizes 1-40 kHz frequency amplitude able to attain hundred micron meters with adaptable setup, however, the effect of is hard to handle. Resonant employs frequency value of more than 20 kHz, amplitude 1-10 micron meters with fix set up, and with booster or horn [1, 34].

5. Conclusion and Potential Industrial Perspectives

This present review illustrates remarkable attention towards the machining performance during Vibration Assisted Turning that needs to be considered for the future aspects in metal cutting Industry. Moreover, the followings outlines reflect the importance in industrial significance for this efficient machining practice.

- The vibration assisted turning practice offers technological feasibility by incorporating improvement of the integrity of the machined surface, material removal rate, minimization of tool wear and cutting time and wide application in metal cutting areas.
- Appropriate selection of Vibration Assisted Turning conditions would be highly significant for obtaining valuable outcomes to be taken into consideration for technological changes leads to stable machining in practical applications and uses of hard to cut materials.
- It would be of immense concern for the machining and scientific communities to further develop initial research work performed in the field of Vibration Assisted Turning by generating analytical or numerical models by adopting developed methodologies and appropriate research to apprehend enhance cutting mechanics principle.
- Thermal behavior of the tool and workpiece along with varying vibration frequency and amplitude may be considered as the most important challenges in the vibration assisted turning for future investigations.

References

[1] Brehl D E, Dow T (2008) *Precision engineering* **32**(3) 153-172
[2] Thoe T B, Aspinwall D K, and Wise M L H (**1998**) *International Journal of Machine Tools and Manufacture* **38**(4) 239-255.
[3] Muhammad R, Ahmed, N, Roy A, and Silberschmidt, V V (2012) **188** *Trans Tech Publications* 277-284
[4] Kumar M N, Subbu S, Krishna P V, and Venugopal, A (2014) *Procedia Engineering* **97** 1577-1586

[5] Zhang S J, To S, Zhang G Q, and Zhu Z W (2015) *International journal of machine tools and manufacture* **91** 34-42

[6] Di Iorio E, Bertolini R, Bruschi S, and hiotti A (2018) *Procedia CIRP* **77** 324-327

[7] Pervaiz S, Kannan S, and Samad, W A A (2018) *International Mechanical Engineering Congress and Exposition ASME* V002T02A014-V002T02A014

[8] Unune D R, Mali H S (2015) *Proceedings of the Institution of Mechanical Engineers, Part B: Journal of Engineering Manufacture* 229 (10) 1681-1693

[9] Chen W, Huo D, Shi Y, and Hale J M (2018) *The International Journal of AdvancedManufacturing Technology*, **97** 5-8 2033-2049

[10] Nosouhi, R, Behbahani, S, Amini S, and Khosrojerdi M (2014) *Proceedings of the Institution of Mechanical Engineers, Part B: Journal of Engineering Manufacture* **228**(6) 837-846

[11] Liu X, Wu D, Zhang J, Hu X, and Cui P. (2019) *Journal of Materials Processing Technology* **267** 186-195

[12] Prasad B S, Rao J U, and Krishna A G (2017*) Proceedings of the Institution of Mechanical Engineers, Part E: Journal of Process Mechanical Engineering* **233** (1) 35-47

[13] Fuchs M., Habersohn C, Kalkan Y, Lechner, C, and Bleicher, F (2015) *FME Transactions* **43**(2) 119- 122

[14] Silberschmidt V V, Mahdy M A, Gouda M A, Naseer A, Maurotto A, and Roy (2014) *Procedia CIRP* **13** 49-54

[15] Nestler A, Schubert A (2014) *Procedia CIRP* **13** 125-130

[16] Amini S, Teimouri R (2017) *Proceedings of the Institution of Mechanical Engineers, Part E: Journal of Process Mechanical Engineering* **231**(5) 978-991

[17] Nath C, Rahman M (2008) *International Journal of Machine Tools and Manufacture* **48** (9) 965-974

[18] Zhou M, Eow Y T, Ngoi, B K A and Lim E N (2003) *Materials and Manufacturing Processes* **18** (5) 825-834

[19] Zhang X, Liu, K, Kumar, A S, and Rahman M (2014) *Journal of Materials Processing Technology* **214**(2) 496-506

[20] Sivareddy, D V, Krishna, P V, Gopal, A V, and Raz, C P (2018) *International Journal of Automotive and Mechanical Engineering* **15** (3) 5400-5420

[21] Adnan A S, Subbiah, S (2010) *International Journal of Machine Tools and Manufacture* **50** (3) 294-302

[22] Yassin M, Hossam M, El-Hofy H. (2018) *Trans Tech Publications* **775** 480- 486

[23] Muhammad R, Mistry A, Khan S W, Ahmed N, Roy A, Silberschmidt, and V V (2016) *Scientia Iranica* **23** (4) 1800-1810

[24] Yan L, Zhang Q, Yu J (2018) *The International Journal of Advanced Manufacturing Technology* **98** 1-4 827-837

[25] Amini S, Kazemiyoun, M (2014) Materials and Manufacturing Processes **29** (5) 627-633

[26] Zhang X, Liu K, Kumar, A S, and Rahman, M A (2014) Journal of Materials Processing Technology **214** (2) 496-506

[27] Amini, S, Khosrojerdi, M R, Nosouhi, R, and Behbahani, S (2014) *Materials and Manufacturing Processes* **29** (3) 331-336

[28] Pervaiz S, Kannan S, and Samad W A (2018) *In ASME 2018 International Mechanical Engineering Congress and Exposition* ASME V002T02A014-V002T02A014

[29] Yan L, Zhang Q, and Yu J (2018) *The International Journal of Advanced Manufacturing Technology* **98** (1-4) 827-837

[30] Zhang X, Kumar A S, Rahman M, Nath C, and Liu K (2011) *Journal of Materials Processing Technology* **211**(11) 1701-1709

[31] Sivareddy, D V, Krishna, P V, Gopal, A V, and Raz, C P (2018) *International Journal of Automotive and Mechanical Engineering* **15** (3) 5400- 5420

[32] Maurotto, A, Roy, A, Babitsky, V. I., and Silberschmidt, V. V. (2010). *In Proceedings of the 4th CIRP International Conference on High Performance Cutting (HPC2010)* 24-26

[33] Astakhov, V P (2011). In Machining of Hard Materials Springer, London 1-32

[34] Kiswanto, G, and Libyawati, W (2019) In IOP Conference Series: Materials Science and Engineering, IOP Publishing **494** 1012095

Pitting and Electrochemical corrosion behaviour of 316L austenitic stainless steel subjected to warm deformation

Viranshu Kumar[1*], Ratnesh Kumar Gupta[2], Ghanshyam Das[2]
[1, 2] Materials and Metallurgical Engineering Department, National Institute of Foundry and Forge Technology, Hatia, Ranchi, 834003, Jharkhand, India

[1*] viranshukumar@gmail.com

Abstract. Electrochemical and pitting corrosion behaviour of undeformed, deformed and annealed after deformation of the 316L stainless steel have been investigated by using electrochemical corrosion test methods. The results indicated that pitting occurred in undeformed sample, while the small pits were observed in deformed and annealed sample after immersion test in 10% $FeCl_3$ solution. Microstructural characterization of undeformed, deformed and corroded sample were analysed by optical microscope and SEM. A Potentiodynamic polarization test in PWR environment was performed to evaluate the electrochemical behaviour of samples after deformation and heat treatment. The results showed that the corrosion rate decrease after heat treatment of deformed samples because heat treatment after deformation form a more protective oxide film on the surface of samples as compared to oxide film form on the surface of undeformed sample. The E_{corr} shifted toward the passive direction with increase in annealing temperature after deformation. The shift of E_{corr} towards the passive direction indicates a more protective passive film formed on the surface of heat treated samples after deformation.

Keywords: Corrosion, Deformation, Electrochemical, Pitting, Polarization

1. Introduction

316L steel exhibit excellent mechanical properties and reasonable corrosion resistance in different environment but it lost their properties for long service period at high temperature [1-3]. For improving the high temperature mechanical strength and corrosion resistance stainless steels are subjected to deformation process like Sever Plastic Deformation, forging and rolling [4-5]. Warm deformation gives optimum mechanical strength and corrosion resistance as compare to cold and hot deformation because homogenised microstructure were observed in warm deformation process [5]. Mechanical and corrosion properties of steel can be improved by producing ultra-fine size grain with the help of deformation and warm deformation gives optimum mechanical and corrosion resistance [6-7]. The present work aims to improve the high temperature mechanical strength and corrosion resistance of 316L steel by warm forging with the help of hydraulic press at 750^0C. In this investigation we produce the ultra-fine grain size with the help of warm deformation and determine the influence of warm deformation and heat treatment parameters on the pitting corrosion and electrochemical corrosion of stainless steel in primary water reactor (PWR) environment. In this study AISI 316L was annealed after warm deformation for 2 hours at different temperatures ranging from

800^{0}C–1100^{0}C with intervals of 100^{0}C. Solution annealing and stress relief treatment were also performed to evaluate the effect of annealing on the microstructure of the warm deformed sample. Mechanical properties and corrosion resistance depend upon the microstructure developed after warm deformation so we investigated the effect of warm deformed microstructure on the mechanical strength and corrosion resistance. Deformations of austenitic stainless steels play a major role in controlling the microstructures and mechanical properties of the steel [8-10]. The present work aims to determine the influence of deformation and annealing parameters on the pitting corrosion and electrochemical corrosion behaviour of austenitic stainless steel in simulated PWR environment.

2. Experimental Procedures

2.1 Materials, deformation and sample preparation.

The 316L stainless steel used for the present study and its composition is given in Table 1. The steel plates were subjected to 50% reduction by warm forging using 150 Ton hydraulic press at 750 ^{0}C. The deformed samples having dimensions of 10 X 10 X 4 mm^3 were cut from the warm forged plate and were subjected to annealing for 2 hours at different temperatures ranging from 800 ^{0}C–1100 ^{0}C with intervals of 100 ^{0}C. Samples prepared for microstructural characterization were etched using solution of 4% Picric acid and ethanol. Pitting morphology and microstructure of undeformed and deformed samples were characterized by SEM and Optical microscope.

Table 1. Chemical composition of 316L stainless steel (wt %)

Elements (Wt %)	C	Si	Mn	P	S	Cr	Mo	Ni	N	Fe
316L	0.02	0.48	1.28	0.035	0.01	16.12	2.17	10.12	-	Bal.

2.2 Pitting corrosion test

The pitting corrosion study of deformed and undeformed sample ware investigated according to ASTM G48-03 by using the 10% $FeCl_3$ solution [5-6]. The samples with a size of 35 X 20 X 4 mm^3, cut from the undeformed and deformed samples were tested in the water bath with a temperature of 50°C for 24 h. Pitting potential and rate of corrosion of the different samples were investigated by potentiodynamic polarization test in PWR environment at room temperature. The 10 X 10 X 4 mm^3 samples were covered with Teflon tap and leaving an exposed area of 1 cm^2 used as working electrode. After the pitting corrosion test electrochemical corrosion test were performed in PWR solution. Composition of simulated primary water (PWR) environment is 1200 ppm boric acid and 2 ppm lithium hydroxide, 30 ppm Cl^- and distilled water. PWR solution was purged in nitrogen gas and cleaned with potentiostat for removing the oxide layer of the sample. OCP were performed before electrochemical test and allowed to stabilize all sample for 1 hour. For the Tafel extrapolation scan rate of 0.182 mV/s used at room temperature. Pitting and electrochemical corrosion resistance and rate of corrosion were calculated with the help of weight loss method and Tafel fitting using Tafel curve.

3. Results and discussions

3.1 *Microstructural Evolution*

Figure 1(a–b) shows the optical microstructure of the undeformed and deformed sample. From figure 1(a) it can be seen that annealing twins with fully austenitic structure is present in the undeformed sample. From figure 1 (b) it can be seen that grain boundaries are not clearly visible. Despite making several attempts at polishing, etching and microstructure observation; grain structure could not be revealed.

Figure 1(b) is at a slightly higher magnification as compare to 1(a), from this we can conclude that formation of ultra-fine grain has occurred due to deformation which is a widely known phenomena during severe plastic deformation.

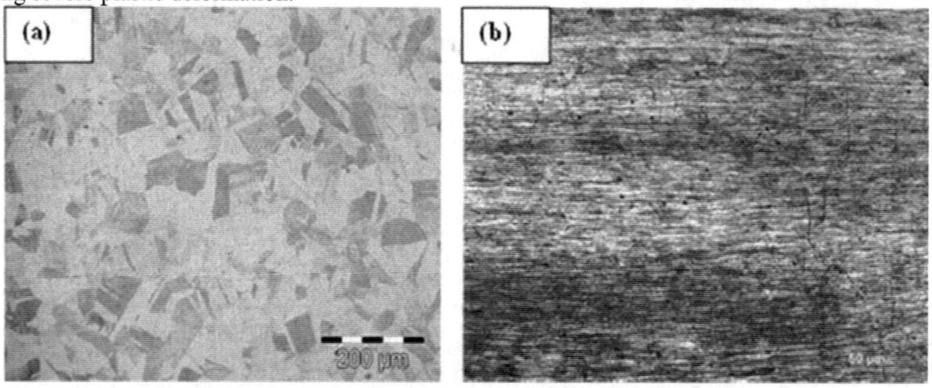

Fig. 1. Optical microstructure of (a) undeformed, (b) warm deformed (50%)

3.2 Pitting corrosion in ferric chloride solution

Figure 2(a-d) shows the surface of undeformed, deformed and annealed samples after the immersion test in 10% $FeCl_3$ solution. As shown in Fig. 2 (a), the small pits are observed in undeformed sample but after deformation number of pits and depth of pits increases which is clearly observed in Fig 2(b). Depth of pit increases after deformation as compare to undeformed sample, which indicates that the pitting is still in growth stage and also indicate that rate of pitting corrosion is higher than as undeformed sample. From figure 2(c) it can be seen that depth of pit and number of pit is very less because after annealing more protective oxide film formed on the surface of the samples as compare to undeformed and deformed samples. From figure 2 (d) it can be observed that very less number of pits and corrosion product formed on the surface which indicate that after annealing at 1000°C repassivation of protective oxide film occurs and pitting corrosion resistance increases as compare to undeformed and deformed samples. Higher pitting potential observed in sample annealed at 1000°C because more protective oxide film formed after annealing.

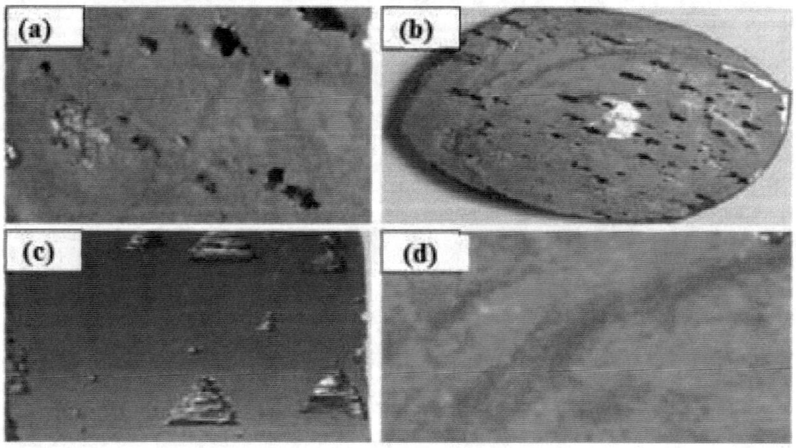

Fig. 2. Pitting test samples (a) undeformed, (b) deformed at 700° C (c) deformed and annealed at 900°C (d) deformed and annealed at 1000° C

3.3 *Microstructural Analysis after pitting corrosion*

Figure 3(a-d) shows SEM microstructure of samples after pitting corrosion test in 10% $FeCl_3$ solution. It can be seen that the undeformed and deformed sample fig. 3(a, b) resulted large number of pits on the surface, whereas deformed samples after annealing resulted less number of pits on the surface. Concentration of pits reduces after annealing at 1000 ^{0}C which indicate that the more protective oxide film formed after annealing. Annealing also helps in reducing metal ion dissolution by improving the repassivation behavior of oxide film. Anealing treatment at 1000°C gives the optimum corrosion resistance and homogenised microstructure because at this temperature chromium carbide gets dissolved in the matrix so that it forms a passive layer of oxide on the surface of the samples. At warm deformation temperature of 750 ^{0}C and annealing at 900 ^{0}C chromium carbides are formed because of chromium and carbon. Formation of chromium carbide at grain boundaries reduces the pitting corrosion resistance at warm deformation temperature and annealing at 900 ^{0}C.

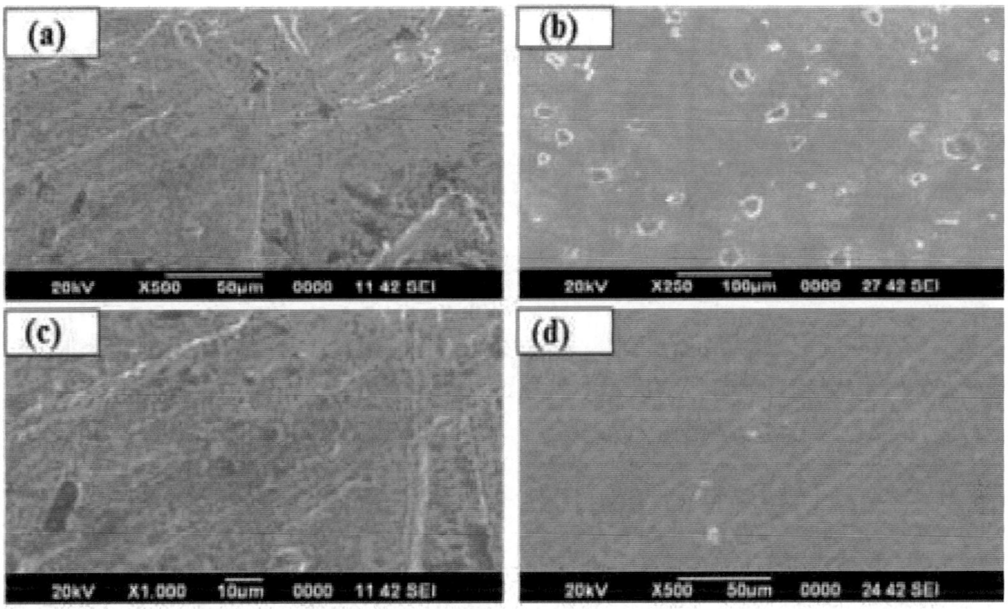

Fig. 3. SEM microstructure of (a) undeformed (b) deformed (c) deformed and annealed at 900°C (d) deformed and annealed at 1000°C

3.4 Potentiodynamic polarization study

Tafel plots of samples in simulated PWR environment shown in figure 4. E_{corr}, corrosion current (I_{corr}) and Tafel slopes are obtained using Tafel plot. From Tafel plot it can observed that the Ecorr shifted toward the passive direction with increase in annealing temperature after deformation which indicates that the formation of protective oxide film of Cr_2O_3 on the surface. Corrosion current suddenly increases at pitting potential which indicate that the formation of stable pits start at this potential. From figure 4 it can be seen that the corrosion rate is drastically increases after deformation as compare to other samples because protective oxide film destroyed after deformation but after annealing at different temperature repassivation capacity improved.

IOP Conf. Series: Materials Science and Engineering **653** (2019) 012038 doi:10.1088/1757-899X/653/1/012038

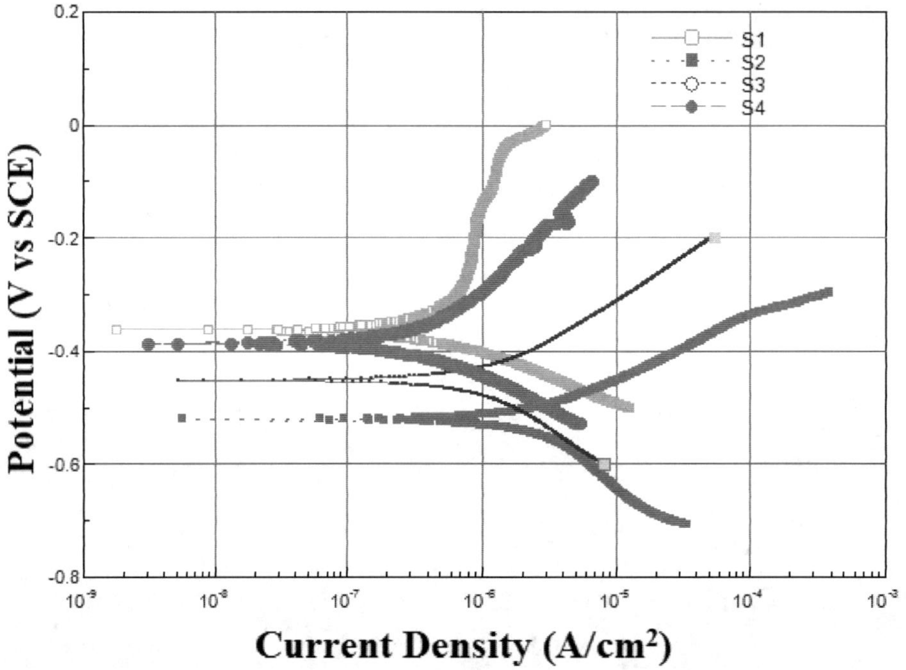

Fig. 4. Tafel plots of samples (a) undeformed (S1), (b) deformed (S2), (c) deformed and annealed at 900° C (S3), (d) deformed and annealed at 1000° C (S4)

4. Conclusions

E_{corr} shifted toward the passive direction with increase in annealing temperature after deformation which indicates that the formation of protective oxide film of Cr_2O_3 on the surface. Corrosion current suddenly increases at pitting potential which indicate that the formation of stable pits at that potential. Less corrosion rate observed in sample annealed sample as compare to deformed and undeformed samples. The corrosion rate drastically increases after deformation as compare to other samples because protective oxide film destroyed after deformation but after annealing at different temperature repassivation capacity improved. It can be seen that the undeformed and deformed samples results in more pits as compare to annealed samples. Heat treatment after deformation enhances the pitting corrosion resistance of 316L stainless steel in 10% $FeCl_3$ solution by forming a more protective oxide film of Cr_2O_3 on the surface. Annealing treatment at 1000^0C gives the optimum corrosion resistance and homogenised microstructure because at this temperature chromium carbide gets dissolved in the matrix so that it forms a passive layer of oxide on the surface of the samples. At warm deformation temperature of 750^0 C and annealing at 900^0C chromium carbide gets precipitated at grain boundaries which destroy the protective layer of chromium hence rate of corrosion increases after deformation. Formation of chromium carbide at grain boundaries reduces the pitting corrosion resistance at warm deformation temperature of 750^0 C and annealing at 900^0C but after annealing at 1000^0C chromium carbides gets dissolved in the matrix hence rate of corrosion decreases after annealing at this temperature.

References

[1] S.J. Zinkle and G.S. was 2013 Materials challenges in nuclear energy *Acta Materials* **6** pp 735-758

[2] T. Allen, J. Busby, M. Meyer and D. Petti 2010 Materials challenges for nuclear systems *Materials Today* **13** 14–23
[3] M.F. McGuire 2008 Stainless Steel for Design Engineers *ASM International* pp 296-301
[4] Junjie Chen , Zhanpeng Lu, Qian Xiao, Xiangkun Ru, Guangdong Han , Zhen Chen , Bangxin Zhou and Tetsuo Shoji 2016 The effects of cold rolling orientation and water chemistry on stress corrosion cracking behavior of 316L stainless steel in simulated PWR water environments *Journal of Nuclear Materials* pp 472-476
[5] Yueling Guo, En-Hou Han and Jianqiu Wang 2015 Effects of Forging and Heat Treatments on the micro structure and oxidation behavior of 316LN Stainless Steel in High Temperature Water *Journal of Materials Science & Technology* pp **31** 403-412
[6] Ankur V. Bansod , Awanikumar P. Patil , Saurabh Suranshe and Ashish Dahiwale 2018 Pitting Corrosion Behavior of Cr–Mn Austenitic Stainless Steel with Addition of Molybdate and Tungstate Under Stagnant and Flow Condition in NaCl Solution *Journal of Failure Analysis and Prevention* ***6*** pp 511-516
[7] Ping Zhu, Xinyuan Cao and Wei Wang, Yonghao Luand Tetsuo Shoji 2017 An investigation on microstructure and pitting corrosion behavior of 316L stainless steel weld joint *Journal of material research* **32** pp 3904-3911
[8] Xingfei Xiea and Dong Ninga 2016 Stress corrosion cracking behavior of cold-drawn 316 austenitic stainless steels in simulated PWR environment *Corrosion Science* **112** pp 576–584
[9] Donghai Dua, Kai Chena, Hui Lua, Lefu Zhanga, Xiuqiang Shib, Xuelian Xub and Peter L. Andresen 2016 Effects of chloride and oxygen on stress corrosion cracking of cold worked 316/316L austenitic stainless steel in high temperature water *Corrosion Science* **110** pp 134–142
[10] S. K. Ghosh, Shikhar Jha, P. Mallick, and P. P. Chattopadhyay 2013 Influence of Mechanical Deformation and Annealing on Kinetics of Martensite in a Stainless Steel *Materials and Manufacturing Processes* **28** pp 249–255

IOP Conf. Series: Materials Science and Engineering **653** (2019) 012039 doi:10.1088/1757-899X/653/1/012039

About the design, development of an automated SMAW setup and the weldment analysis in air and wet working environment

S BReyaz, S KParida*

Manufacturing Engineering Department, NIFFT, Hatia, Ranchi, India-834003

*Corresponding author sambitparida@gmail.com

Abstract. The aim of this research is to design and to develop a computer controlled low-cost underwater welding setup for shielded metal arc welding (SMAW) process to work in dry as well as in wet welding conditions. Investigations of the microstructure and mechanical properties of the welded joints made from AISI 1010 steel plates using this setup, in air, tap water, and saltwater working environment, respectively, have been carried out. Weld bead geometry, microstructure, ultimate tensile strength (UTS), ductility, impact strength and Vickers microhardness of air and wet welded joints have been characterized and compared. The general weld bead size is quite dissimilar for air, tap water and salt water welded joints. The tap water and salt water welds seem to be wider with a slightly higher weld penetration with lower reinforcement. The maximum value of the the average UTS of the welded sample as per ASTM was 485.74 MPa. The experiments showed that the maximum percentage of elongation of samples welded in different conditions is 8.06 % and the maximum value of the average impact strength of 25.64 J was found in case of air welds. The microhardness value of fusion zone (FZ) was greater than base metal (BM) in all the welded samples. It was noted maximum up to 310.8 HV10 at coarse grain heat affected zone (CGHAZ) of tap water welds. The mechanical properties of the salt water welds deteriorated due to porosity, slag inclusion, undercut, increase in hydrogen and oxygen content in the welds.

1. Introduction

Underwater welding is essentially employed to repair offshore structures such as submerged pipelines, offshore oil drilling structure, nuclear power plants, marine structures, etc. [1-2]. Two underwater welding methods is mainly carried out, i.e. wet and dry methods. Wet welding is conducted directly in water environment as it is required for repairing geometrically complex structures many times in the submerged conditions. But, in dry welding, a hyperbaric vessel under the water is required to do the welding [3]. As welding is carried out in a wet environment in submerged conditions, there is a significant effect of tap water and salt water, their flow rate on the stability of the welding arc and metal transfer to the weld pool. Therefore, the quality of the welding bead is being significantly affected. In addition, the dissociation of water molecules into oxygen and hydrogen during welding changes the surface morphology, a mechanical property greatly due to oxide inclusions and hydrogen cracking phenomenon [4]. A number of experimental research works have been reported in recent years based on the metallurgical investigation and the mechanical properties of different arc welded joints in dry and wet conditions. Many of them focus on the favorable metallurgical investigation by customized electrodes or using several auxiliary energy methods. Rowe et al.[5] have shown the importance of ferro-alloy additions by the coating of the electrode in wet SMAW welds. Their results represent a significant improvement in tensile strength, weld metal chemical composition and acicular ferrite microstructures (60 to 90 vol-%). Santos et al. [6] (2012) described the experimental development of an oxyrutile electrode for the wet welding of structural ship steels. They observed lower porosity and superior toughness and ductility of wet welds. Fydrych et al. [7] correlated different welding parameters such as (welding current, welding polarity, water salinity) by using a mercury displacement method for determining diffusible hydrogen amount in deposited metal during the wet welding process. However, there is little information available in research works about the design and development of automated SMAW system and the effect of different water environment on the microstructure and mechanical performance of low carbon steels.

So on the basis of research gap, in the present study (i) design and development of low cost automated SMAW setup to do welding in air and wet environment and (ii) low carbon AISI 1010 steel plates were welded under the different working environments such as air, tap water, and saltwater environment respectively.

2. Design and Development of Automated SMAW Setup

The important requirement of the automated SMAW setup is to maintain the constant arc gap and constant welding speed during the welding operation. The present setup is assembled with two stepper motors, two lead screws, a guided and movable welding tray and electrode holder, an SMPS, a reprogrammable microcontroller system (Arduino Uno), a water tank along with submersible water pump and a welding machine etc. To avoid electric shock and leakage of welding current to the motor, control computer and Aurdino microcontroller during welding, 6000V isolation has provided. In this arrangement the speed of the motors can be controlled in two-stage i.e. at software and also by adjusting the microswitches of the controller TB6600. One stepper motor and lead screw control the movement of the tray while the other controls the movement of the electrode. TB6600 Toshiba makes stepper motor driver having a current capacity of 4A, 9-42V DC voltage has been used. Several microswitches in the controller make control the speed as well as the torque of the motor. The computer control of this welding system allows welding to be started and stopped conveniently and provide synchronous movement of the weld plate and the electrode. The flux electrodes are specially coated with materials which provide shielding environment even when the welding is done in the water. The flux coats the welded surface with a waxy type substance by dispelling the water around it during the welding simultaneously. The motions generated by the stepper motors are converted to the linear motion through the lead screw during SMAW welding operation. Figure 1(a) shows the actual and figure 1 (b) shows the schematic representation of the automated SMAW system.

The present SMAW setup can have a maximum speed of 800 rpm with 20 microns in one step. Cold rolled low carbon AISI 1010 steel plate of size 185 × 100 × 6mm was selected as a base metal because of its wide application in marine structures. The carbon equivalent (CE) is 1.9 according to IIW. The electrode is a commercial E7014 rutile underwater electrode with a diameter of

3.2 mm and length is 350 mm. The chemical analysis of the base metal and electrode were obtained by optical emission spectrometer are listed in Table 1.

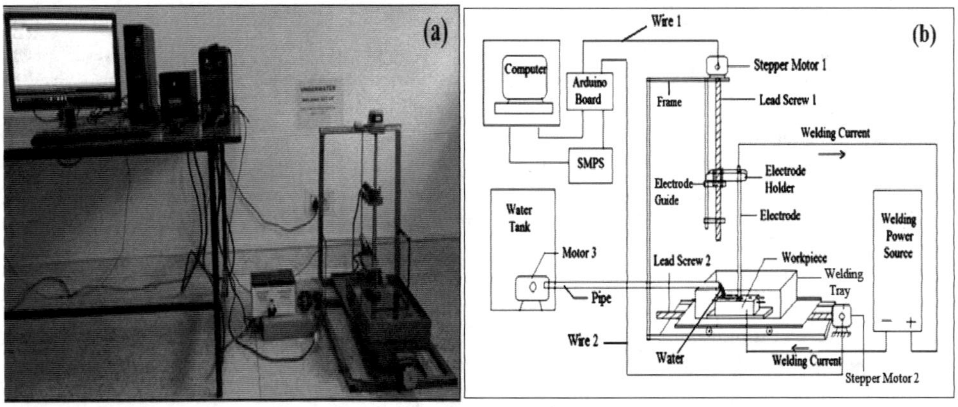

Figure 1. (a) Actual and (b) schematic representation of the automated underwater SMAW system

Table 1.Chemical composition of base metal and electrode (Spectrometer) (wt. %)

Elements	C	Mn	P	S	Si	Cu	Cr	Al	Ti	Ni	*Ceq.
Base metal	0.09	0.54	0.036	0.052	0.17	0.003	0.005	0.011	-	0.112	1.9
Electrode	0.07	0.46	0.056	0.035	0.32	0.03	0.06	0.006	0.011	0.009	-

Air and underwater welding experiments were conducted by using developed automatic welding setup on the plates with prepared edge apriori. Direct current electrode positive (DCEP) or reverse polarity was used in all experiments because of the ease in initiating and maintaining welding arc, and the lesser melting rate of the electrode, with deeper weld penetration. The welding condition and process parameters are shown in Table 2. After all, weldings were performed, the welded joints were subjected to microstructural examination, hardness, tensile and impact tests as per ASTM standard. For each weld, six standard samples were prepared as per ASTM.

Table 2. Welding condition and process parameters of wet SMAW

Working environment	Temperature (°C)	Average welding current (A)	Average arc voltage (V)	Average welding speed (mm/min)	Number of passes	Water flow rate (L/min)	Salinity (ppm)
Air	27	110.02	24.17	135.36	2	-	-
Tap water	23	180.02	31.13	163.27	2	10	500
Salt water	23	140.71	27.14	146.56	2	10	35000

3. Results and discussions

3.1. *Weld bead appearance* In the air welded joint, the globular mode of metal transfer across has taken place resulting in large droplets, usually larger than the diameter of the electrode. Hence, bead appearance was good in dry welding as shown in figure 2 (a). In tap water and salt, water welding is carried out under the combined effect of surface tension, gravity, and the interaction of water and water vapour. Hence, a repelled globular mode of metal transfer has taken place. The weld pool is distorted and the spattering of molten metal has taken place due and hence bead appearance was comparatively poor (as shown in figure 2 (b) and Figure 2 (c)).

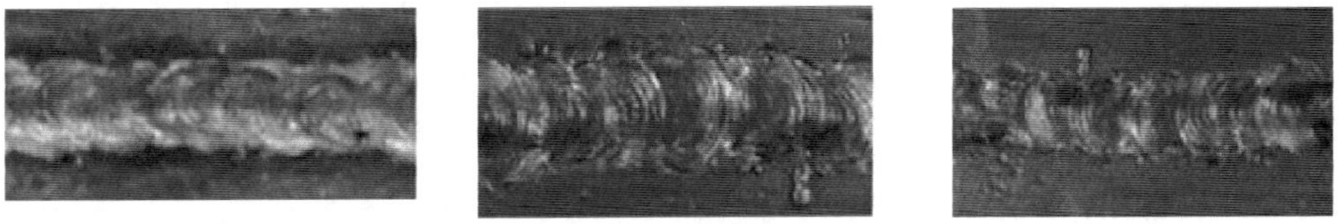

(a) (b) (c)

Figure 2. Weld bead appearance (a) air (b) tap water and (c) salt water

3.2. *Weld beadgeometry* The tap water and salt water welds seem to be wider with a slightly higher penetration and have lower weld reinforcement as shown in Figure 3 (a) to Figure 3 (c). The dissimilarity of the weld bead shape is due to the existence of pressure of water column and water vapour, convection of the water current apart from the gravity force, surface tension force and, the recoil pressure in the weld pool. In addition, the differential cooling rate of the weld pool in each case severely influences the weld bead geometry, the weld bead microstate, and mechanical properties. The increase in weld width and penetration is due to increased welding power input and decrease in reinforcement is due to lesser time available for weld pool is to spread out resulting from the fast cooling. Some welding defects like slag inclusion, porosity and undercuts are also observed in tap water and salt water weld bead as shown in figure 3 (b) and figure 3 (c).

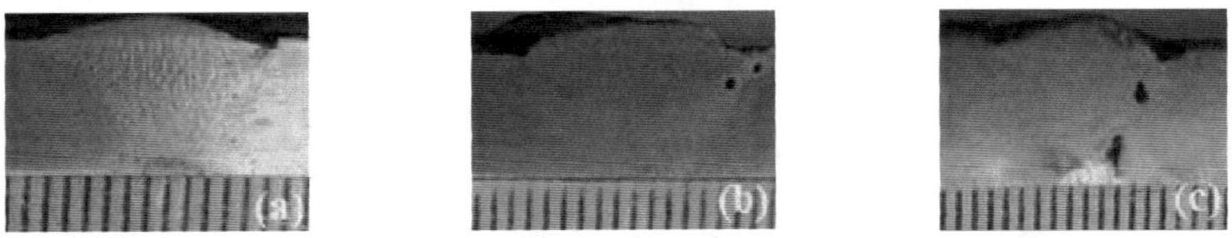

Figure 3. Weld bead cross-section of (a) air joint; (b) tap water joint; (c) salt water joint and (d) air joint with complete geometry

3.3. Microstructure studies

Figure 4 shows the microstructures of BM (base metal) and the heat affected zones (HAZs) corresponding to the three welding environments air, salt water, and tap water, respectively.

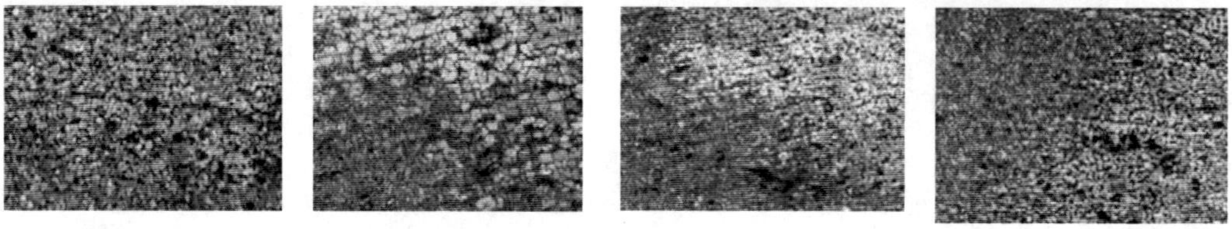

Figure 4. Microstructures of (a) BM; (b) HAZ of air joint; (c) HAZ of tap water joint and (d) HAZ of salt water joint

BM consists mostly of ferrite and some pearlite microstructures with light phase being the ferrite (a) and dark phase (p) being pearlite as shown in figure 4 (a).The microstructural changes of weld joint based on the rate of heat dissipation from the fusion zone (FZ) to the BM. Hence, near the FZ coarse grain and near the unaffected BM refined grain structures have appeared in all the joints. Due to slow cooling rates near the fusion line of air welded joint equiaxed grains have appeared with pearlite within a ferrite matrix structure. Further, due to the very fast cooling rate near the fusion line of tap water and salt water joints, nonequilibrium transformed microstructures are observed. HAZ widths is maximum in air welded joints and minimum in tap water welded joints as air joint undergoes slowest heat dissipation rate from FZ to BMas shown infigure 4 (b). tofigure 6 (d).

3.3. *Tensile strength* In order to compare the tensile strength, six samples of BM (Figure 5 (a)) and the welded joint in air, tap water and salt water have been conducted at room temperature. In air welded joints, most of the tensile samples have fractured in the base metal zone whereas in tap water and salt water welded joints they are fractured mostly in the weld zone as shown in figure 5 (b) to figure 5 (d). Subsequently, air welded joint shows highest UTS (up to 640 MPa) and saltwater welded joint shows the lowest UTS (up to 394.36 MPa). The low UTS of tap water and salt water welded joint may be due to the presence of undercut, porosity, slag, inclusion and increase in hydrogen content and oxygen content in the welds. The tensile samples of base metal show cup-cone failure while all welded joints show brittle failure. Also, the highest elongation has been found in air welded joint (up to 8.06 %) whereas the lowest tensile elongation presented in salt water joint (up to 2.13 %). UTS and ductility (in terms of percent elongation) are determined and the data have been presented in figure 6 (a) and figure 8 (b).

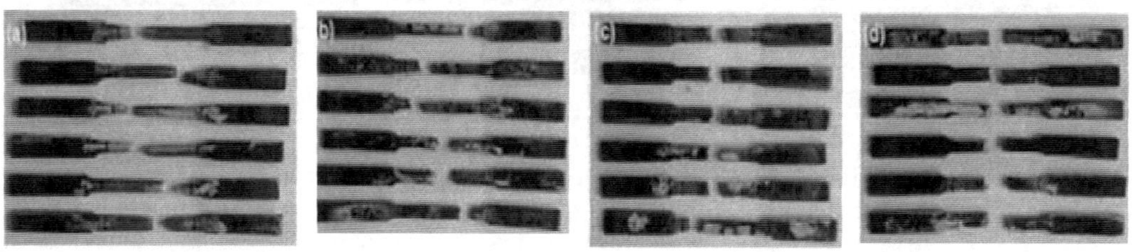

Figure 5. Broken transverse tensile test samples of (a) BM; (b) air welded joint; (c) tap water welded joint; (d) saltwater welded joint

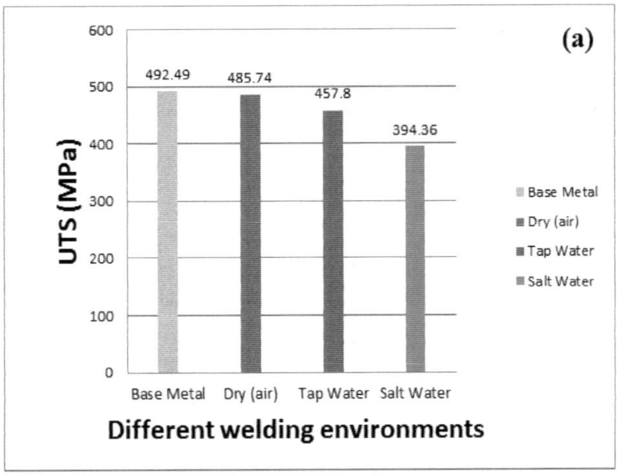

Figure 6 . (a) Ultimate Tensile Strength of samples welded in different conditions

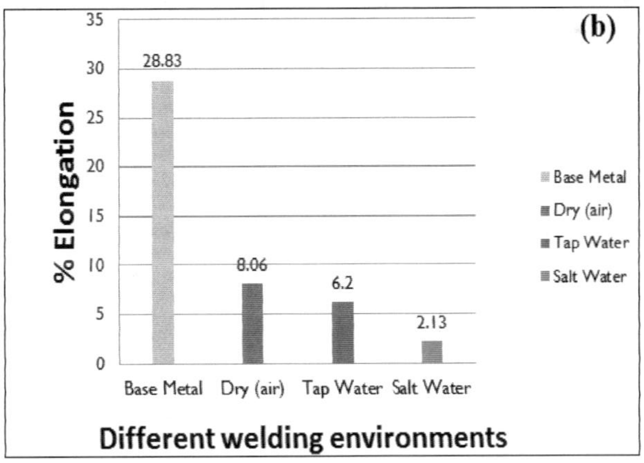

Figure 6 (b) Percentage of elongation samples welded in different conditions

3.4. *Impact strength* Charpy impact strength tests are conducted as per ASTM E-23 at room temperature of six samples of BM and for each different welded joint. Figure 7 shows the variation of the impact strength of the welded joints of the different working environment. Air welded joint shows highest impact strength (up to 32.89 J) and saltwater welded joint shows the lowest UTS (up to 13 J. The low impact strength of tap water and salt water welded joint is mainly due to the presence of slag, porosity and undercut, etc.

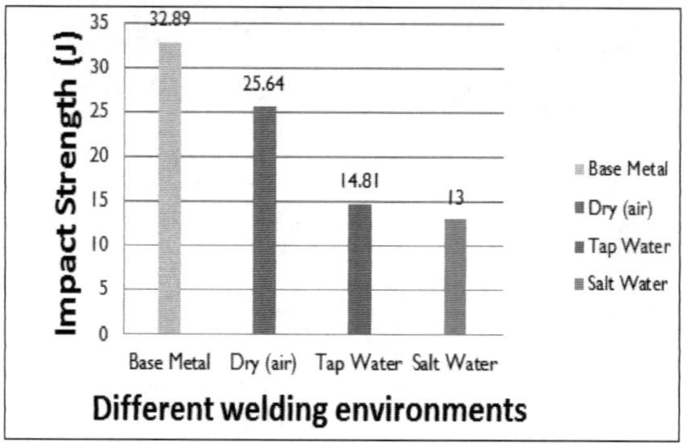

Figure 7. The impact strength of BM and different welded joints

3.5. Vickers microhardness distribution

Figure 8represents the Vickers hardness values (indenter load: 981 N) along the line at the distance of 1 mm from the fusion zone towards the base metal of the cross-section of the welded joint. The value of microhardness value is more in the fusion zone than the base metal in all the welded conditions. This may be due to the fine microstructure in HAZ. The highest microhardness value 310.8 HV_{10} is measured in the AZ of the tap water welded joint and it decreased up to 226.04 HV_{10} air welded joint at CGHAZ close to the fusion zone, where very fast cooling has occurred.

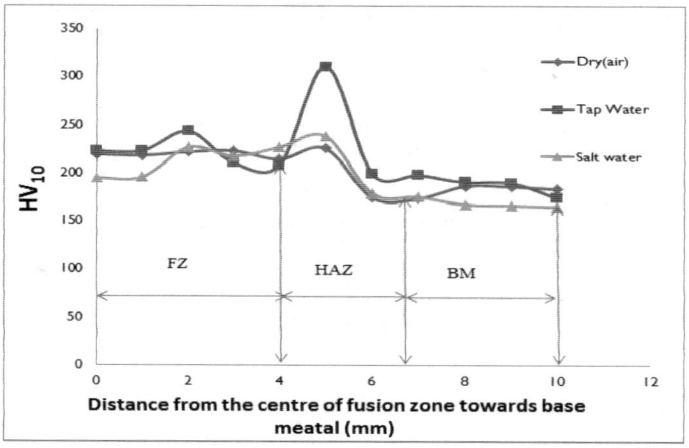

Figure 8: Vickers hardness distribution of the joint at different welding environment.

4.0 Conclusions

Based on the experimental results obtained from the different welding conditions using the automated welding setup, the following conclusions have been drawn.

With this set up more uniform welding could be possible compared with manual and semi-automatic welding system because of the visibility issue a very high skilled welder is required to get good quality of welded joint. The general weld bead size is quite dissimilar similar in air, tap water, and salt water welded joints for same welding parameters. The tap water and salt water welds seem to be wider with slightly higher weld penetration and lower reinforcement. The maximum Ultimate Tensile Strengthof the samples welded in atmospheric air is 640 MPa for AISI 1010 steel whereas for the saltwater welded sample the uts is the lowest (up to 394.36 MPa). The maximum elongation of the air welded joint is 8.06 % while the lowest value occur(i.e. 2.13%) in the saltwater welded joint. The air weld shows the better impact performance (i.e. up to 32.89 J) among all the welded joints whereas salt water weld shows poor impact performance (i.e. up to 32.89 J).

Equiaxed crystal microstructure of pearlite within a ferrite matrix is observed in air welded joints. Due to the rapid cooling rate and solidification of the joint in tap water and salt water environment, non-equilibrium transformed microstructures have been observed near the fusion line of FZ. HAZ widths are maximum in air welded joints and less in case tap water welded joints, as air joint undergoes slowest heat dissipation rate from FZ to BM. For almost all the welded joints, the increased value of micro hardness in the fusion zone than the base metal has been measured. The highest micro hardness value 310.8 HV_{10} is found in tap waterwelded joint and is 226.04 HV_{10} air welded joint at CGHAZ close to the fusion zone. Smaller grains haveformed due to the rapid solidification in HAZ zone which caused increased in hardness in this region. Air welds give the best mechanical properties while it gets deteriorated in salt water welded joint due to defects like porosity, undercut, slag, increase in hydrogen and oxygen content in the welds.

References

[1]. Labanowski J 2011 *Weld Int.* 933–937

[2]. Rowe MD, Liu S 2013*Sci Technol. Weld Join* 387–396 .

[3]. Guo N, Wang MR, Du YP, Guo W, Feng, JC 2015*Metallography, Microstructure, and Analysis*90–92

[4]. Santos VR, Monteiro MJ, Rizzo FC, Bracarense, AQ.*2012 Weld J.*319–328

[5]. RoweM.D, Liu S, Reynolds TJ2002*Weld. J.* 156–166

[6]. SantosX.R, Monteiro MJ, Rizzo FC, Bracarense, AQ, Pessoa E, MarinhoRR,Vieira LA., 2012 *Weld. J.*319–328

[7]. Fydrych D, Swierczynska A., Rogalski G 2015 *Metall. Ital.* 47–52.

Free vibration analysis of adhesive bonded composite T-joints using finite element method

S K Panda, P K Mishra
Department of Mechanical Engineering, CAPGS, BPUT, Rourkela, India

Abstract. This study presents the responses of natural frequency on adhesive bonded T-joint structure. For this purpose, free vibration analysis in the T-joint structure have been performed with a three-dimensional finite element method using ANSYS parametric design language (APDL) code. Using Solid 185 element type the structure has been modelled. To verify the accuracy and stability of the present formulation, intensive comparisons are made with existing results available in the literature and good agreements are obtained. Additionally, Influence of various geometrical and material parameters on the dynamic responses of joint structure has been investigated and discussed in detail. It is observed that the natural frequency increases with increase in aspect ratio of the joint structure. Also, the joint made up of Glass fiber/epoxy indicates more resistance towards free vibration.

Keywords: ANSYS APDL, Adhesive bonded joints, Free vibration, Glass fiber/epoxy, T-joint

1. Introduction

In today's world as the technology advances the demand for composites also increases. The demand for lightweight materials is going higher because of its greater performance capabilities. With the use of composite materials the weight of the structure is reduced by 10-50% and the manufacturing cost also reduced by 10-20%, with equal or higher performance. The demand for composite materials is higher in automobile sector and aerospace industry. Composites have many advantages over usual materials like steel, aluminum, as it offer weight reduction, fuel saving, higher acceleration, high specific strength and stiffness, increased stability, high fatigue life and corrosion resistance.

Adhesive bonded joints show improved performance with reduced material and manufacturing cost, which permit to use these products widely in many high technological applications such as automotive, marine industry, wind turbines, space and aeronautics, etc. One of the technique of joining composites are through the use of an adhesively bonded structural T-joint.

For complete use of bonding technology in joint structures, it is essential to determine the mechanical properties and mechanical behavior of both adhesive and adherend connection. The mechanical properties depend upon various parameters like geometrical shape, environmental effects, boundary and loading conditions. Adhesive bonded T-joints with mechanical behaviors are typically more complex in nature compared to the other joints. The T-joint structures are generally subjected to tension, flexure, and out of

Corresponding author: S K Panda
Email: subhransupanda3@gmail.com

plane type normal and shear loadings. Theotokoglou [2] experimentally studied the composite T-joint made of sandwich structures subjected to out-of-plane loading. He found the presence of both geometrical and material nonlinearity effects on the joint structure and the geometrical nonlinearities occur due to the out-of-plane loading. T-joints strength affected by varying geometries at its bonding region, which are examined by researchers [3-5].

The natural frequencies and loss factors associated with coupled longitudinal and flexural vibrations are the important factors while analyzing adhesive joints. This has been investigated by Saito and Tani [5]. Finite element Analysis is basically used for the calculation of free vibration. Ko et al. [6], Reddy [7], and Lin et al. [8] have used this method for evaluating vibration response of bonded composite plates. Oyadiji [9] analyzed the effect of transverse free vibration on structural adhesives in single-lap cantilever beam joint structures.

Free vibration analysis of the delaminated cantilever composite plate was also conducted. This has been done by Tenek et al. [10], who followed finite element method (FEM) and 3D-elasticity theory in conjunction and carried out a comparison with the experimental data. The analysis to identify the effect by varying dimensions and position on the laminated composite plate's frequency observed by Ju et al. [11]. The response of natural frequencies on composite plates having different fiber orientations subjected to different boundary conditions are analysed.

To understand the dynamic behavior of adhesive bonded joints using both numerical and analytical methods has been completed in the above reviewed papers. Some of the papers has done analysis to study the effect of material parameters and varying boundary conditions. It is found that the geometry, loading and boundary affect the frequency responses of the adhesive bonded joint structure considerably. However, none of the above literature have done extensive work on dynamic response of adhesive bonded T-joint structures. Evaluating the frequency responses of adhesive bonded T-joints, is authors' purpose, in order to bridge the gap for which Finite element analysis method is used. Lastly, the response of natural frequency for different aspect ratio of the joint structure is computed. The free vibration behaviour is also calculated for different material properties of adhesively bonded T-joints.

2. Numerical modeling

The schematic geometrical configuration of adhesive bonded T-joint made with Glass fiber reinforced polymer (GFRP)/epoxy composite material is shown in figure 1 and table 1 shows the dimensions of the sample joint. In the preprocessing part the material properties, material type, and laminate shell properties are assigned and the structure is modelled. The eight noded three dimensional structural volume elements termed as SOLID 185 of Finite element package software ANSYS 19.1 and material properties as shown in table 2 have been used for modelling of the adhesive bonded T-joint. Much better-quality finite elements have been used near the overlap regions to confirm convergence of the solution. Here the model is discretized into 1280 elements and 2856 number of nodes. The meshed view of the adhesive bonded T-joint along with the boundary conditions is shown in figure 2. Here one side of the flange part is fixed (i.e. degrees of freedom in all directions is zero). The system has a built-in edge at the left horizontal end and free at the right and top end.

In case structure having constant stiffness, mass effects and no time varying boundary conditions i.e. displacements, forces, equation motion for an undamped system is represented by

$$[M]\{\ddot{u}\}+[k]\{u\}=\{0\} \tag{1}$$

Where, nodal $\{\ddot{u}\}$ and $\{u\}$ are acceleration and displacement vectors, $[M]$ is mass matrix and $[k]$ is stiffness matrix. Finite element method yields

$$([K]-\lambda[M])\{\bar{D}\}=\{0\} \tag{2}$$

IOP Conf. Series: Materials Science and Engineering **653** (2019) 012040 doi:10.1088/1757-899X/653/1/012040

Where, $\lambda = \omega^2$

Solution of the above eigenvalue problem is given by

$$\det([K] - \lambda[M]) = \{0\} \tag{3}$$

Each of eigenvalue λ_i related to an eigenvector $\{\bar{D}\}_i$, which is defined as natural mode. Eigenvalues and eigenvectors are calculated using the block Lanczos eigenvalue extraction method, as models have large degree of freedoms [12][13].

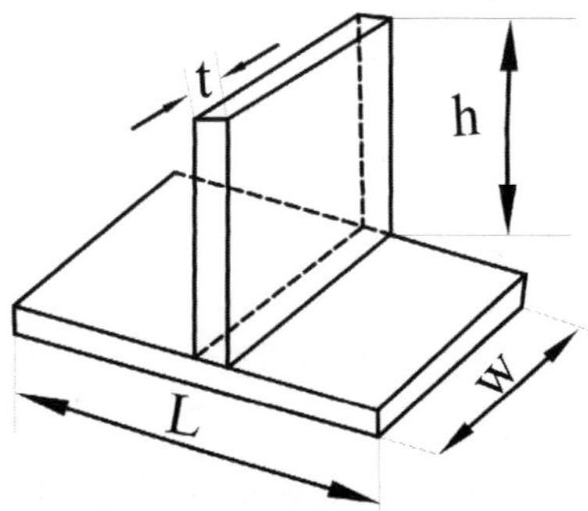

Figure 1. Schematic representation of adhesive condition of bonded T-joint.

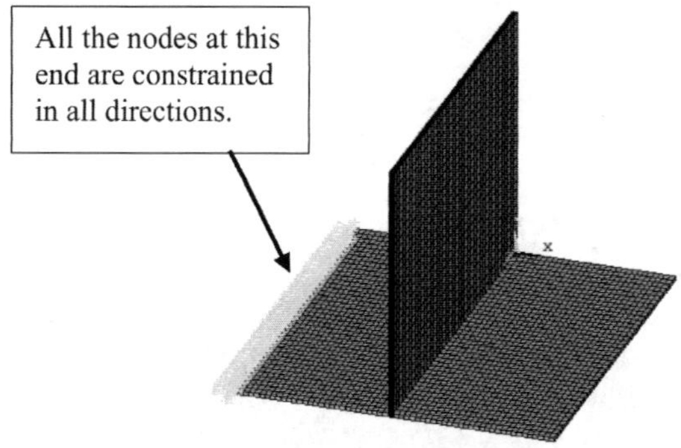

Figure 2. FE mesh and boundary adhesive bonded T-joint.

Table 1: Dimensions of the adhesive bonded T-joint.

Length (L) in meter	Width (w) in meter	Height (h) in meter	Thickness (t) in meter
0.16	0.18	0.16	0.0025

Table 2: Mechanical properties of the adhesive bonded T-joint. [14]

Material	E_x (GPa)	E_y (GPa)	E_z (GPa)	ν	G_{xy} (GPa)	G_{yz} (GPa)	G_{xz} (GPa)	ρ (kg/m^3)
Glass Fiber	8.739	7.926	7.926	0.17	3.75	1.875	3.75	14.276

3. Results and discussion

3.1. Validation study

For validation and comparison study, the natural frequencies of laminated composite having different stacking orientation and CCCC (four edges of plate are clamped) boundary condition, has been analysed [11].

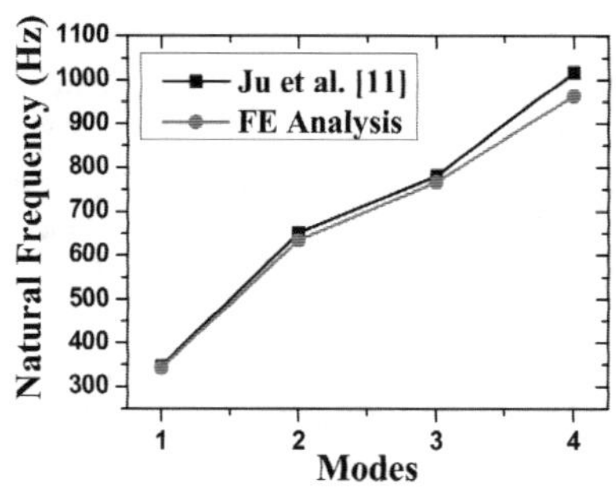

Figure 3. Comparison between current finite element analysis and Ju et al. [11]

The necessary input parameters i.e. geometry, material properties and boundary conditions etc. required for the validation purpose are taken as same as the reference [11]. The first four natural frequencies are obtained and the comparison presented in figure 3. It is found that the results of present finite element analysis shows agreement with the available literature.

The free vibration response of the T-joint structure for different modes are shown in figure 4. First mode shows, the displacement in the x-direction is quite less vibrant than that in y-direction. This could cause an impact on the dimensions of the system. It is identified that the adhesive layer has slopes in two different directions in the second mode. If no tearing and peering occur in the bonding and the bonding has sufficient strength, then identical results can be seen in other modes. In-plane bending modes resulted in the first three modes, whereas fourth one is longitudinal.

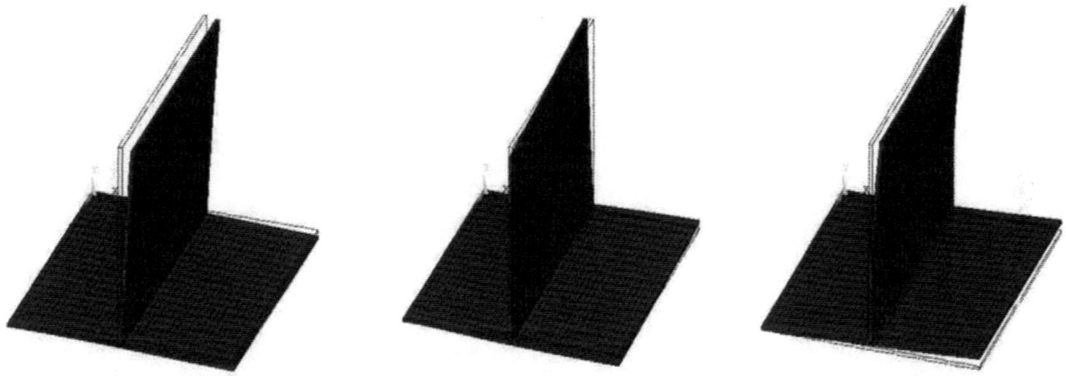

(a) Mode-1
Natural Frequency-19.571Hz

(b) Mode-2
Natural Frequency- 47.943Hz

(c) Mode-3
Natural Frequency- 56.774Hz

(e) Mode-4
Natural Frequency- 88.501Hz

(f) Mode-5
Natural Frequency -163.4Hz

Figure 4. Mode shapes of the adhesive bonded laminated composite T-joint for free vibration.

4. Parametric Study

4.1. Natural frequencies of adhesive bonded laminated composite T-joint with varied aspect ratio

Natural frequencies of T-bonded layered structure have been evaluated by varying aspect ratio (L/h) i.e. 0.5 to 2. Here, the length remains constant whereas height varies. Figure 5 shows five different modes of natural frequency. It is observed that natural frequency varies directly with the aspect ratio (i.e. *L/h*). This is due to bulky nature of the joint structure. It can be seen from the graph after the aspect ratio of 1.0, the free vibration becomes more dominant factor. Hence the aspect ratio for the T-joint structure should be less than 1.

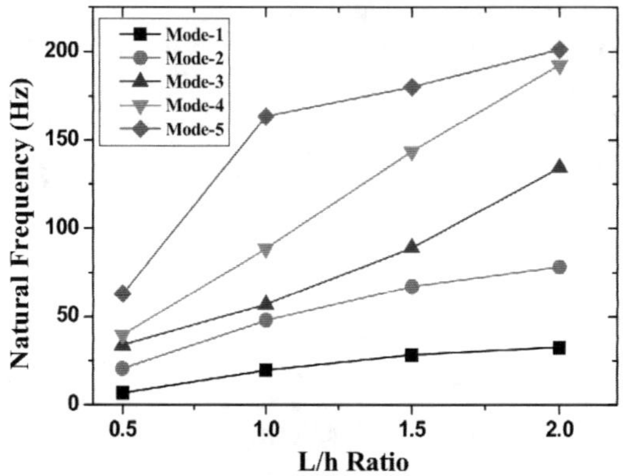

Figure 5: Variation of natural frequencies of adhesive bonded composite T-joint with different L/h ratio.

4.2. Natural frequencies of adhesive bonded T-joint with varied material properties (degree of anisotropy)
In this case, an attempt has been made to study the vibration behavior of adhesive bonded T-joint made with Glass fiber, Graphite fiber, Carbon fiber and Boron fiber materials with varying degrees of anisotropy as shown in table 3.

Table 3: Material properties of T-joint with varying degree of anisotropy. [14]

Fibre	E_x (GPa)	E_y (GPa)	E_z (GPa)	ν	G_{xy} (GPa)	G_{yz} (GPa)	G_{xz} (GPa)	ρ (kg/m³)	Order of Anisotropy $\left(E_x/E_z\right)$
Glass	8.74	7.93	7.93	0.17	3.75	1.86	3.75	14.28	1.10
Boron	207.00	21.00	21.00	0.30	7.00	7.00	7.00	25.20	9.86
Carbon	181.00	10.30	10.30	0.28	7.71	7.17	7.71	13.80	17.57
Graphite	26.20	1.49	1.49	0.28	1.04	1.04	1.04	13.80	17.58

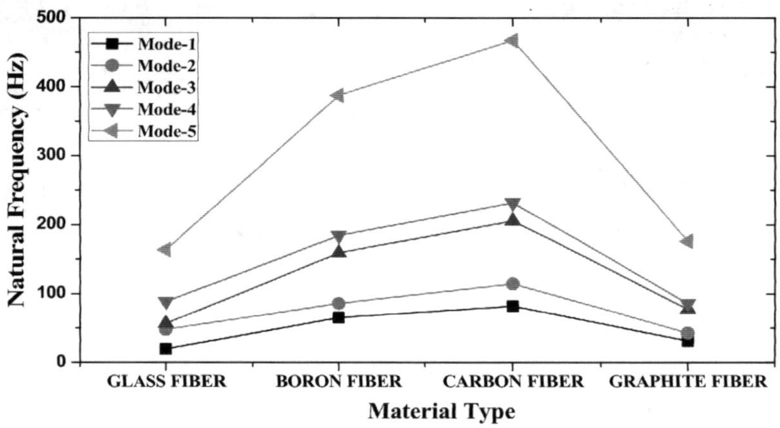

Figure 6: Natural frequency (Hz) of T-bonded structure with different materials.

The free vibration response for first five modes has been potted in the graph as shown in Fig. 6. Here the material properties are taken in increased value of order of anisotropy.It is found that the joint structure with Glass/epoxy fiber laminate exhibits less vibration in compared to other three types of fiber laminate. This behavior attributed due to the simultaneous effects of both mass and degree of anisotropy of the joint structure.

5. Conclusion

In this present study, both experimental and FE free vibration analyses of adhesive bonded laminated composite T-joint have been conducted. The modal test has been conducted using PXIe-1071.The vibration responses are recorded for laminated composite T-joint under CFFF support through the accelerometer mounted on the joint. The finite element analyses have been carried out using ANSYS (APDL) 19.1. Influence of various geometrical and material parameters on the dynamic responses of joint structure has been investigated by varying aspect ratio (L/h), and different lamination schemes. The salient observations are given as follows:

- It is observed that natural frequency increases with increase in aspect ratio (*L/h*). This is due to bulky nature of the joint structure.
- It is found that the joint structure with Glass/epoxy fiber laminate exhibits less vibration in compared to other three types of fiber laminate of the joint structure. This may be due to the simultaneous effects of both mass and degree of anisotropy of the joint structure.

References

[1] Shenoi RA, Violette FL. 1990 *Journal of composite materials.* **24(6)** 644-66
[2] Phillips HJ, Shenoi RA. 1998 *Composites Part A: Applied Science and Manufacturing.* **29(4)** 465-78
[3] Grant LD, Adams RD, da Silva LF. 2009 *International journal of adhesion and adhesives.* **29(4)** 405-13
[4] Grant LD, Adams RD, da Silva LF. 2009 *Journal of Adhesion Science and Technology.* **23(2)** 317-38
[5] Saito H, Tani H. 1984 *Journal of Sound and Vibration.* **92(2)** 299-309
[6] Ko TC, Lin CC, Chu RC. *Journal of Sound and Vibration.* **184(4)** 567-83
[7] Reddy JN. 1979 *Journal of Sound and Vibration.* **66(4)** 565-76
[8] Lin CC, Ko TC. 1997 *Computers & structures.* **64(1-4)** 441-52
[9] He X, Oyadiji SO. 2001 *Journal of Materials Processing Technology.* **119(1-3)** 366-73
[10] Tenek LH, Henneke II EG, Gunzburger MD. 1993 *Composite Structures.* **23(3)** 253-62
[11] Ju F, Lee HP, Lee KH. 1995 *Composites engineering.* **5(2)** 195-209
[12] Cook RD. 2007 *John Wiley & Sons*
[13] Apalak MK, Ekici R, Yildirim M. 2008 *Journal of Scientific and industrial research.* **67**
[14] Jones RM. 2014 *CRC press*

Tribological Characteristics of Epoxide based Bio lubricant – commercial gear oil blends

Jagadhi Priyadharshan SP[1] ,V Srinivas[1*] Kodanda Rama Rao Chebattina[1] , V Vandana[2] ,Pentakota varaaha subrahamanyam[1]

[1] Dept of Mechanical Engineering, GITAM, Visakhapatnam, India.
2 Department of Chemistry, GITAM, Visakhapatnam, India.

[1*]vsrinivas@gitam.edu, [1] kodandaram.nitw@gmail.com

Abstract

This study investigates the effect of epoxide based bio-lubricant derived from Calophyllum Inophyllum blended in commercial gear oil. The tribological properties of the blends are investigated on a four-ball wear tester. The base lubricant used is EP-140 gear oil and study is carried out by blends in 5%, 10%, 15%, 20% volume percentages of Calophyllum Inophyllum oil. Both wear test and friction test were carried out as per ASTM standards for various blends and the results were studied. The experimental results show that the base oil wear scar diameter increased with an increase in load and reduced by the addition of Calophyllum Inophyllum concentrations. The frictional torque is analyzed and the coefficient of friction is calculated. The results shows that 15% of bio-oil concentration is relatively more efficient compared to the other three. Despite a fall in seizure load, the overall results reveal that the addition of 15% Calophyllum Inophyllum to EP-140 gear oil performs better thereby improving anti-wear characteristics of the commercially available lubricants .

Keywords: Bio lubricant, Coefficient of friction, Tribological properties, Epoxide.

1. Introduction

The dependency of lubricants in the field of engineering is unarguable and the importance of its efficiency in multitudes of directions grows every day and rightfully. With concerns over the effects of mineral lubricants over its ecological impacts, usage of bio-lubricants is suggested by many researchers[1-10]. They exhibit excellent properties to be deemed as lubricants. For instance, Canola biodiesel-derived bio-lubricant performed better than methyl oleate- derived bio-lubricant, exhibiting high viscosity indices, good pour point, and moderate oxidative stabilities [1]. But bio-lubricants show inferior properties compared to that of mineral lubricants like low flash points and poor thermal stability, thereby making it impossible to be a replacement of the latter [2, 3]. This phenomenon is attributed to the hydrogen atoms on the ß-carbon atom of the alcohol fragment in ester molecule leading to poor oxidative and thermal stability [4]. In order to find middle ground bio lubricants blended with mineral oils are preferred. Hence, mineral lubricant and bio-based lubricant Calophyllum Inophyllum is blended at different Volume per cent of 5%, 10%, 15%, 20% with the base oil commercially available EP140. The change in its properties and improvement is further studied for possible commercialization.

2. Materials and Methodology

2.1 Materials used

The major resource for this project viz. Calophyllum seeds were collected from GITAM, Visakhapatnam. The Analytical grade supporting reagents and solutions like Basic aluminium oxide, hexane, toluene, and ethyl acetate were purchased from Finar, India. Basic aluminium oxide, hexane, toluene, and ethyl acetate were purchased from Finar, India. P-TSA (para-toluene sulphonic acid) was purchased from AVRA labs, Hyderabad, India.

2.2 Methodology

2.2.1 Oil Extraction method

The Calophyllum nuts were collected form GITAM, Visakhapatnam campus, dried and finely powdered. The oil was extracted using hexane as a solvent. Later, hexane was evaporated and crude oil was dried at low-pressure conditions.

2.2.2 Preparation of calophyllum fatty acid (Undecylenic acid)

The TLC equipment was used for the preparation of Calophyllum fatty acid. Calophyllum oil (300 g) and aqueous sodium hydroxide solution (54 g in 675 ml of water) were stirred mechanically for 4 h at 80-90oC using hexane and ethyl acetate (90:10 v/v) as the solvent. The process was carried out until the last appearance of triglyceride in TLC. Concentrated HCl was used to neutralize the mixture and was cooled to room temperature. Then it was extracted with ethyl acetate and washed twice with water. Finally, 270 g of the product was obtained by drying ethyl acetate layer over anhydrous sodium sulphate and concentrated under vacuum.

2.2.3 Epoxidation of Calophyllum Inophyllum

Undecylenic acid is treated with TME (trimethylolethane) in presence of pTSA (paraTouluenesulfonic acid) as a catalyst and toluene as a solvent to obtain TME ester. Esterification is carried out in the presence of pTSA (1% based on reactant). Later TME ester is treated in the presence of Mcpba (meta-Chloroperoxybenzoic acid) as a catalyst at 0^0C to obtain TME epoxide.

Table 1: Physico-Chemical Properties of calophyllum inophyllum Epoxides and EP 140

Properties	**ASTM methods**	**calophyllum inophyllum epoxide**	**Ep 140**
Density (g/cc)	-	1.0446	1.032
Specific gravity	-	1.0446	1.032

Moisture	-	0.0021	0.0018
Viscosity at 400C(cSt)	ASTM-D-445	38.84	24 – 32.5
Viscosity at 1000C(cSt)	ASTM-D-445	6.64	5
Viscosity index	ASTM D2270	126	93
Cloud point(^{0}C)	ASTM D2500	-1.8	-6
Pour point(0C)	ASTM D97	-25	-9
Flash point(0C)	ASTM D92	270	240
Fire point(0C)	ASTM D92	285	255
Copper strip Corrosion	ASTM-D-130	1a	3

All properties were done thrice and an average was tabulated.

2.3 Experimental methods

2.3.1 Bio lubricant sample preparation

Multipurpose gear oil, EP140 was used as the base lubricant. The base oil is then blended with 5%, 10%, 15%, 20% by volume of Epoxide based Calophyllum Inophyllum bio lubricant. An Ultrasonic sonicator and a bath sonicator were used to blend the composition of base oil and bio oil properly.

2.3.2 Four ball setup

As per the American Society of Testing Materials, the standard test method for measurement of Extreme Pressure (E.P.) properties of lubricating oils is by using Four Ball Extreme Pressure Oil Testing Machine. The four-ball wear tester is used to evaluate the friction and wear properties of the lubricant. Three balls are fixed in the bottom pot called as ball pot and a ball is fixed in the chunk above the three balls firmly held together and immersed in lubricant under test. It is controlled by changing parameters such as test load, temperature, speed and time duration based on the test schedule.

2.4 Test procedures

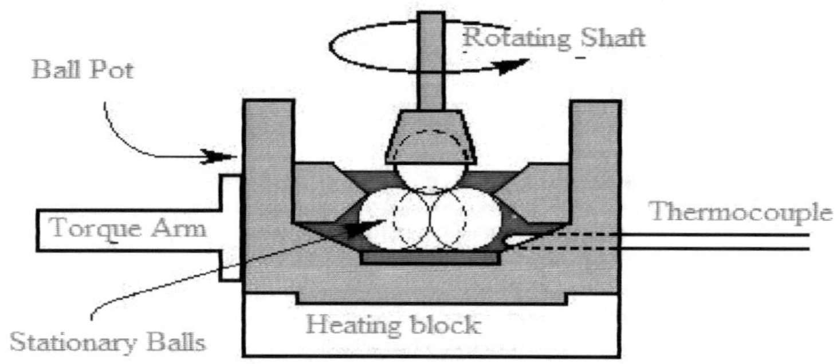

Fig. 1. Schematic diagram of four ball wear tester

2.4.1 Anti-Wear test

Tested based on ASTM D 4172 – 94 standard test procedure. The test load is maintained at 392 and 60 Kgf (588 N), duration was 1hr, the temperature is maintained at 75 0C and rotational speed is maintained at 1200±60 RPM in set in accordance with the standard test schedule. A schematic view of four ball wear tester is shown in figure-1

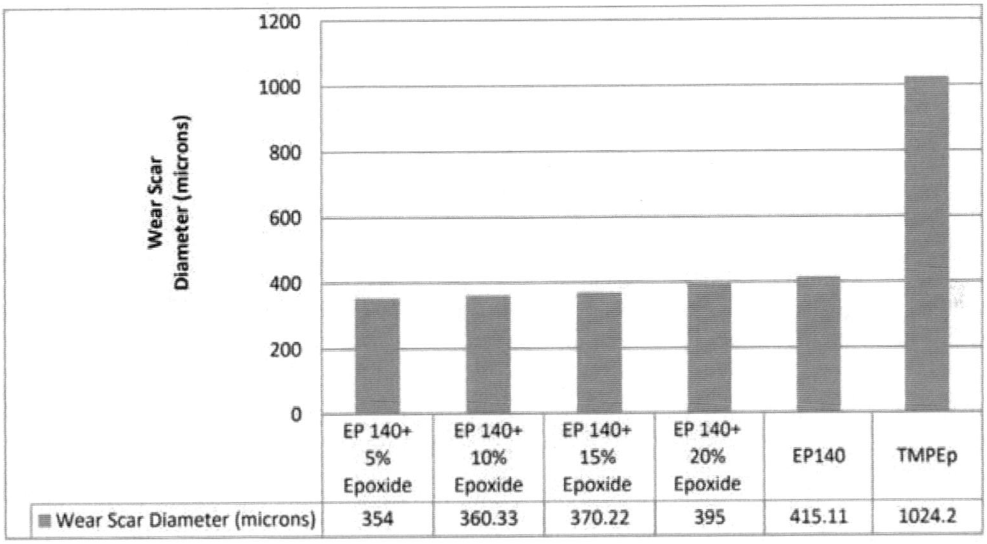

Fig. 2. Comparison of wear scar diameters of different test samples

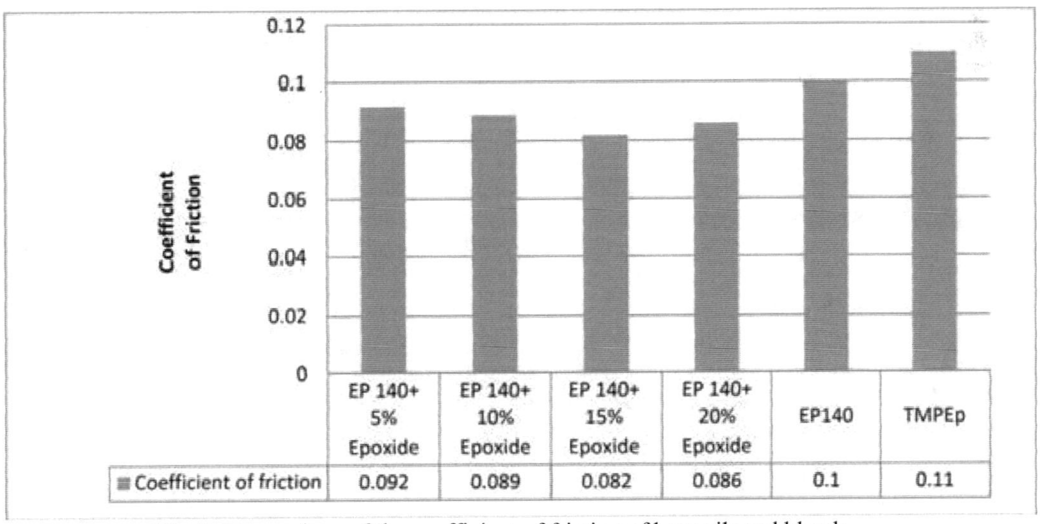

Fig. 3. Comparison of the coefficient of friction of base oils and blends

2.4.2 Friction test

Following the ASTM D 5183 – 95, four 12.7 mm (0.5 in.) diameter steel balls are clamped together and covered with 10 mL of the wear-in lubricant with the fourth ball a 'Top ball' with a force of 392 N (40 kgf) into the cavity formed by the three bottom balls for three-point contact. The top ball is rotated at 600 rpm for 60 min with a lubricant temperature of 75°C (167°F). The fluid is discarded and the balls are cleaned, a fresh oil of 10 mL is added to the ball pot with the worn-in test balls. The top ball is rotated at 600 rpm at 98.1 N (10 kgf) for 10 min with increasing orders of the load by 10 kgf at the end of each successive 10 min interval until the seizure occurs. The friction coefficient of the oil is precise for every 10 min.

3. Results and Discussions

The wear scar diameter of test oils is shown in Figure- 2. Analysis of the wear scar diameter of the test balls at different blends shows that the wear scar diameter for a commercial lubricant is high as with the bio lubricant. But blends of the same show improvement in wear bearing capacities. The coefficient of friction of test oil is presented in Figure- 3. From the graph above, Figure- 3, it is observed that blends of bio-lubricant with commercial lubricant could decrease the friction coefficient of commercial lubricant with improvement in seizure loads too. This is due to the presence of epoxies which slides in rubbing motion causing lesser friction.
Further inference of the graph shows that pure EP-140 shows high friction co-efficient compared to the blends with 15% blend showing the best results. At 20 % The increase in epoxy increases the viscosity and other properties of the oil and thus increase the friction coefficient Similarly, a study of Figure- 4 shows that pure TMP epoxide offers lesser coefficient of friction but has a very low seizure load. EP-140 has superior load bearing capacities. But friction coefficient value is slightly higher. Further inference of the graph proves that improvement in both seizures bearing capabilities and friction coefficient values like 15% blend show very good friction coefficient values with almost the same load-bearing properties as commercial lubricants, thereby making 15% blends the optimal choice for commercialization

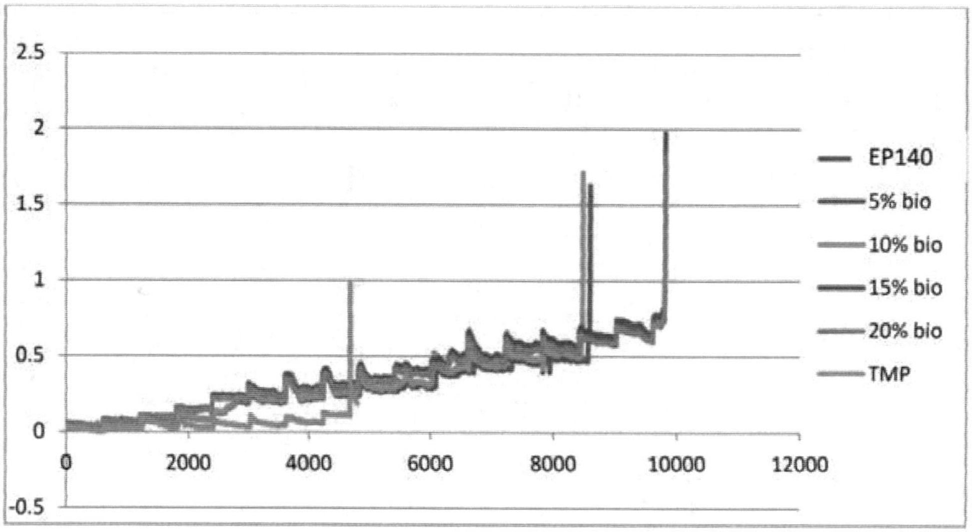

Fig. 4. Frictional Torque vs data points

Conclusions

Fatty acid epoxide based bio-lubricant is derived from Calophyllum Inophyllum seeds. The blends of EP-140 and TMP epoxide showed lower friction coefficients and better wear bearing capabilities. While showing close to similar scar diameters, the 15% blend showed a better coefficient of friction and load bearing capabilities. The increase in epoxy increases the viscosity and other properties of the oil and thus increase the friction coefficient and the fatty acid content helped in reducing the wear and friction rate, while the oleic acid and stearic acid controlled abrasion rates.

Acknowledgment

The Authors sincerely thank the management of GITAM (Deemed to be University) for their Enhanced support

References

[1]. Sripada P K, Sharma R V and Dalai A K, 2013 *Industrial Crops and Products* **50** 95– 103.
[2]. Imran A, Masjuki H H, Kalam M A, Varman M, Hasmelidin M, Al Mahmud K A H, Shahir S and Habibullah M 2013 *Procedia Engineering* **68** 78 – 85.
[3]. Zulkifli N W M, Azman S S N, Kalam M A, Masjuki H H, Yunus R and Gulzar M 2016 *Tribology International* B **93** 555-62.
[4]. Havet L, Blouet J, Valloire F R, Brasseur E and Slomka D 2001 *Wear* **248** 140–46.
[5]. Syahrullail S, Nuraliza N, Izhan M I, Abdul H M K and Razaka Md D 2013 *Procedia Engineering* **68** 158 – 165.
[6]. Mobarak H M, Mohamad N E, Masjuki H H, Kalam M A, Mahmud Al K A H, Habibullah M and Ashraful A M 2014 *Renewable and Sustainable Energy Reviews* **33** 34–43.
[7]. Fox N J and Stachowiak G W 2007 *Tribology International* **40** 1035–46.
[8]. Habibullah M, Masjuki H H, Kalam M A, Ashraf A M 2014 *Procedia Engineering* **90** 740-45
[9]. wang H S, Erhan S Z 2006 *Industrial Crops and Products* **23** 311–317
[10]. Heikal E K, Elmelawy M S, Salah A K and N M Elbasuny 2017 *Egyptian Journal of Petroleum* 53–59.

Optimization and Modeling Approaches in Machining of High Temperature Super Alloy Steels: An Overview

Ramai Ranjan Panigrahi, Ramanuj Kumar*, Amlana Panda, and Ashok Kumar Sahoo

School of Mechanical Engineering,
Kalinga Institute of Industrial Technology (KIIT),
Deemed to be University, Bhubaneshwar-751024, Odisha, India

*ramanujkumar22@gmail.com

Abstract. Over the years, the need for appropriate selection and implementation of optimal cutting conditions has been a major issue among the metal cutting researchers. The implementations of multicriteria decision tools in modern machining were highly valued due to a better capacity to select the optimum parametric combinations. In the optimization method, an objective function is developed with or without mathematical model whereas a mathematical equation based model is formulated, based on input and response parameter in modeling technique to obtain the best efficient cutting environment. In this review paper, various modeling techniques like artificial neural networks, fuzzy modeling, and regression modeling analysis along with Taguchi technique, genetic algorithm and response methodology for application in turning have been focused. Moreover, some new investigations and observations are focused on the advancement of new strategies like (PSO) Particle Swarm Optimization, as it is simple and easy to utilize. In addition, multi-objective optimization is an essentially needed when a single optimization technique can't produce a satisfactory result. This overview states the significance of optimization and modeling technique in the field of high-temperature material and discloses the critical issues related to it.

Keywords: High Temperature Super Alloy, Mathematical Model, Optimisation

1. Introduction

Every manufacturing industry intends to produce a number of machined parts in a short time period. However, a reduction in the production time may cause serious loss of product quality. Therefore, to achieve these valuable criteria, it is important to evaluate the quality of the component. The characteristics might be quantitative or qualitative [1]. In the rapidly expanding manufacturing sector, utilization of optimization and modeling strategies are needed to reduce production, tooling and maintenance costs and to enhance the overall productivity [2].

For any design and modeling reason, the focal goal is to provide progressively exact expectations and better methodology. Hence, optimization, simulation, and modeling computations are an integral part of the modern design practice [3, 4]. Many challenging issues are still to be explored despite of advances in the cutting industry [5]. The decision making methods are mainly focused on appropriately establishing the relationship between input cutting conditions and measured responses. The improvement of process parameters provides the ideal conditions signifying at a lower manufacturing cost [3-5].

Optimization is one of the foremost approaches where design engineering is focused on achieving the best possible outcomes under certain circumstances. Consequently, the Bacteria Foraging Optimization (BFO) and Particle swarm optimization (PSO) was implemented to optimize the feed rate, depth-of-cut and cutting speed to reduce tool flank wear. Despite the fact that the two models (PSO and BFO) performed well, the established results are the predominance of PSO technique. These techniques are recognized for the determination of non-linear response inside a short time period [6].

2. Review of Literature

Superalloys have complexity in the material that is caused by massive strength at the higher temperature, rapid hardening, and low thermal conductivity. [7-8]. Palanisamyan and Selvaraj [9] optimized the process parameters and examine the turning behavior of Incoloy800H superalloy utilizing uncoated carbide inserts. Chakraborty et al. [10] applied ANOVA analysis to evaluate the impacts of machining factors, to recognize the importance of response variables of MRR and Ra. Furthermore, the relation between responses and process parameters are presented and examined by regression analysis [11].

Maiyar et al. [4] examined the influence of various machining responses during milling operation of Inconel718 super alloy. Taguchi design is implemented and results are investigated to accomplish the rate of material removal and ideal surface roughness condition. To analyze the ideal machining parameter settings, grey relation analysis was performed. In the research performed by Muthu et al. [12] established the mathematical models that can be utilized in assessing the responses without conducting trials. Thakur et al. [13] optimized the pulse frequency, coolant quantity, cutting speed, nozzle direction and feed rate to obtain a decreased tool wear, force, and temperature during the turning process.

Jafarian et al. [14] observed that during machining of superalloys, the modeling and optimization intelligent system is used to demonstrate the most suitable surface integrity factor. Krain et al. [15] used optimization methods to evaluate the efficiency and tool life during the milling process of Inconel718 superalloy. Senthilkumar et al. [16] utilized both GA and ANN technique to optimize surface roughness and tool flank wear parameters during the machining process of Inconel718 super alloy. Gupta and Sood [17] observed the sustainable machinability investigation of Incoloy800 super alloy where the critical parameters are they optimized.

Zhou et al. [18] described the fruitful usage of PSO technique during machining of Inconel super alloy. Pimenov et al. [19] utilized artificial intelligence techniques for modeling of surface quality relying upon the tool flank wear. Thirumalai et al. [20] observed the wear design for the turning procedure of Inconel718 under few machining conditions dependent on the Taguchi optimization technique. Satyanarayana et al. [21] implemented the Taguchi optimization technique to acquire the process parameters at high-speed machining of Inconel 718 superalloy.

Kumar et al. [22] implemented modeling and optimization strategy to anticipate the parameters influencing mechanical wear of Incoloy-800 using Taguchi's methodology. Paul et al. [23] used RSM method to optimize the MRR of die-sinking electrical discharge machining on Inconel800 super alloy. The overall summary of optimization techniques used in application of the high-temperature metal as presented in table 1.

Table 1. Summary of optimization and modeling techniques used in machining of High temperature super alloy steel

Year and Author (s)	Work material	Optimization/modeling techniques	Process parameters	Responses
2016 Gupta and Sood [17]	Inconel800	Regression analysis	cutting speed, feed rate, and approach angle.	surface roughness, cutting forces, , tool wear, and chip control
2012 Senthilkumar et al.[16]	Inconel718	Taguchi	Speed, depth of cut, feed rate	Surface roughness, flank wear

2017 Zhou et al.[18]	Inconel718	Particle swarm optimization, radial basis function neural network	Cutting speed, feed rate, inclination angle	compressive residuals tress, surface roughness
2012 Thirumalai et al.[20]	Inconel718	Taguchi method, response surface methodology	Feed rate, depth of cut, speed	flank wear, surface quality
2015 Paul et al.[23]	Inconel 800	RSM	Speed, depth of cut, feed rate	Surface quality

3. Optimization Techniques

Optimization of cutting parameters for differentiable and non-differentiable objective function is dependent on input variables and the optimal conditions are observed. Super alloys are identified as a major challenge in a high-temperature working environment [46]. Design of Experiment and soft-computing methods are the two frequently used optimization methods [24]. Different techniques like Taguchi method, Genetic algorithm, Response Surface Methodology are generally used which generates optimum cutting condition whereas an objective function is developed to determine local optimal solution in soft computing [25]. Saini et al. [26] discussed about the various applications optimization and modeling techniques.

DOE Based Technique

Design of experiment techniques has been utilized for investigational parameter and optimization of chosen factors, to find out the better optimal solution [44].

Taguchi Methodology

It is a design and optimization technique where experiments are performed by orthogonal array (OA), the ideal condition is obtained and a confirmation run is conducted at the ideal condition. To quantify the performance characteristics, Taguchi method changes the objective function values to (S/N) ratio [27]. S/N ratio demonstrates the quality attributes and is utilized as the objective functions. Further, the use of Analysis of variance (ANOVA) is predicted to found the importance of process parameters [28]. It is visualized that when a single quality attribute is optimized, this leads to the failure of other quality attributes. Hence, a multi-objective optimization is needed [29].

Response Surface Methodology (RSM)

RSM is a method in which the relation between response and machining parameter is created. The aim is to measure the impacts of process parameter on the optimization of responses [30]. Firstly, the RSM is utilized to map the response with the independent variable.RSM technique is principally appropriate for applications where a response or a dependent variable simultaneously rely on various input parameters [1]. Generally, RSM is used in various metal cutting problems by various researchers. During invigilating the surface roughness using the RSM method, it was found that the influence of feed on surface quality is more than speed and depth of cut [31]. RSM model enhances the surface roughness under dry cutting condition [32]. Also, RSM modeling technique indicates the possibility for defining tool life with less experimental tests.RSM technique determines experimental data by implementing regression modeling and statistical experimental design principles to establish an efficient relationship between the process parameters and the response variables.RSM technique combines both optimization and modeling methods to get the attainable responses. Numerical optimization, point prediction and graphical optimization are namely the three optimization tools available. Among the three, much frequently utilized tool in metal cutting applications is numerical optimization [1].Some disadvantages of the RSM model is, RSM model can work for maximum three responses. Sometimes the RSM model is not cost effective due to a series of experimentation for many manufacturing situations [33].

Genetic Algorithm (GA)

Genetic Algorithm is a technique where complex issues are solved by the natural evolution procedure. Genetic Algorithm originates from the group of evolutionary algorithms which are employed for the optimization process separated from modeling. In machining, generally, optimization of complex cutting operations is achieved by GA [1]. This method provides a solution space compassing of the high proportion of good solution. GA is very efficient for a single as well as a multi optimization technique and has the capability to solve multi-model function, multi-objective non-linear discreet and continuous functional problems [34]. The GA includes three stages that are selection, cross-over, and mutation which make new generation from the initial population [35]. The application of GA is an exceptionally helpful technique for acquiring process parameters so as to get the required surface completion [36]. GA model shows a certain limitation also.GA is required to merge with some neural network technique to get optimize cutting condition [4].

Multi-objective Optimization

In few cases when a single optimization technique can't give the adequate arrangement, frequently a combination of various optimization and modelling techniques procedures is made to accomplish better outcomes, called Multi-Objective Optimization [1]. Research dependent on multi-optimization techniques are generally, Taguchi based grey relational analysis (TGRA), RSM with TM, ANN with regression technique, ANN with GA [5]. The study conveyed that radial basis function optimization approaching the best outcomes. Predictions made by N-Ns and regression analysis were looked through different statistical techniques. MOO was utilized to limit flank wear, surface roughness and machining force using high-speed dry turning of Inconel-718 super alloy [21].

4. Modeling techniques

To understand the process behavior, numerical condition based models like the analytical model and experimental models are mainly applied [37]. The analytical modeling approach applied for establishing the relationship between cutting inputs and responses [39,43].

Linear regression modeling

In regression-based modeling, a useful relationship between dependent variables and independent variables are accepted. Regression method helps to know, how dependent variables change with the change in the independent variables [4]. For parameter estimation and control, Statistical regression might be applied [1]. Feng and Wang [38] found that, by using regression analysis, the estimated value predicted from ANN model compared with the predicted surface roughness during turning operation. Uthayakumar et al. [40] modeled speed as independent variables and cutting forces as a function of feed, depth of cut by utilizing multiple linear regressions method. Hossainy et al. [41] implemented the regression analysis method to model the machining forces and surface quality in terms of different input responses. Regression-based modeling is used to set up the dependency of cutting force and surface roughness on machining parameters [42]. In regression modeling, a prior assumption is an essential need for a functional relationship; the function may be linear, quadratic, polynomial or exponential. Regression technique is substantial in the region of the regression variable of cutting data and not for whole universe data. For successful utilization of the regression model, the error segment of the equation must have commonly perpendicular and normally distributed with constant variance [4]

Artificial Neural Network

ANN is a numerical condition based model that reproduces the human nervous system in which the information dealing occurs in the form of neurons [44]. Utilization of ANN in turning process is very helpful as it avoids mathematical complexity and no prior assumption is required in the functional form of relationship.ANN model is much useful in machining process for controlling surface roughness characteristics [45]. The ANN model is commonly used when regression analysis fails to provide a solution. The major issues in the modeling of ANN are that the model parameters are unpredictable for a non-linear relationship. A large set of data is needed for training the data and sometimes the model is over fitted [4].

Fuzzy modeling

Fuzzy logic modeling technique can be learned through human intelligence, decision making, and different prospects are commonly utilized when objective function is constrained and decision variable is not exactly measured [1,4]. This method additionally stipulates features like tool selection and optimizing cutting environments [1]. It was demonstrated that fuzzy modeling with ANN technique is required in the manufacturing process to control different metal cutting issues [4]. It was anticipated that the surface roughness is enhanced by implementing fuzzy model during machining process [10]. Fuzzy modeling technique finds its applications where obvious objectives are not characterized [1]. Chakrobarty et al. [10] found that a combination of fuzzy modeling with ANN is much preferable to prevent complex machining procedure. A fuzzy rule is depended on subjective knowledge and expert opinion to set the objective function [4].

5. Conclusion

Appropriate selection of geometrical parameters setting is very essential for high temperature material machining processes which may significantly affect the cutting performance. This review paper has presented the application of modeling and optimization approaches in machining of high temperature superalloy to select the optimal parameter setting for responses. In manufacturing firms, the main objectives are to produce superior quality parts with less cost and within time limits. In today's rapidly expanding of production sector, the applications of multi criteria decision making is to maximize the overall productivity and subsequently contributing to the overall cost of the manufacturing product.. The ongoing tendency in computational optimization method move far away from conventional strategies, however, newly developed methods can be effectively utilized for solving problems. Nevertheless, some new investigations and observations have been focused on the new strategies like Particle Swarm Optimization (PSO) due to its simplicity and easy utilization.

References

[1] Dureja J S, Gupta V K, Sharma V S, Dogra M, and Bhatti, M S, (2016) *Proceedings of the Institution of Mechanical Engineers, Part B: Journal of Engineering Manufacture* **230(3)** 389-404

[2] Asal V D, Prajapati C A, and Patel P K (2013) *Indian journal of research* **2** 4.

[3] Yang A, Koziel X S, and L. Leifsson (2013) *Procedia Computer Science* **18** 855-860

[4] Maiyar L M, Ramanujam R, Venkatesan K, and Jerald J (2013) *Procedia engineering* **64** 1276-1282

[5] Kumar S, Rizvi Y and Kumar R (2018) *International Journal of Mechanical Engineering and Technology* **9(3)** 1146–1156

[6] Singh GR, Gupta M K, Mia M, and Sharma V S (2018) *The International Journal of Advanced Manufacturing Technology* **97** 1-14 481–494

[7] Guo Y B, Li W, and Jawahir I S (2009) *Machining Science and Technology* **13(4)** 437-470

[8] Pan Z, Feng Y, Liang S Y (2017) *Manufacturing Review* **4** 5

[9] Palanisamy A, and Selvaraj T (2018) *Materials Today: Proceedings* **5** (2), 7708-7715

[10] Chakraborty S, Bhattacharyya B, and Diyaley S (2019) *Management Science Letters* **9**(3) 467-494.

[11] Suresh R, Basavarajappa S, Gaitonde V N, Samuel, G L, and Davim J P (2013) *Proceedings of the Institution of Mechanical Engineers, Part B: Journal of Engineering Manufacture* **227**(2) 191-209.

[12] Muthu K V, Suresh B A, Venkatasamy R, and Raajenthiren M (2010) *International Journal of Engineering Science and Technology* **2**(6) 1538-1547

[13] Thakur D G, Ramamoorthy B, and Vijayaraghavan L (2009) *Materials and Manufacturing Processes* **24** (4) 497-503

[14] Jafarian F, Amirabadi, H and Fattahi, M (2014) *The International Journal of Advanced*

Manufacturing Technology **71**(5-8) 817-827
[15] Krain H R, Sharman, A R C, and Ridgway K (2007) *Journal of Materials Processing Technology* **189** (1-3) 153-161.
[16] Senthilkumar J S, Selvarani, P, and Arunachalam, R M (2012) *The International Journal of Advanced Manufacturing Technology* **58** (9-12) 885-894
[17] Gupta M K, and Sood, P K (2017) *Proceedings of the Institution of Mechanical Engineers, Part C: Journal of Mechanical Engineering Science* 231(8) 1445-1464
[18] Zhou J, Ren J, and Yao, C (2017) *Measurement* **102** 271-285
[19] Pimenov D Y, Bustillo A, and Mikolajczyk, T (2018) *Journal of Intelligent Manufacturing* **29** (5), 1045-1061
[20] Thirumalai R, Senthilkumaar J S, Selvarani P, Arunachalam R M, and Senthilkumaar K M (2012) *Australian Journal of Mechanical Engineering* **10**(2) 157-168
[21] Satyanarayana B, Janardhana, G R, and Rao, D H (2013) *Indian Journal of Engineering and Material Sciences* **20** 269-275
[22] Kumar M, Bose S C, and Kotiveerachari, E P B (2013) *International Journal of Engineering Development and Research*
[23] Paul T R, Majumder H, Dey V, and Dutta P (2015) *Journal of Material Science and Mechanical Engineering* **2** 27-31.
[24] Mukherjee I, and Ray P K (2006) *Computers & Industrial Engineering* **50** (1-2) 15-34.
[25] Goldberg D E and Holland J H (1998) *Machine learning*, 3(2) 95-99
[26] Saini S, Ahuja I S, and Sharma V S (2012) *Materials and Manufacturing Processes* **27**(6) 583-598.
[27] Phadke M S, *Prentice-Hall* New-Jersy 1989.
[28] Gunay M, Kaçal A, and Turgut, Y (2011) *e-Journal of New World Sciences Academy* **6**(1) 1A0165
[29] Aggarwal A and Singh H (2005) *Sadhana* **30**(6) 699-711(2005)
[30] Myers R H, Montgomery D C, and Anderson-Cook, C M (1995) *Applied Probability and Statistics*
[31] Choudhury I A, and El-Baradie, M A (1997) *Journal of Materials Processing Technology* **67**(1-3) 55-61
[32] Aruna M, and Dhanalakshmi, V (2012) *International Journal of Manufacturing Technology and Management* **25**(1-3) 113-134.
[33] Singh, N, Routara, B C, and Das, D (2018) *Materials Today: Proceedings* **5**(2) 3438-3449. (2018).
[34] Aszkowicz, W (2013) *Materials and Manufacturing Processes* **28** (7) 708-725
[35] Zitzler, E, and Thiele, L (1998) Springer Berlin, Heidelberg 292-301
[36] Suresh, P V S, Rao, P V, and Deshmukh, S G. (2002) *International Journal of Machine Tools and Manufacture* **42**(6) 675-680
[37] G. E. P. Box and N. R. Draper, Empirical model-building and response surface, John Willey and sons New-York (1987)
[38] Wang, X, and Feng, C X (2002) *The International Journal of Advanced Manufacturing Technology* **20** (5) 348-356
[39] Luong L H, and Spedding T A (1995) *Journal of Materials Processing Technology* **52**(2-4), 585-591
[40] Uthayakumar M, Prabhakaran, G, Aravindan, S, and Sivaprasad, J V (2012) *Materials and Manufacturing Processes*, **27**(10) 1078-1083
[41] El-Hossainy T M, El-Zoghby A A, Badr M A, Maalawi, K Y, and Nasr, M F (2010)*Materials and Manufacturing* Processes **25**(10) 1101-1114
[42] Bartarya, G, and Choudhury, S K (2014) *Part B: Journal of Engineering Manufacture* **228**(9) 1068-1080(2014)
[43] E. A. Peak and D. C. Montgomery, *Introduction to linear regression analysis*, Second

edition, Wiley, New-York(1992).

[44] Fausett, *Fundamental of neural network*, Prentice- Hall International Editions, NewJersy (1994).

[45] Grzesik, W and Brol S (2003) *Journal of Materials Processing Technology* **134**(2), 265-272

[46] Roy S, Kumar R, Panda A, and Das R K (2018) *Materials Today: Proceedings* **5**(9) 18664-18673

An Experimental Investigation of surface roughness of FDM Build parts by Chemical Misting

S.B. Mishra[1*], Eeshan Acharya [2], Dipabrata Banerjee[2] and M.S. Khan[3]

[1] Assistant Professor, School of Mechanical Engineering KIIT University, Bhubaneswar, Odisha 751024

[2] Research Scholar, School of Mechanical Engineering KIIT University, Bhubaneswar, Odisha 751024

[3] Associate Professor, Faculty, Dept. Industrial Engineering, King Khalid University, Abha, Saudi Arabia

swayambikash86@gmail.com

Abstract. Rapid Prototyping or additive manufacturing is a process of creating 3 dimensional objects by adding materials layer by layer. It is a modern manufacturing process of producing more complex shaped materials which might be difficult or impossible in traditional machining processes. This technique has application in all fields of life and many tools artefacts circuit boards etc. have been produced by this process. It also produces new bones to be replaced by older ones. In this treatise an experiment has been conducted to determine the surface roughness before and after chemical treatment thereby comparing both values. Design of Experiments was conducted in Minitab by Taguchi analysis and six parameters, three-part building parameters Raster Angle, Raster Width, Air Gap, and three-part smoothing parameters Temperature, Time, Cycle have been found out. Analysis of Variance (ANOVA)has been carried out to determine the relationship of each parameter along with regression equation. Also surface plots have been drawn to examine the effect of parameters on surface roughness.

...

1 Introduction

Rapid prototyping technologies are growing very rapidly and are in great demand due to cheap cost easy and quick manufacturing of large complex shaped parts [7, 8, 11 & 12]. There are numerous 3d printers and they differ in the way material is deposited or cured. Among them Fused Deposition Modelling is the most widely used technology. It uses Acrylonitrile Butadiene Styrene (ABS) plastic to produce material. ABS is a widely used plastic, used for manufacturing of low cost end use parts having high strength and stiffness. The appraisal shows that ABS plastic is mainly used for production of prototypes due to its exceptional dimensional stability and for easy machinability. The powder is deposited on the model tray. The movement of build platform is over the z-axis (i.e. up and down) and the material extruder movement is over x and y axis. The head applies resin in shape of model. After 1st layer is formed model tray moves down distance of layer and another layer is deposited on model tray. The print head again applies resin in shape of model. This process continues until the total material is formed. Plastic components have major advantages over metal built up parts in terms of weight reduction, ease fabrication of complex geometry and design flexibilities. However, some of the major disadvantages of plastic parts are, its limited wear resistance and structural weakness. Also the surface roughness of parts manufactured by FDM process is one of the main concerns in the industries. The poor surface roughness is due to the stair-stepping effect resulting from the layer by layer deposition of 3D parts. Other factors like raster angle, raster width & air gap also affect surface roughness. Chemical misting is one of the post processing process by which the surface texture of the ABS-FDM 3D printing parts are enhanced.

2 Experimental approach

Our prime objective is to develop relationship between surface roughness and process parameters. In order to do this Taguchi method [3] is adopted in Minitab software by going to the Design of Experiment option. Using a design of experiment (DOE) approach (i.e. Taguchi L orthogonal array, L27) test specimens are fabricated by controlling the process parameters at their different levels. Taguchi design is adopted due to its easiness in data handling and low experimental run order as compared to other DOE approach. Specimens are fabricated using FDM Fortus 250mc in a controlled environment as recommended by ASTM D618 standards. (Layer Thickness: 0.254mm fixed). To obtain uniform result ASTM D618 standards is adopted since different results will be produced in different environment. From various literature survey, few parameters are listed those have significance effect on the response. Process parameters along with their values at different levels is presented in the table (Table 1) below.

1.

Table 1. Process parameters and their levels

Name	Low Level	Centre Level	High level
Raster Angle (degree)	0	45	90
Raster Width (mm)	0.3556	0.5306	0.7306
Air Gap (mm)	0	0.05	0.1
Temperature (^{0}C)	50	60	70
Time (Sec)	10	20	30
Cycle	1	2	3

In order to determine the surface roughness of ABS parts at first the parts are created using Fortus 250 mc (Stratasys Inc.) FDM machine. The material used for preparing the test specimen is acrylonitrile butadiene styrene (ABS M30) as it is a combination of monomeric chemical acrylonitrile butadiene and the styrene in presence of carbon hydrogen and nitrogen. ABS is a carbon chain copolymer and belongs to styrene ter-polymer chemical family. It is made by dissolving butadiene-styrene copolymer in a mixture of acrylonitrile and styrene monomers and then polymerizing the monomers with free-radical initiators. It contains 90-100% acrylonitrile/butadiene/ styrene resin and may also contain mineral oil (0-2%), tallow (0-2%) and wax (0-2%). ABS is a widely used plastic, used for manufacturing of low cost end use parts having high strength and stiffness. The material is added one over another in a sequential manner as defined by the machine software (Insight 10.2). Since it is an additive manufacturing process material waste is minimum than other conventional manufacturing processes. The movement of build platform is over the z-axis (i.e. up and down) and the material extruder movement is over x and y axis. The post processing technique is most important step in rapid prototyping process as parts made have comparatively rough surface. To reduce the surface roughness chemical vapor treatment using acetone is carried out which is the simplest easiest and environment friendly technique. It has other advantages like enhancement of aesthic beauty.

3 Fabrication of specimens

To study surface roughness of ABS parts, specimens are manufacture using Fortus 250 mc (Stratasys Inc.) FDM machine (Fig. 1). The part specifications are 50×20×15 mm (Fig. 2). The parameters set are raster angle (0°, 45° and 90°), raster width (0.3556, 0.5306 and 0.7306) and Air gap (0, .05 and

IOP Conf. Series: Materials Science and Engineering **653** (2019) 012043 doi:10.1088/1757-899X/653/1/012043

0.1mm). The numbers of experimental runs are 27, for each experiment two specimens are fabricated. Hence, a total of 54 numbers of specimens are fabricated.

Fig. 1. FDM 250mc (Stratasys Inc.) with a specimen

The material used for fabrication of test specimen is acrylonitrile butadiene styrene (ABS M30). ABS is a combination of monomeric chemical acrylonitrile butadiene and the styrene in presence of carbon hydrogen and nitrogen. ABS is a carbon chain copolymer and belongs to styrene ter-polymer chemical family. It is made by dissolving butadiene-styrene copolymer in a mixture of acrylonitrile and styrene monomers and then polymerizing the monomers with free-radical initiators. It contains 90-100% acrylonitrile/butadiene/ styrene resin and may also contain mineral oil (0-2%), tallow (0-2%) and wax (0-2%). Its three structural units provide a balance of properties with the acrylonitrile providing heat resistance, butadiene imparting good impact strength and the styrene gives the copolymer its rigidity.

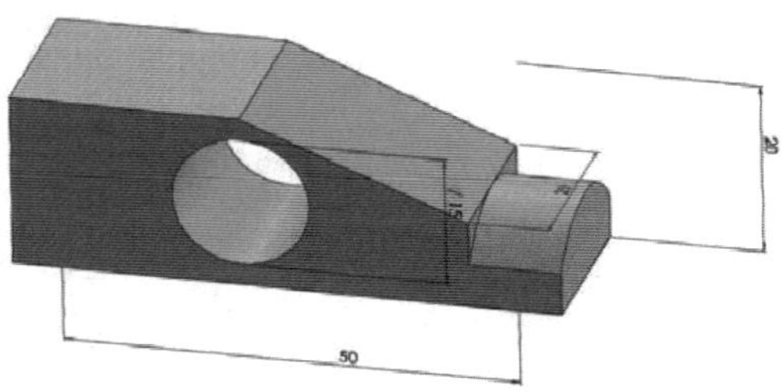

Fig. 2. CAD Model of Specimen

4 Smoothening of abs plastics

For enhancing the surface roughness of ABS plastics, a chemical post processing has been conducted [1, 2, 4, 5 & 9]. As it is well known that ABS plastics are soluble in acetone liquid. Therefore, in this method, Acetone in vaporized form is used to enhance the surface of the ABS specimen [10]. In order to know the effect of chemical process on the specimen, the specimen is so selected that it contains all the three surface, i.e., flat surface, inclined surface and curved surface [9 & 10]. First of all, an electrical hot plate with temperature control is taken. A tin container with a glass lid is kept on it. There are two slots on the lid of the container. One for digital thermometer and another for hanging the specimen inside the container. The container is heated to various temperatures such as 50, 60 and 70 degrees Celsius with the help of the temperature controller provided on the hot plate. Then 20-25 ml of acetone is poured inside the container. The acetone now becomes in a vaporous state. At this time, the specimen, tied with a thread is introduced inside the container for various time period like 10, 20 and 30 seconds. The samples are also exposed to the acetone vapour for one, two or three cycles. In one cycle, the specimen is exposed only once. In the two cycle method, the specimen is exposed for certain period of time at a particular temperature. Then it is kept outside for a period of 10 minutes to cool. Again it is exposed to the acetone vapour with the same constraint. Similarly, for three cycle experiment, the specimen is cooled in air for two times in between three exposures. The smoothening process is shown in the following figure (Fig. 3.)

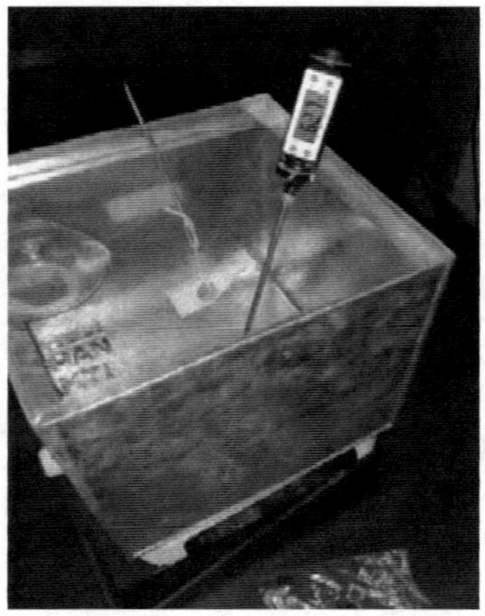

Fig. 3. Smoothening Process of ABS Specimen

5 Surface roughness photographs

The photos of surface texture [5, 6 & 9] of ABS-FDM parts were taken using metallurgical microscope OLYMPUS STM6 measuring microscope (Fig. 4.1) for curved surface before and after chemical treatment. The specimen was kept on the base of microscope and image was seen alongside the computer attached to the microscope by focusing the microscope metallurgical controller. When the relevant image was formed, snapshot of it was immediately captured on the attached computer (Fig 4.2).

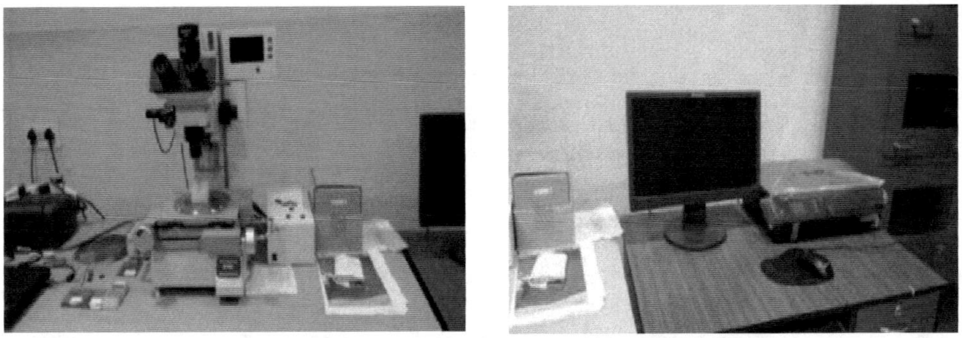

Fig. 4.1. Metallurgical microscope (OLYMPUS STM6)

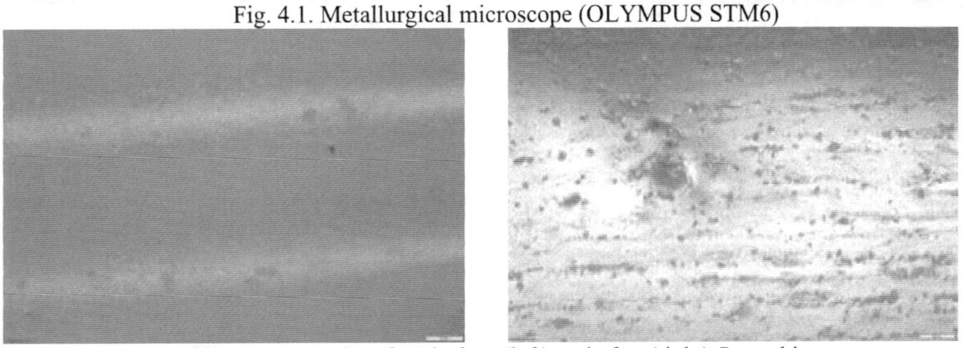

Fig. 4.2. Curved surface before (left) and after (right) Smoothing

The Fig. 4.2. shows surface textures curved surface before and after smoothing. In the before smoothing sections of curved surface the rasters are clearly visible and they are in oblique direction. Inside the raster of curved surface some marks or globules are spotted. The small spots are also visible in before smoothing section of curved surface. Those spots indicate that those areas are rough. In the after smoothing section small spots are seen which are uniformly scattered in straight line path which shows the surface roughness is reduced after chemical treatment by acetone and the surface has been smoothed. There is also a small hole like image is observed which indicates that particular area has been over smoothened.

6 Results

For better understanding of relationship between surface roughness and process parameters, S/N Plot regression equations and ANOVA table has been presented. The S/N plot shows maximum and minimum values of process parameters. The regression equation gives relationship of process parameters with surface roughness. The ANOVA table shows significant influence of parameters over surface roughness when significant level is less than 0.05. The coefficient of determination (R2), which indicates the percentage of variation explained by the terms in the model to the total variation in the response is 0.9818 for Surface roughness of Copper plated FDM Parts. It is to be noted from the table that lack of fit is not significant. Residual analysis has been carried out and found that residuals are normally distributed.

IOP Conf. Series: Materials Science and Engineering **653** (2019) 012043 doi:10.1088/1757-899X/653/1/012043

Table 2. Result Table

Sl. No.	FDM Parameters (Part Building)			Smoothing Parameters (Part Testing)			Surface Roughness of Curved Surface		
	Raster Angle	Raster width	Air Gap	Temp	Time	Cycle	Before Smoothening	After Smoothening	% Error
1	0	0.3556	0	50	10	1	3.207	2.992	6.70
2	0	0.3556	0	50	20	2	3.104	3.043	1.96
3	0	0.3556	0	50	30	3	2.335	2.147	8.05
4	0	0.5306	0.05	60	10	1	2.571	2.242	12.79
5	0	0.5306	0.05	60	20	2	2.922	2.557	12.19
6	0	0.5306	0.05	60	30	3	2.701	2.455	9.10
7	0	0.7306	0.1	70	10	1	1.927	1.662	13.7
8	0	0.7306	0.1	70	20	2	2.994	2.713	9.38
9	0	0.7306	0.1	70	30	3	2.475	2.360	4.64
10	45	0.3556	0.05	70	10	2	2.456	2.295	6.55
11	45	0.3556	0.05	70	20	3	2.485	2.264	8.89
12	45	0.3556	0.05	70	30	1	2.242	1.997	10.92
13	45	0.5306	0.1	50	10	2	2.263	2.109	6.80
14	45	0.5306	0.1	50	20	3	1.551	1.352	12.83
15	45	0.5306	0.1	50	30	1	2.183	2.041	6.50
16	45	0.7306	0	60	10	2	2.180	2.310	5.6
17	45	0.7306	0	60	20	3	2.111	1.835	13.07
18	45	0.7306	0	60	30	1	1.743	1.503	13.76
19	90	0.3556	0.1	60	10	3	2.231	2.154	3.45
20	90	0.3556	0.1	60	20	1	2.562	2.241	12.52
21	90	0.3556	0.1	60	30	2	1.666	1.455	12.66
22	90	0.5306	0	70	10	3	2.325	2.134	8.21
23	90	0.5306	0	70	20	1	1.712	1.489	13.02
24	90	0.5306	0	70	30	2	2.191	1.996	8.90
25	90	0.7306	0.05	50	10	3	1.650	1.488	9.81
26	90	0.7306	0.05	50	20	1	2.036	1.851	9.08
27	90	0.7306	0.05	50	30	2	1.776	1.617	8.95

ROUGHNESS REGRESSION EQUATION =-2.30567 + 1.87407A + 0.80375B + 0.165028C - 0.183444D + 0.773742E + 0.269687F - 0.559833A×B - 0.34616A×C - 0.0166429A×E + 0.00169048A×F + 0.181417B×C - 0.0483889B×E - 0.0349444B×F - 0.0164444C×E + 0.135944C×F - 0.0791667D×E - 0.233083E×F.

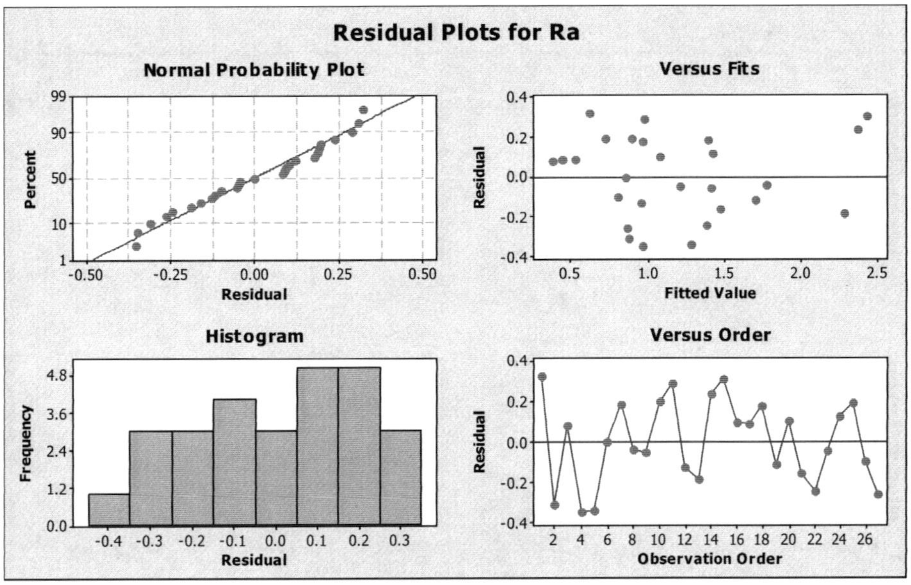

Fig. 5. Residual Plot for Surface roughness

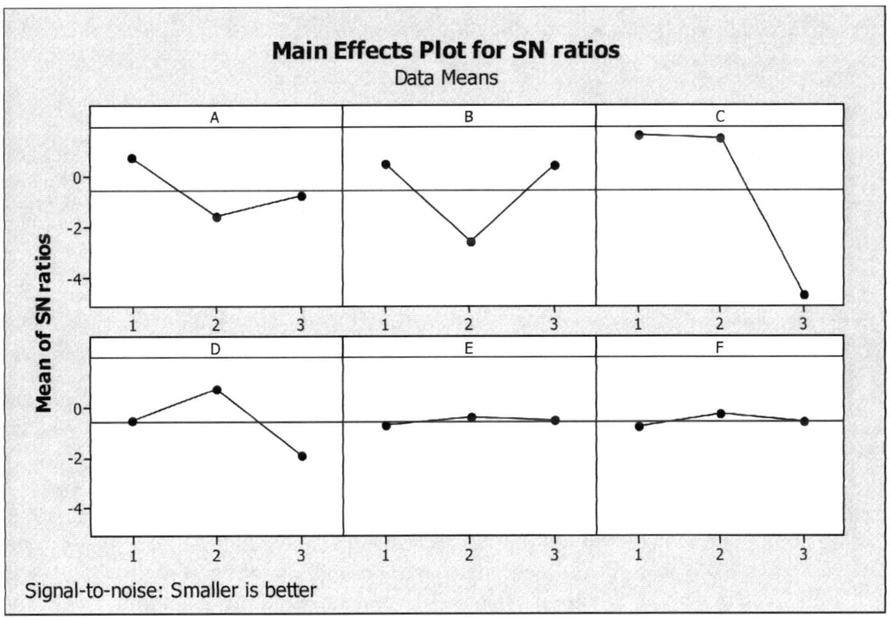

Fig. 6. Main Effect plot for S/N ratios

IOP Conf. Series: Materials Science and Engineering **653** (2019) 012043 doi:10.1088/1757-899X/653/1/012043

Table 3. ANOVA Table

Source Regression	DF	Seq SS	Adj SS	Adj MS	F	P
	17	3.82069	3.82069	0.224747	8.8951	0.001083
A	1	0.06055	0.56187	0.561872	22.2379	0.001096
B	1	0.05120	0.25445	0.254448	10.0706	0.011304
C	1	2.26136	0.00378	0.003780	0.1496	0.707913
D	1	0.22267	0.02423	0.024229	0.9590	0.353038
E	1	0.00381	0.13121	0.131211	5.1931	0.048655
F	1	0.00534	0.00884	0.008838	0.3498	0.568787
A×B	1	0.12649	0.52236	0.522356	20.6739	0.001392
A×C	1	0.49093	0.19972	0.199719	7.9045	0.020327
A×E	1	0.08943	0.00088	0.000881	0.0349	0.855988
A×F	1	0.15545	0.00001	0.000009	0.0004	0.985279
B×C	1	0.02326	0.09403	0.094034	3.7217	0.085786
B×E	1	0.00460	0.01686	0.016859	0.6672	0.435105
B×F	1	0.08217	0.00550	0.005495	0.2175	0.652046
C×E	1	0.00089	0.00195	0.001947	0.0771	0.787594
C×F	1	0.06225	0.08316	0.083164	3.2915	0.103039
D×E	1	0.02507	0.02507	0.025069	0.9922	0.345230
E×F	1	0.15522	0.15522	0.155222	6.1434	0.035072
Error	9	0.22740	0.22740	0.025266		
Total	26	4.04809				

Optimum parameter to reduce surface roughness of ABS-FDM parts is given below. This optimum parameter is obtained by firefly algorithm which is based on movement of fireflies in the atmosphere.

Table 4. Optimum Parameter Setting for Curved Surface

Raster Angle (A) (Degree)	Raster Width (B) (mm)	Air Gap (C) (mm)	Temperature (D) (^{0}C)	Time Period (E) (Sec)	Cycle No. (F) (Magnitude)	Surface Roughness (µm)
37	0.5102	0.02	53	18	1	5.74

7 Conclusion

It has been observed that surface roughness is largely influenced by the air-gap and less influenced by other process parameters. Comparison of surface roughness before and after chemical treatment was successfully carried out and it has been noted that surface roughness is reduced after chemical vapour smoothing and surface textures of part are enhanced. From Analysis of Variance (ANOVA), the significance as well as the influence of each process parameters have been checked. Regression equations has been generated considering all process parameters and with the performance characteristics. Since FDM is a process dependent on parameters so the performance characteristics are largely influenced by the selection of parameters. To understand the effect of individual parametric effect on process characteristics design of experiments were conducted based on Taguchi L 27 orthogonal array. In this research, vapour smoothing with acetone has been considered for post processing of ABS printed parts. The reason being, it is cheaper, quicker and gives better finish. Also, all the three smoothing parameters such as temperature, time and number of cycle have substantial influence on the surface roughness of ABS build parts. The optimal parameter settings have been calculated using these regression equations.

References

1. Espalin D, Medina F, Wicker R (2009) Vapor Smoothing, A Method for Improving FDM-Manufactured Part Surface Finish, Int. Rep. of the W.M. Keck Center for 3D Innovation, Univ. of Texas at El Paso, provided by Mrs. C. Gerling, Fortus Marketing Association, Stratasys Inc.
2. Pandey PM, Reddy NV, Sanjay. Dhande G (2003) Improvement of Surface Finish by Staircase Machining in Fused Deposition Modelling. Journal of Material Processing Technology 132:323–331.
3. Sood, A.K.; Ohdar, R.K.; Mahapatra, S.S. Parametric appraisal of fused deposition modelling process using the grey Taguchi method. Proceedings of the Institution of Mechanical Engineers Part B: Journal of Engineering Manufacture 2010, 224 (1), 135–145.
4. Galantucci, L.M.; Lavecchia, F.; Percoco, G. Experimental study aiming to enhance the surface finish of fused deposition modeled parts. CIRP Annals—ManufacturingTechnology2009, 58, 189–192.
5. Boschetto, A.; Giordano, V.; Veniali, F. Surface roughness prediction in fused deposition modelling by neural networks. The International Journal of Advanced Manufacturing Technology 2013, 67, 2727-2742.
6. A. Boschetto, L. Bottini, Roughness prediction in coupled operations of fused deposition modeling and barrel finishing, J. Mater. Process. Technol. 219 (2015) 181–192.
7. S. H.Masood (2014), "Advances in Fused Deposition Modeling" Comprehensive Materials Processing, Volume 10, pp. 69-91.
8. C.K.Chua, S.H.The and R.K.L.Gay, “Rapid Prototyping and Virtual Manufacturing in Product Design and Manufacturing,” International Journal of Advanced Manufacturing Technology, vol.15, 1999,pp. 597-603.
9. A. Garg, A. Bhattacharya, A. Batish, On surface finish and dimensional accuracy of FDM parts after cold vapor treatment, Mater. Manuf. Process. (2015).
10. C.C. Kuo, R.C. Mao, Development of a precision surface polishing system for parts fabricated by fused deposition modeling, Mater. Manuf. Process. (2015).
11. Kruth JP, Leu MC, Nakagawa T (1998) Progress in Additive Manufacturing and Rapid Prototyping. CIRP Annals - Manufacturing Technology 47(2):525–540.
12. T. Wohler, “Additive Manufacturing State of the Industry Annual Worldwide Progress Report,” *Wohler associates*, Colorado, USA, 2010.

Effect of cenosphere filler on the hardness and impact strength of wild cane fibre mat reinforced laminated composites

Pruthwiraj Sahu[1] and Sambit Kumar Parida[2]

[1] **KIIT (Deemed to be University), Bhubaneswar 751024, INDIA**

[2] **NIFFT, Ranchi 834003, INDIA**

[1]**psahumech.fme@kiit.ac.in**

Abstract. This study deals with the development of laminated composites consisting of epoxy resin and cellulosic wild cane fibre (Arundo Donax L.) woven mat (cross-ply), and the effect of cenosphere filler on the microhardness and impact strength. Cenosphere used as a filler material, produced during burning of coal in thermal power plants. The conventional hand lay-up practice is applied to prepare different composite samples with varying weight percentage of filler content. The wild cane fibre is treated with alkali (5% NaOH) to improve its surface properties. It is observed that, mechanical properties, such as microhardness and impact strength are significantly influenced by incorporation of this waste-based ceramic filler for treated fibre composite samples. The laminated composites developed from this fibre mats can have potential application in automobile interiors and lightly loaded structural applications.

Keywords: Cenosphere filler, hardness and impact strength, wild cane fibre mat, etc.

1. Introduction

The synthetic fibres such as glass, carbon and amide fibres are not sufficiently eco-friendly because of their non-biodegradability and disposal problem. The natural fibres such as pineapple, flax, abaca, sisal, bamboo, cane, coir etc. as reinforcement in composites is now becoming a popular area of research due to its great potentiality. Abundantly available wild cane fibre has higher specific strength, stiffness, low processing cost, non-corrosive, eco-friendly and bi-degradability and possesses very good tensile and flexure strength, because of the presence of high volume of longitudinal oriented cellulosic fibre (α-cellulose) in them.

The treatment of fibre by suitable chemical reagents like NaOH, tri-methoxysilane removes the oily and sticky materials from the natural fibres. Thus, the treatment of fibres can improve the physico-mechanical, triobological and thermal stability properties. Some of the authors reported the advantage of alkali treatment (5 % NaOH solution at 20^0 C for 30 minutes) of natural fibres such as sisal, bamboo, cane and coir which develops the adhesion characteristics of composites [1-4]. After treatment the tensile and flexural strength increases substantially. Further, impact strength and hardness of treated bamboo and coconut fibre composite improves, has been studied [5, 6]. The impact energy absorbed of giant bamboo epoxy composites increases with the amount of fibre content has been noted [7].

However, Biswas *et al.* [8] found that tensile strength decreases, flexural and impact strength increases linearly by the incorporation of ceramic nano fillers. Rao *et al.* [9] investigates that the average tensile strength and tensile modulus, and impact strength of bamboo fibre with fly ash filler reinforced hybrid composite is increased as the increase of volume fraction of fibre for certain value up to 0.5 and then decreased slowly. Jena *et al.* [10] studied the effect of cenosphere on the mechanical properties such as microhardness, tensile, flexural, interlaminar and impact strength and found that the increase of cenosphere increases the strength up to certain limit after which it is decreased on further addition. Dalbehera *et al.* [11] studied the mechanical properties of hybrid composites are significantly influenced by the addition of cenosphere up to 15 wt. % and Kushnoore *et al.* [12] reported impact strengths decreases with the increase of cenosphere as reinforcement and epoxy as matrix.

Moreover, tailor made properties can be obtained, depending on the application by suitable combination of the parameters such as fibre volume fraction, number of layers and the weight percentage of the filler. Here, cenosphere is used as filler material produced during burning of coal in thermal power plants. The composite laminates made from wild cane fibres and suitable filler cenosphere can find applications in the places like in automobile dashboard, corroding roofs, furniture, household appliances.

2. Materials and methods

Wild Cane (Spanish cane, river cane, giant reed or Arundo Donax L.) is cultivated in north eastern state of India. These fibre mats are woven from the thin strips taken out from the woody stem part of the seasoned cane plants. The property of this fibre is almost in the same range that of the bamboo fibre. So, the laminated composite made from wild cane fibre is thought to be potential candidate for the making laminates having structural usefulness.

The matrix material utilized in the fabrication is *epoxy resin* (Lapox-12). A comparatively lower viscous hardener K-6 is added with moderate viscous epoxy in 10:1 proportion by weight. The epoxy and hardener used here have a good melt flow index, thermal stability, longer working time and high cure shrinkage stability.

Cenosphere made largely of silica and alumina and filled with air or inert gas, typically produced as a by-product of thermal power plants during combustion of coals. They are produced through complicated chemical and physical transformation at temperature of 1500 to 1750 °C. The SEM image of cenosphere is shown in fig. 1(a). The detail specifications and properties of all the raw materials are given in Table 1.

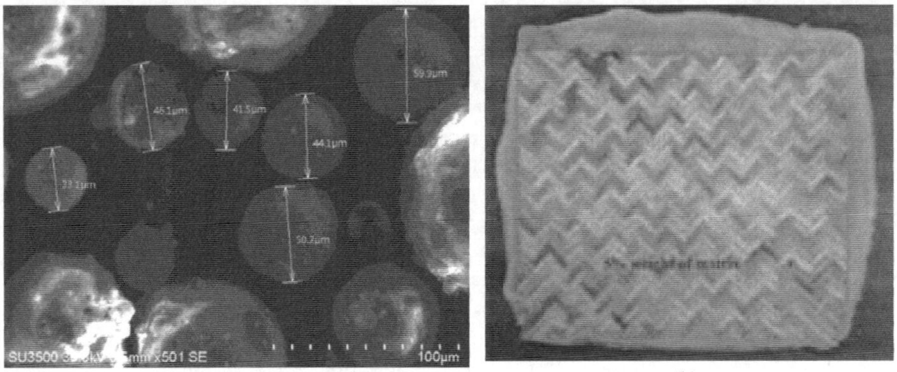

(a) (b)

Fig. 1. (a) SEM image of Cenosphere and (b) Fabricated composite sample

2.1. Fabrication procedure of laminate composite

The woven wild cane fibre mats cleaned and dried before treated with 5% NaOH solution for 6 hours maintained at 40^0C. The fibre mats are then dried for 12 hours in drying oven at 70^0C. The fillers in four different weight proportions (i.e., 2.5, 5, 10 and 15) and the hardener (K-6) are added (10:1 by weight) to the epoxy resin matrix (Lapox 12) and mixed thoroughly. The mixture is stirred till it becomes little warm. The composite laminate is fabricated by hand lay-up technique and was cured under a load of 50 kg for 48 hrs. Samples of suitable dimensions were cut in a diamond cutter for mechanical tests. Utmost care has been taken to maintain uniformity and homogeneity of the laminate composites. One of the sample has been fabricated is shown in fig. 1(b). The composite samples so made with varying weight percentage of cenosphere filler are categorised as shown in Table 2.

Table 1. Specifications and properties of raw materials used for making composites [11].

Materials	Grade	Characterstics
Epoxy	LAPOX L 12	Viscosity 9000-12,000 mPa-s, Density 1.1 g/cc
Hardner	K 6	Refractive index: 1.494-1.50 (at 25^0C)
Wild cane fibre	--	Weave type [$+45^0/-45^0$] , Density :0.95gm/cc Width: 3.7-4.0 mm, Thickness: 0.3-0.4mm, Specific gravity= 0.5-0.7
Cenosphere	CS 300	Density: 0.45-0.80 g/ml, size: 60-94 mm, Colour: Light grey, thermal conductivity: low Insoluble in water, water absorption: low Melting temp. 1300-1500 ^{0}C

Table 2. Designations of composites made from wild cane fibre with different wt. % of filler.

Group	Designation	Composite with varying wt.% of cenosphere
A (Untreated fibre)	A0	0
	A1	2.5
	A2	5
	A3	10
	A4	15
B (Treated fibre)	B0	0
	B1	2.5
	B2	5
	B3	10
	B4	15

3. Experiments

The different mechanical tests such as SEM, micro-hardness and impact test of developed laminated composite samples have been carried out as per the ASTM standards.

(a) For Scanning Electron Microscopy (JSM-6390LV) the samples are fixed on car-bon stubs. Before the photomicrographs are taken, a thin film of gold is plated on them by vacuum evaporator. Then the surfaces of the composite samples are scanned directly under the scanning electron microscopy.

(b) For Vicker's micro-hardness test the composite samples are tested in Zwick Roell (ZHV 30) hardness tester as per ASTM E92 test procedure. Pyramid type diamond indenter is used with load of 200 gram force (gf). The hardness values are measured at various locations randomly on the surface of the composite laminate.

(c) For Izod impact strength (Impact meter 6545, Ceast, Italy), a sample has to be prepared as per ASTM D 256 standard of dimension: length = 63.5 mm, width = 12.7 mm and thickness = 3 mm. A V-notch is made at the centre of the sample (notch depth: 2.54 mm and notch angle: 450). The hemispherical head impacts the test sample at a velocity of 1.06 m/s with energy of 15.26 J. The impact energy of various sample are taken directly from the dial indicator.

4. Results and discussion

The morphological study of the composite reinforced with untreated and treated fibre surfaces has been carried out using an SEM. From the SEM micrograph it has been seen that the untreated fibre composite surfaces are appear to be rough compared to treated one as shown in Fig. 2(a, b). This may be due to the occurrence of lignin, wax and other impurities which can be removed by the alkali treatment.

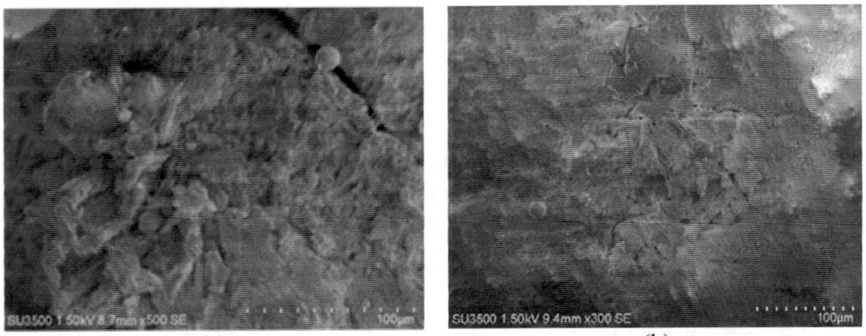

(a) (b)

Fig. 2. SEM micrograph of (a) untreated and (b) treated composites with filler.

The images are captured by 20X magnification metallurgical microscope attached to the machine. The hardness is measured at various locations randomly on the composite surface. The average values of hardness for different samples (both untreated and treated fibre composite) for different wt. % of filler were presented in Fig. 3(a). From the graph, it has been seen that for both untreated and treated fibre composite samples, the hardness value is slightly differs by 1 HV. With the addition of cenosphere the bulk hardness of the sample substantially increases. The increasing trend in hardness values in all the samples is attributed due to inclusion of uniformly distributed harder ceramic particles in the continuous resin matrix as per the rules of mixture.

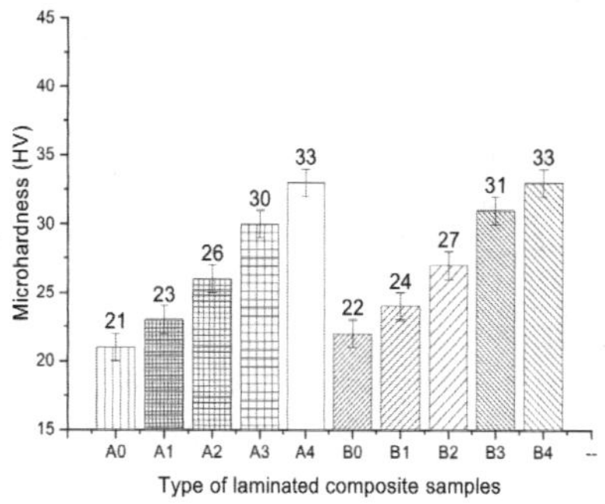

(a)

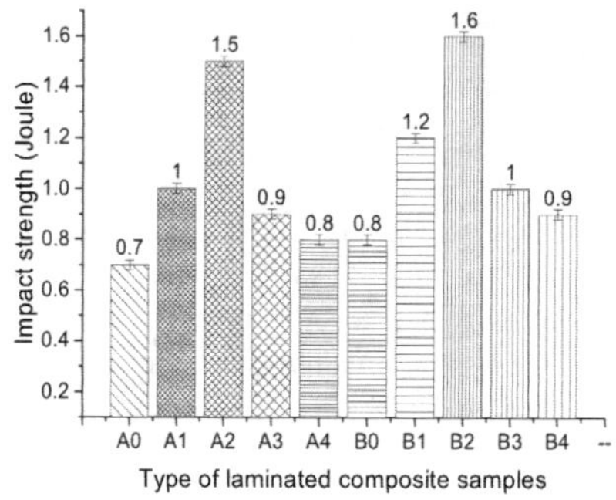

(b)

Fig. 3. (a) Micro-hardness and (b) Impact strength of different composite samples

Fig. 3(b) shows the impact strength values of the composite samples made from both untreated and treated fibre mats. At least five tests for each category (described in Table 2) of the samples are carried out. It can be observed from Fig. 3(b) that the impact strength increases by the addition of filler content up to a certain limit (i.e., up to 5 wt. % of filler), and decreases thereafter. This may be due to the presence of higher volume fractions of the voids and problem related to non-uniform dispersion of filler in the wild cane fibre based composites. The composite samples made with treated fibre mats have better impact strength than the corresponding samples made with untreated one. More weighting area acts as better stress transfer media and hence possess better energy absorption capacity due to impact.

5. Conclusion

The wild cane fibre reinforced epoxy composites filled with cenosphere were successfully fabricated using hand lay-up technique. The mechanical properties i.e., micro-hardness and impact strength of the composite laminates were evaluated experimentally by following the appropriate ASTM standards. The experimental observation reveals that a limited inclusion of the cenosphere as filler in the wild cane fibre based composite improves the micro-hardness and impact strength. Thus, the composite laminates can be customized depending on the structural requirement by varying weight fractions of treated wild cane fibre mat, cenosphere, and number of layers. However, the impact strength of laminates of treated fibres is higher. The micro-hardness value gradually increases with an increasing filler percentages and has found a maximum value of 33HV for 15 wt. % of filer. Further, the use of cenosphere reduces the load on the environment which otherwise may has been disposed to the environment causing pollution. Finally, the research inferred that cenosphere filled composite laminate can be adopted for the low strength type structural material, can find the application in industries like furniture, automotive, sports, recreation, farming etc. .

References

[1] Bisanda, E.T.N: The effect of alkali treatment on the adhesion characteristics of sisal fibres. Applied Composite Materials 7, 331–339 (2000).

[2] Manalo A.C, Wani, E., Zukarnain, N. A, Karunasena, W: Effect of alkali treatment and elevated temp. on the mechanical properties of bamboo fiber-polyester composite, Composites :part B 80, 73-83(2015).

[3] Chikouche, M.D.L, Merrouche, A, Azizi, A, Rokbi, M and Walter, S: Influence of alkali treatment on the mechanical properties of new cane fibre/polyester composites. Journal of Reinforced Plastics and Composites 34(16) 1329–1339 (2015).

[4] Yan, L, Chouw, N, Huang, L, Kasal, B: Effect of alkali treatment on microstructure and mechanical properties of coir fibres, coir fibre reinforced-polymer composites and reinforced-cementitious composites. Construction and Building Materials 112, 168–182 (2016).

[5] Nuthonga, W, Uawongsuwanb, P, Pivsa-Arta, W, Hamadab, H: Impact property of flexible epoxy treated natural fiber reinforced PLA C composites. Energy Procedia. 34, 839 – 847 (2013).

[6] Srinivasa, C.V, Bharath, K.N: Impact and hardness properties of areca fiber-epoxy reinforced composites, J. Mater. Environ. Sci. 2 (4), 351-356 (2011).

[7] Gloria, G.O, Margem, F.M, Vieira F, Moraes, Y.M, Ribeiro, C.G.D, Gomes, L.E.R, Monteiro, S.N: Charpy impact tests of epoxy composites reinfoced with bamboo fibers of the specimen dendrocalmus giganteus. Congresso Brasileiro de Engenharia e Ciência dos Materiais. 9(13), (2014).

[8] Biswas S, Satapathy A and Patnaik A: Effect of ceramic fillers on mechanical properties of bamboo fibre rein-forced epoxy composites: a comparative study. Adv Mater Res. 123-125, 1031–1034(2010).

[9] Rao, V, Venkatarao, K, Lakshmikanth: Mechanical Properties of Bamboo Fibre filled with Fly Ash filler Reinforced Hybrid Composites. International Journal of Engineering Research & Technology (IJERT) 3(9), (2014).

[10] Jena, H, Pandit, M.K, and Pradhan, A.K: Effect of cenosphere on mechanical properties of bamboo–epoxy composites. Journal of Reinforced Plastics and Composites 32(11) 794–801 (2013).

[11] Dalbehera, S and Acharya, S.K: Effect of cenosphere addition on the mechanical properties of jute-glass fiber hybrid epoxy composites. J. Industrial textiles 46(1), 177–188 (2016).

[12] Kushnoore, S, Varma, G.A, Akhil, K, Jathin, Ch, Reddy, M.M: Experimental investigation on mechanical behaviour of cenosphere reinforced epoxy composites. International Journal of Mechanical Engineering and Technology (IJMET), 9(4), 73–81(2018).

Tribo-performance of Thixoformed A356-5TiB_2 *in-situ* Composites

S. Deepak Kumar[1,*], Shailesh Dewangan[1], S.K. Jha[1], A. Mandal[2]
[1]Department of Production Engineering,
Birla Institute of Technology Mesra, Ranchi - 835215, Jharkhand, India
[2]School of Minerals, Metallurgical and Materials Engineering,
Indian Institute of Technology Bhubaneswar, Bhubaneswar - 751007, Odisha, India

[*]E-mail: dks10@iitbbs.ac.in

Abstract. In the current work, thixoformedA356-5wt%TiB_2in-situ composites were subjected to dry sliding wear. Further, the wear surface topography and wear debris was analyzed. The results show that wear rate is significantly influenced by redistribution of TiB_2 particles in the matrix, refinement of fine globular α-Al grains and modification of eutectic Si. As expected, the thixoformed composites show better wear resistance than the alloy. Moreover TiB_2 particles lower the coefficient of friction when reinforced in A356 alloy, in both gravity cast and thixoformed conditions. Also, it is interesting to note that there exist a linear relationship between wear rate and hardness of composites, irrespective of the processing history. The wear rate in the thixoformed composite is reduced by 48% as compared to gravity cast alloy. The improvement in wear resistance in the present thixoformed composite is not only due to the presence of fine TiB_2 particles but also due to refinement of α-Al and eutectic Si caused due to the plastic deformation during thixoforming. In addition, high hardness and strength of thixoformed specimens compared with gravity cast parts has resulted in lower coefficient of friction in thixoformed A356-5TiB_2 in-situ composite.

Keywords: Dry sliding wear; Thixoforming; A356; TiB_2; in-situ composites;

1. Introduction

Aluminum metal matrix composites (AMMCs) are considered as a potential material for automotive, aerospace and other engineering applications [1]. Thixoforming of commercial alloys and MMCs has received significant attention in recent years [2-4]. Tribological properties are one of the important considerations for application of Al based alloys and composites. The tribological behaviour determines the suitability of a material for engineering applications [5].

The wear properties of Al-Si-Mg alloys have been studied by various researchers. Das et al. [6] investigated the wear properties of rheocast A356 alloy. Birol et al. [7] investigated the same for thixoformed Al-Si-Cu-Fe alloys. They reported that the formation of globular α-Al grains in rheocast and thixoformed alloys resulted in improvement of wear resistance compared to their conventional counterparts. The influence of the thixoforming process on dry sliding wear behaviour of A319 alloy was investigated by Alhawari et al. [8]. The wear rate and coefficient of friction of thixoformed samples were found to be lower than that of as-cast alloy. They ascribed it to the increased hardness, finer grain size and uniform distribution of silicon particles in comparison to the cast alloy.

Kumar et al. [9] have extensively investigated the influence of in-situ formed TiB_2 particles on wear behavior of Al-4Cu and Al-Si alloys. They stated that hard in-situ TiB_2 particles resist plastic deformation in subsurface region and thereby improve the wear resistance. In-situ TiB_2 particles are attractive for use in wear applications by virtue of its higher hardness compared to SiC particles. This is due to the fact that TiB_2 is stiff, hard and have cleaner interfaces aided by strong particle-matrix

IOP Conf. Series: Materials Science and Engineering **653** (2019) 012045 doi:10.1088/1757-899X/653/1/012045

bonding [10-12]. Accordingly, Al-based MMCs reinforced with TiB_2 are expected to exhibit better wear performance than those reinforced with SiC, Al_2O_3 or any other ex-situ particles. Based on the previous literature on wear properties of semi-solid metal processing (SSMP) parts, it appears that there is definitely an improvement in the wear properties of thixoformed products. Thus, there is no information is available on the tribological properties of thixoformed AMMCs reinforced with in-situ particles. Therefore, the present investigation was undertaken to study the dry sliding wear behavior of A356-5wt%TiB_2 in-situ composite subjected to thixoforming.

2. Materials and Methods

2.1. Synthesis of in-situ composites

In the present work, A356-5TiB_2 in-situ composites were fabricated by mixed salt route. At a reaction temperature of 800°C the halide salts, K_2TiF6 and KBF_4, undergo an exothermic reaction with molten Al-7Si alloy and form titanium-diboride (TiB_2) particles in the melt. Further, details of composite synthesis can be found elsewhere [11-12].

2.2. Thixoforging of composites

Thixoforging was carried using a 80-ton hydraulic press (Flowmech Engineers Private limited, India, Model No.-HP 50) on both alloy and composite specimens. The load applied was 150 Kg/cm^2 at a ram speed 20 mm/sec. The forming dies were preheated up to 200 °C to prevent the forged specimens from cooling. The heating process was monitored using K-type thermocouples to control the temperature fluctuation in the range of ± 2 °C. The thixoforging process was described in detail earlier [12]. The gravity cast alloys and composite specimens have been designated as Gravity-cast alloy (GCA) and Gravity-cast composite (GCC) respectively, while the thixoforged specimens were designated as thixoforged alloy (TFA) and thixoforged composite (TFC) respectively.

2.3. Dry sliding wear testing

Dry sliding wear tests were carried out on the gravity cast and thixoformed alloy and composites at ambient temperature, using a pin-on-disc type wear-testing machine (DUCOM, India) following ASTM G99-05 standards. For the purpose of wear testing, cuboidal specimens having dimensions of 6 x 6 x 9 mm were cut from gravity cast and thixoformed alloys and composites and attached to cylindrical brass bar of 20 mm height to be used as pin. These specimens were then subjected to sliding wear tests under normal loads of 20 and 40 N against a disc of hardened chromium steel (Rc 64). Each test was carried out for a sliding distance of 1800 m in contact with the disc rotating at 245 rpm. The process of sliding was carried out on a track radius of 40 mm such that the linear circumferential speed was 1 m/s. A computer aided data acquisition system was used to record both height loss and frictional force. For each condition, at least three samples were tested and the average height loss was calculated. Wear surfaces and debris of selected specimens were characterized using a Field Emission Scanning Electron Microscope (FESEM).

3. Results and discussion

3.1. Worn surface morphology

The micrographs showing the worn surfaces at normal loads of 40 N are presented in figure 1. The variation of surface roughness values is shown in figure 2. It was observed that the worn surfaces of gravity cast and thixoformed alloys and composites at a low load of 20 N are less damaged than at high loads of 40 N. This clearly shows that the volume loss increases as the load increases in both alloys and composites, as expected. It is noted that the extent of wear in thixoformed samples is lower than that of gravity cast alloy and composites. Small craters, adhesive pits and abrasive marks were observed in the alloy and composites. However, in comparison to the gravity cast samples, the thixoformed samples under the same condition have fewer grooves and smaller craters. At a high load of 40 N the worn surfaces of both alloys and composites show adhesive marks which are indicative of material flow. The worn surface of thixoformed A356-5wt%TiB_2 composite (figure 1c) reveals narrow groves and almost flat surface. The average surface roughness (R_a) of the worn surface of TFC is 2.96 µm, which is about 35% lower than that of gravity cast alloy. This is due to the presence of TiB_2 particles which restrict the flow of the material under the applied load. In addition, adhesion and ploughing are predominant at lower loads and delamination is predominant at higher loads. Therefore, the wear mechanism of the thixoformed A356-5TiB_2 at high load is dominated by delamination of mechanical mixed layer.

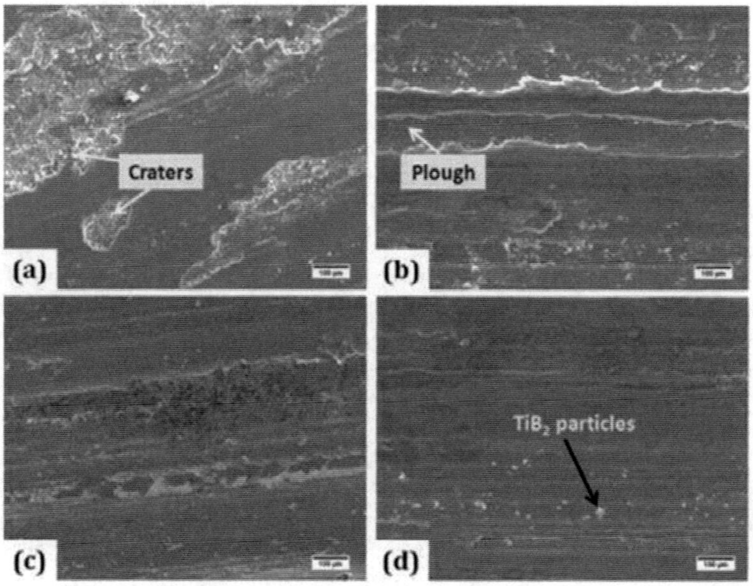

Figure 1. FESEM of worn surfaces of alloy and composites at applied load of 40 N (a)GCA (b) GCC (c) TFA (d) TFC

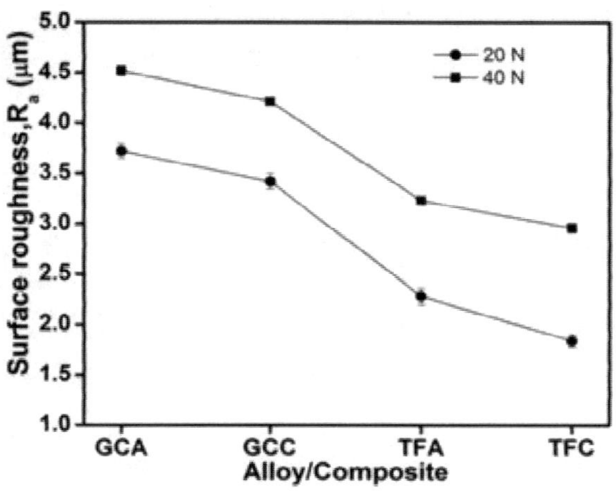

Figure 2. Variation of surface roughness

IOP Conf. Series: Materials Science and Engineering **653** (2019) 012045 doi:10.1088/1757-899X/653/1/012045

3.2. Wear rates

The wear rates of alloy and composite specimens as a function of normal load are shown using bar charts in figure 3. The wear rate of alloy and composites in both gravity cast and thixoformed conditions increases with increasing normal load from 20 N to 40 N. It is interesting to note that the extent of wear is lowered in thixoformed alloy and composites compared to gravity cast specimens. On the other hand, the wear rates are further lowered on addition of TiB_2 particles in both gravity cast and thixoformed conditions. This may be attributed to the presence of secondary phases such as Si and TiB_2 which restrict the flow of metal during sliding. The wear rate of the gravity cast alloy (GCA) is reduced by about 13% as compared to gravity cast composite (GCC) which is further reduced to about 48% in thixoformed composites (TFC). Thus, the improvement in the wear resistance in the present thixoformed in-situ composite (TFC) is not only due to the presence of fine TiB_2 particles but also due to refinement in grain size of α-Al and eutectic Si caused due to the plastic deformation during thixoforming.

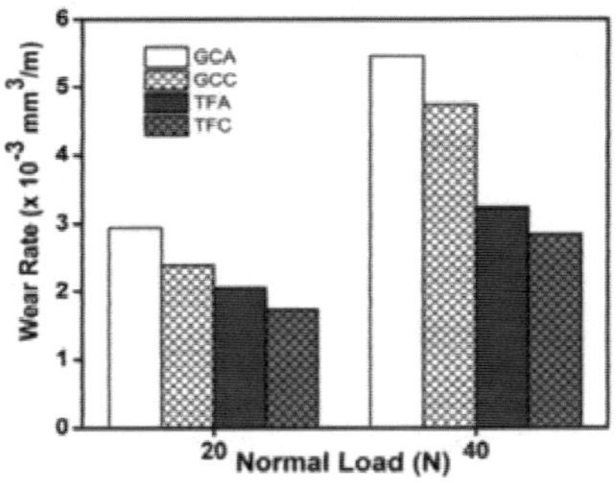

Figure 3. Wear rates of the alloys and composites

3.3. Coefficient of Friction

The bar charts depicting the Coefficient of friction (COF) for loads of 20 N and 40 N applied during steady state wear of alloy and composites in both gravity cast and thixoformed conditions are presented in figure 4. It is noted that the coefficient of friction is very high at lower loads for both alloy and composites. However, thixoformed alloy and composites showed a lower COF value compared to gravity cast specimens. Further, it is noted that with addition of TiB_2 particles to A356 alloy, the COF decreases irrespective of applied load in both gravity cast and thixoformed conditions. The average COF at 20 N normal load for the thixoformed composite (TFC) is about 0.24 which is significantly lower than that of gravity cast alloy (GCA) which is about 0.49.The better wear performance of thixoformed alloy and composites compared to gravity cast parts can be attributed to favorable microstructural features, relatively fine α-Al globules and a uniform dispersion of fine primary Si particles. In addition, high hardness and strength of the thixoformed alloy and composites compared with gravity-cast parts may reduce the occurrence of fracture and deformation of surface asperities and consequently real contact area may reduce. Hence, the above may be attributed to low friction coefficient in sliding of thixoformed alloy and composites compared to gravity-cast parts.

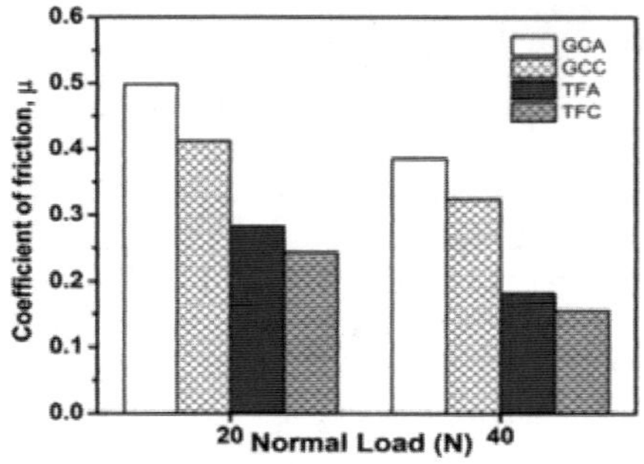

Figure 4. Coefficient of friction of alloys and composites

4. Conclusions

The detailed study on dry sliding wear behavior of thixoformed A356-5wt%TiB_2 in-situ composite led to following important conclusions.

- Thixoforming of alloy and composites led to significant increase in wear resistance. The improvement was attributed to combination of factors such as redistribution of fine TiB_2 particles, refinement of α-Al and eutectic Si due to plastic deformation during thixoforming.
- The thixoformed alloy and composites showed a lower coefficient of friction than that of gravity cast specimens.
- The wear rate in the thixoformed composite is reduced by 48% as compared to the gravity cast alloy.
- The average coefficient of friction at 20 N normal load for the thixoformed composite is about 0.24 which is about ~ 51% lower than that of gravity cast alloy.
- The improvement in wear resistance in present thixoformed in-situ composite is not only due to the presence of fine TiB_2 particles but also due to refinement of α-Al and eutectic Si caused due to the plastic deformation during thixoforming.

Acknowledgment

The authors gratefully acknowledge the financial support received from Birla Institute of Technology Mesra Ranchi under the SEED project grant for performing the Thixoforming and Tribology experiments.

References

[1] Surappa M K 2003 *Sadhana.* **28** 319
[2] Fan Z 2002 *Int. Mater. Rev.* **47** 49
[3] Fadavi Boostani A and Tahamtan S 2010 *Mater. Des.* **31** 3769
[4] Dodangeh A, Kazeminezhad M and Aashuri H 2012 *Mater. Sci. & Eng. A* **558** 371
[5] Kori S A and Chandrashekharaiah T M 2007 *Wear* **263** 745
[6] Das P, Show B K, Rathore A and Samanta S K 2015 *Trib. Trans.* **58** 1054
[7] Birol Y and Birol F 2008 *Int. J Mater. Form.* **1** 981
[8] Alhawari K S, Omar M Z, Ghazali M J, Salleh M S and Mohammed M N 2015 *Mater. Des.* **76** 169
[9] Kumar S, Sarma V S and Murty B S 2007 *Mater. Sci. & Eng. A* **465** 160
[10] Mandal A, Murty B S and Chakraborty M 2009 *Wear* **266** 865
[11] Deepak Kumar S, Pandu Ranga V, Mantry S, Mandal A and Chakraborty M 2014 *Proc. Mater. Sci.* **5** 232
[12] Deepak Kumar S, Mandal A and Chakraborty M 2015 *Mater. Sci. & Eng. A* **636** 254

IOP Conf. Series: Materials Science and Engineering **653** (2019) 012046 doi:10.1088/1757-899X/653/1/012046

Investigations on the effect of electrode gap variation over pulse-electrodeposition profile

Hrudaya Jyoti Biswal[1], Pandu R Vundavilli[1], Ankur Gupta[2,*]

[1]School of Mechanical Sciences, Indian Institute of Technology Bhubaneswar, India
[2]Department of Mechanical Engineering, Indian Institute of Technology Jodhpur, India
*Corresponding Author: hjb10@iitbbs.ac.in, ankurgupta@iitj.ac.in

Abstract. The influence of gap between the cathode and anode over electrodeposition profile was investigated using both the experimental and simulation techniques. Experiments were performed through deposition of copper on stainless steel substrates while varying the electrode distance with a constant applied voltage. Pulse current waveform with 70% duty cycle was applied for better uniform deposition. Complementary 3D model was developed to simulate the effect of electrode distance on current density distribution and thickness of deposition with COMSOL Multiphysics 5.3a. For the chosen experimental conditions, an optimum inter-electrode distance of 60 mm was identified for maximum deposition and good surface morphology. In spite of the application of pulsed current waveform, some amount of non-uniformity was encountered due to edge effect. Inter-electrode gap was demonstrated to be an important factor influencing the deposition profile by correlating experiments with simulation.

1. Introduction

Electroplating, a bottom-up approach in the field of manufacturing has enormous potential to be utilized for micro-fabrication. In many cases electroplating stands as the only feasible technique for surface property improvement in a comparatively cost-effective manner. It is also a suitable process for the plating of various functional composites in the micro and the nanoscale. The coatings not only increase the aesthetic values, they also enhance properties like high-temperature endurance, impact strength, resistance against friction and corrosion. Free-standing rigid components are also produced by 'Electroforming' which acts on the same principle as that electrodeposition. Both electroplating and electroforming have been extensively being used in the fabrication of fuel cells and battery electrodes **[1][2]**. In the case of electroplating, the current passed causes the dissolution of metal ions from the anode at an appropriate pH of the electrolyte and migrate towards the cathode. The type of current waveform applied influences the coating properties along with other process parameters like current density, pH and conductivity of the electrolyte, type of additive used. Pulse electrodeposition (PED) is a distinct technique for producing nanostructured coatings on metallic substrates with the use of pulse current waveform. Reduction in porosity, low level of inclusions, and a higher rate of deposition are the many advantages associated with PED as compared to conventional direct current (DC) electro-deposition process. There is much more flexibility involved in varying three basic parameters such as pulse peak current density, pulse frequency, on time and off time in pulse electro-deposition resulting in unique composition and microstructure of the coating being deposited. In the PED method, the current at TON (ON time) has a certain value, but at TOFF (OFF time) it is zero. In DC electroplating, a negatively charged layer of defined thickness is created around the cathode which obstructs the ions from reaching the mandrel. But in PED when the output is periodically switched off the layer discharges which facilitates the passage of ions onto the parts. In addition to that because of the pulse waveform more evenly distributed ions are available in the solution for deposition **[3]**.

Pulse current has been found to facilitate coatings of copper with better mechanical and electrical properties **[4]**. Rosa et al. **[5]** studied the effect of current density, electrolyte charge density, electrolyte concentration and the stirring of electrolyte on the deposition of copper upon titanium wire. Effect of pulse waveform parameters like frequency, duty cycle were studied for material depositions other than copper also **[6]**. But very limited research was found regarding the effect of distance between the electrodes on the process of electroplating. Some investigations were carried out for electrocoagulation which is the reversal of the electrodeposition process and varied responses were found. In one study the removal efficiency was increased with decreasing electrode distance **[7]** while in another an optimum distance was noticed corresponding to maximum removal **[8]**. Hence this paper aims to evaluate the effect of distance between the electrode on the plating thickness and properties both by experiment and simulation. Firstly experiments are done to study the impact of distance between the electrodes on the deposited thickness along with the thickness uniformity. The complexity of current distribution when the electrode distances change and the non-uniform potential field distribution can be studied by finite element analysis of COMSOL Multiphysics 5.3a software.

2. Modelling and Simulation

The physics of current distribution with varying electrode distance was chosen to model with COMSOL Multiphysics 5.3a software using the electrochemical principle of deposition. Based on the experimental setup, a 3D model was constructed consisting of model domains like anode, cathode, electrolyte cell and electrolyte. Charge balance along with Butler-Volmer equation was taken into consideration to account for the current distribution and electrochemical kinetics. The model was built taking into account the following assumptions:

1. Both the electrodes are homogeneous and isotropic.
2. The concentration gradient in the electrolyte is negligible.
3. The double-layer formation on the cathode and subsequent change in potential has not been taken into account.
4. The applied voltage for electroforming is very small (in the range of 1 V for the experiment), hence the species transport due to electrical migration can be neglected.
5. 100 % current efficiency is assumed for the experimental conditions.
6. All the chemical reactions involving all the species in the electrolyte are complicated. Hence to simplify the equation only reduction reaction of the depositing species is considered.

3. Experimental Study

Stainless steel substrates of size 47 mm × 11 mm × 3 mm were cut into sizes, ground with the abrasive belt grinder and subsequently polished with emery papers of different grades. The plates were cleaned with the acetone and ethanol before being used as the cathode. The plates were dipped up to 33 mm inside the electrolytic solution for deposition to take place. The anode was taken as copper plates of dimension 41 mm × 41 mm. The electrolytic solution was prepared using 220g/L of Cupric sulfate pentahydrate (CDH, 99% purity), 60 ml/L of Sulphuric acid and 50 ppm of Hydrochloric acid. The electrolytic ion concentration was found to be 880 mol/m^3. The solution was maintained at a temperature of 400 C with stirring action provided by a magnetic stirrer. The experimental parameters were decided based on the experiments carried out before. Optimum values were chosen as constant parameters with the distance between the electrodes as the variable parameter. The anode and cathode were placed face to face and pulse current waveform was applied with a frequency of 500 Hz and a duty cycle of 70 % with a cycle time of 2 ms. The initial distance between the electrodes was chosen to be 35 mm keeping in view the smooth flow of electrolyte through space and it was varied up to 105 mm. A constant time of 3 hours was maintained for all the experiments.

4. Results and Discussions

4.1 Simulation results

A 3D model was developed for simulating the copper deposition process with the application of pulse current waveform. The values of parameters like frequency, duty cycle for the pulse waveform were chosen based on some preliminary experiments conducted earlier. The information regarding current density distribution both during forward and reverse cycle and thickness profile of copper deposition on the stainless steel substrate has been obtained through simulations and presented in the figures below.

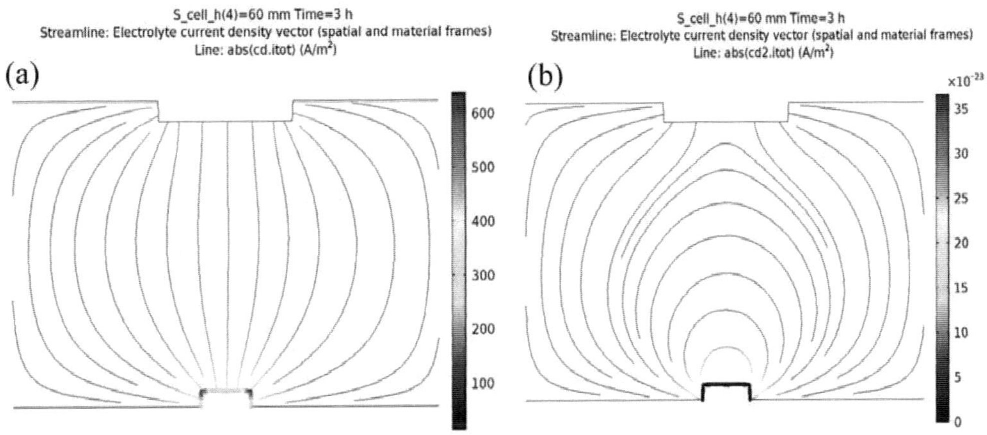

Figure 1: Current density distribution (a) Forward cycle (b) Reverse cycle.

The copper deposition thickness distributes non-uniformly over the substrate decreasing from the outside towards the center of the substrate. This is due to the edge effect which is caused by the uneven distribution of current density. The intensity of the potential field is more at the edges as compared to the middle of the substrate due to a path of lower resistance as shown by current density plots in **Figure 1**. **Figure 2** depicts the thickness profile across the stainless steel substrate clearly demonstrating higher deposition at the edges. Also, the simulation results show the deposition thickness to be higher at the front side (exposed towards the anode) than the rear side due to the shelter effect. The distribution of current is less uniform at the rear due to the shelter effect caused by the front side. The maximum deposition from the simulation result was obtained at an electrode gap of 60 mm which gradually decreased up to the gap of 105 mm. The simulation results harmonized with the experimental findings with a maximum thickness at the optimum distance of 60 mm. Hence the simulation results show that electrode distance has an important effect on the current density which in turn influences the deposition thickness.

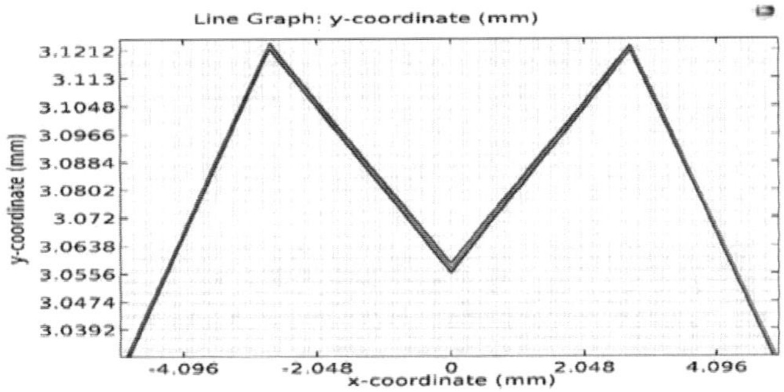

Figure 2: Deposition profile along the cathode surface.

IOP Conf. Series: Materials Science and Engineering **653** (2019) 012046 doi:10.1088/1757-899X/653/1/012046

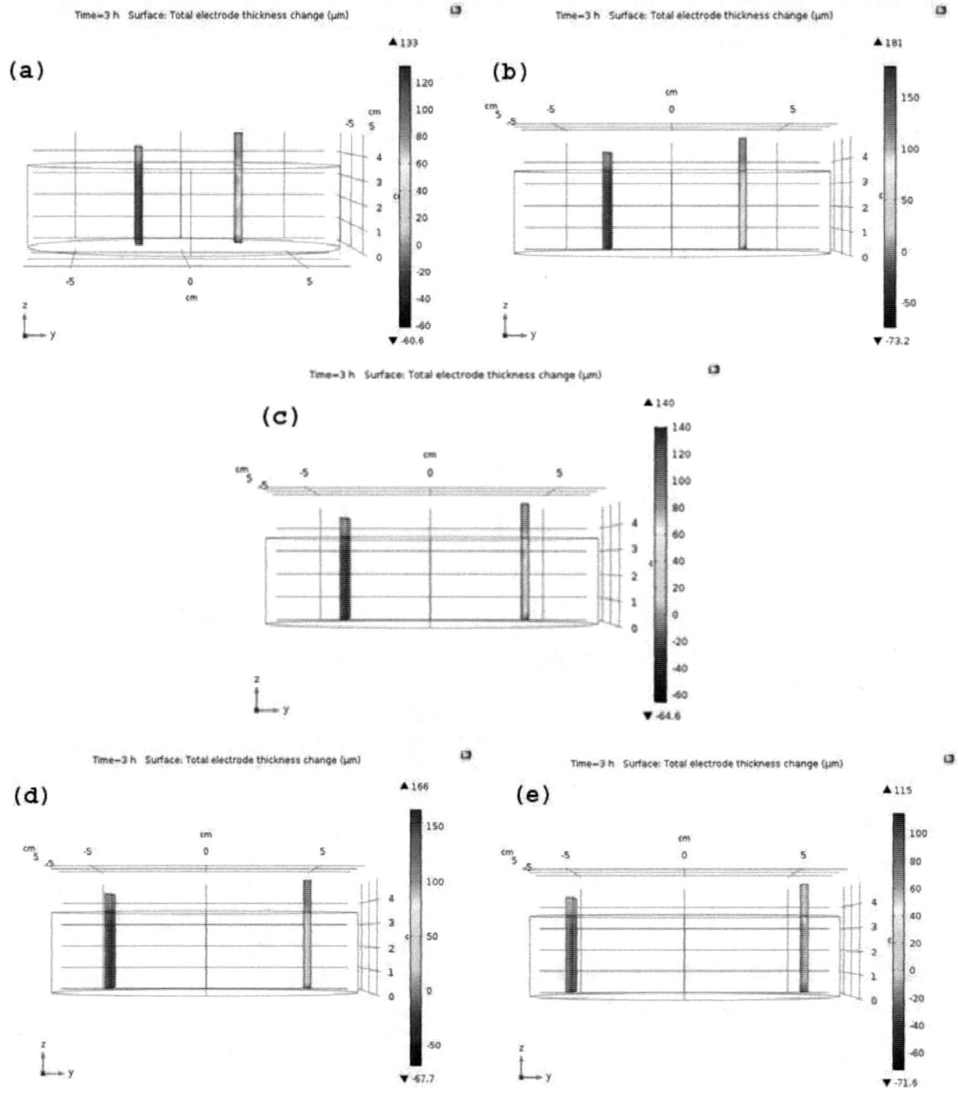

Figure 3: Thickness variations with electrode distance of (a) 45 mm (b) 60 mm (c) 75 mm (d) 90 mm (e) 105 mm.

4.2 Experimental Results

Experiments were conducted according to the model developed and results were matched with simulation results obtained. The copper deposition efficiency ascended at first with an increase in electrode distance and reached an optimum at an inter-electrode gap of 60 mm. Thereafter, the thickness of deposition decreased with a further increase in the gap. Short electrode distances inhibited the diffusion of ions due to restriction to the free flow of electrolytes. A maximum thickness deposition of 115 μm was obtained when the inter-electrode gap was 60 mm (**Figure 5**). On the other hand, the electrode reaction rate decreased with an increase in the gap beyond 60 mm due to the increase in resistance. Also, it can be noticed in **Figure 4** that the quality of deposition is superior at the gap of 60 mm.

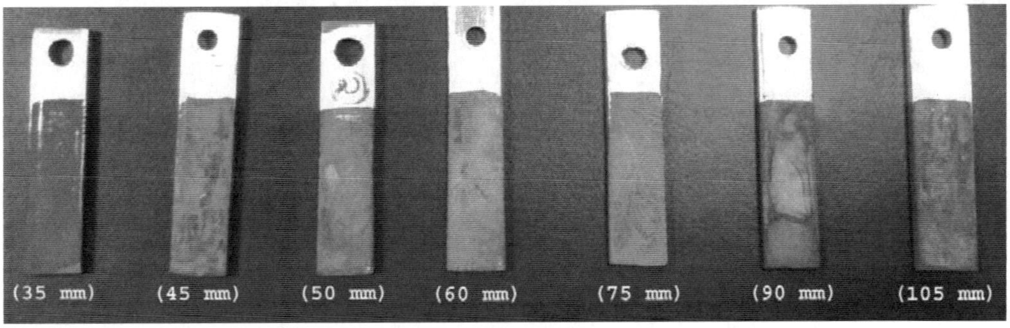

Figure 4: Copper deposited stainless steel substrates with varying distance between electrodes. Value inside bracket shows the electrode gap maintained.

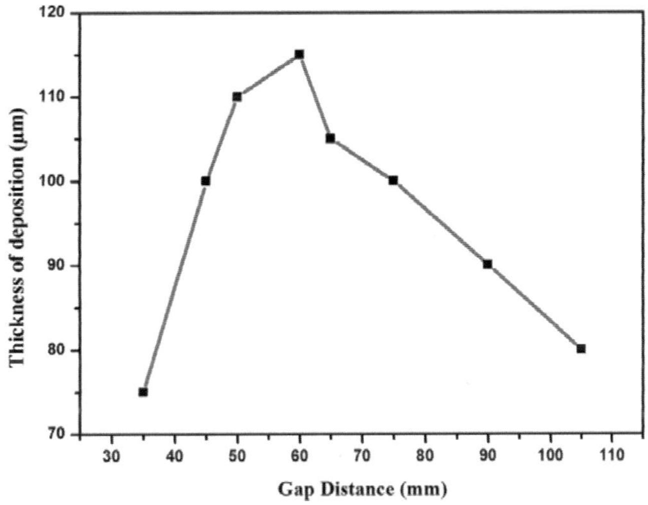

Figure 5: Plot of thickness variation with varying gap between electrodes.

5. Conclusion

The experimental, as well as simulation results, show that the surface morphology and the thickness of copper deposition on the stainless steel substrates are influenced by the gap maintained between electrodes. The results indicate that the distance influences the current density which in turn affects the thickness of deposition. The higher density of potential lines around the edges gives rise to non-uniformity in the deposition. The resistance increase due to the higher electrode gap leads to higher energy consumption. Hence varying electrode distance is not a factor that can be neglected in the process of electrodeposition or electroforming. In fact, it is a vital factor that can influence the deposition properties by regulating the average current density and electrolyte resistance. An optimum inter-electrode gap of 60 mm was suitable for the above copper deposition with increased deposition efficiency and optimum energy consumption.

References

[1] Sörgel Ş, Kesten O, Wengel A, Soergel T. Nickel/sulfur composite electroplated nickel foams for the use as 3D cathode in lithium/sulfur batteries–A proof of concept 2018 *Energy Storage Materials* 10:223-32

[2]Zhang X, Zhang HL, Yang XG, Zeng CL. Evaluation of electroplated Co-Cu metallic coatings for intermediate-temperature solid oxide fuel cells 2018 *International Journal of Hydrogen Energy* 43(49):22458-66

[3] Chandrasekar MS, Pushpavanam M. Pulse and pulse reverse plating—Conceptual, advantages and applications 2008 *Electrochimica Acta* 53(8):3313-22

[4]Pena EM, Roy S. Electrodeposited copper using direct and pulse currents from electrolytes containing low concentration of additives 2018 *Surface and Coatings Technology* 339:101-10

[5]Rosa JL, Robin A, Silva MB, Baldan CA, Peres MP Electrodeposition of copper on titanium wires: Taguchi experimental design approach 2009 *Journal of Materials Processing Technology* 209(3):1181-8

[6] Lajevardi SA, Shahrabi T. Effects of pulse electrodeposition parameters on the properties of Ni–TiO2 nanocomposite coatings 2010 *Applied Surface Science* 256(22):6775-81

[7] Modirshahla N, Behnajady MA, Mohammadi-Aghdam S. Investigation of the effect of different electrodes and their connections on the removal efficiency of 4-nitrophenol from aqueous solution by electrocoagulation 2008 *Journal of Hazardous materials* 154(1-3):778-86

[8] Aoudj S, Khelifa A, Drouiche N, Hecini M, Hamitouche H. Electrocoagulation process applied to wastewater containing dyes from textile industry 2010 *Chemical Engineering and Processing: Process Intensification* 49(11):1176-82

Investigation of impact performance of aluminum metal matrix composites by stir casting

Priyaranjan Samal*, Pandu R. Vundavilli

School of Mechanical Sciences, IIT Bhubaneswar, Odisha, India

*ps23@iitbbs.ac.in

Abstract.

In the present study, the impact behavior of in-situ synthesized aluminum alloy 5052 reinforced with titanium carbide (TiC) was investigated successfully by stir casting process. The Charpy test was conducted to evaluate the impact strength of the composites. Field emission scanning electron microscope was utilized to examine the fracture mechanism of the impact test. With increase in the TiC content, the impact strength was found to be decreased due to the brittleness of the composites. The fracture surface of the composites was characterized by cracks, voids, delamination confirming the brittle nature.

Keywords: AA5052, TiC, stir casting, impact.

1. Introduction

Aluminum metal matrix composites (AMMCs) are preferred as suitable materials for automotive and aerospace application for its high strength to weight ratio, good wear resistance, and good machinability. Therefore, AMMCs are considered good replacement of the conventional material for the above properties. By regulating the processing parameters, types of reinforcement and suitable fabrication method, the properties can be improved as per the requirement. The in-situ synthesis process is favorable over the ex-situ process because it offers advantages such as uniform dispersion, grain refinement, and very good interfacial bonding. Many researchers had developed AMMCs by choosing different series of aluminum and different types of reinforcement such as borides [1, 2], oxides [3], carbides [4, 5], etc. Chen et al. [1] fabricated Al-TiB_2 composites using stir casting method and reported that uniform distribution of reinforcement particles was achieved as a result of enhances stir casting method. A clear phase of TiB_2 particle was formed by an increase in the duration of the stirring. More ever, the tensile and yield strength of the composites was found to be increased by 260% and 180% respectively upon addition of the TiC particles. Titanium diboride reinforced with AA7075 composites was synthesized by in-situ liquid metallurgy process [2]. The TiB_2 particles were uniformly distributed throughout the composites with a clear interface between the matrix and the reinforcement. The interfacial bonding resulted from the in-situ process resists the detachment of TiB_2 particles from aluminum thereby increasing the mechanical strength. Silicon magnesium based Al alloy reinforced with Al_2O_3 particulate composites were reported by Yilmaz and Buytoz [3]. The porosity content in the composites depends upon the stirring speed and time. The optical micrographs

IOP Conf. Series: Materials Science and Engineering **653** (2019) 012047 doi:10.1088/1757-899X/653/1/012047

show the Al_2O_3 formation in the composites was found to be uniform. Samal et al. [4] reported the in-situ TiC composites of AA5052 alloy by stir casting method. The TiC particles were dispersed uniformly and the mechanical strength was increased significantly as a result of the clear interfacial bonding. Both the hardness and tensile strength increased considerably with the addition of titanium carbide. Hardness and ultimate tensile strength was enhanced upon addition of boron carbide and silicon carbide differently in aluminum 6061 alloy [5]. Commercially available pure aluminum composites reinforced with B_4C composites were fabricated through the liquid stirring method [6]. With increase in the stirring time, better uniformity was achieved of the boron carbide particles. Also, the intermittent secondary product was eliminated with an increase in the reaction time which provides a complete in-situ chemical reaction. SiC particles enhanced the impact strength of AA6068 particulate composites [7]. There was observed an increase of 75% in the impact strength on addition of SiC particles into AA6068 alloy. The increase in the impact strength resulted by lack of debonding due to better interfacial characteristics in the composites.

From the literature it was observed that liquid metallurgy i.e. stir casting is the most economical and productive fabrication method for the particulate reinforced aluminum composites. The objective of the present work is to study the impact behavior of the 5052 alloy composite reinforced with titanium carbide particles prepared by stir casting. The Charpy test was carried out to calculate the impact strength of the composites. The morphological behavior of the impact tested fracture surface was also studied.

2. Materials and Method

Selection of the suitable material for the matrix and reinforcement is necessary for the productive fabrication of the aluminum composites. The commercially available AA5052 alloy ingots were chosen for matrix and TiC as reinforcement material. AA5052 is primarily alloyed with magnesium. The chemical composition of the AA5052 matrix is given in **Table 1**. The schematic diagram of the stir casting set up used for the fabrication of the composite is shown in **Figure 1**. Titanium and graphite powder (purity >99%) was used for the synthesis of the reinforcement particles.

Table 1. Chemical composition of AA5052 alloy matrix.

Element	Mg	Si	Cu	Fe	Zn	Cu	Al
Weight %	2.62	0.11	0.02	0.23	0.07	0.26	Balance

The aluminum base alloy 5052 was completely melted at a temperature of 700°C in a carbide crucible. The preheated titanium and graphite particles were introduced into the melt with a continuous supply of argon gas to prevent any oxide reaction. For the complete in-situ formation of TiC particles to take place, the melt was heated up to 1000°C. Simultaneously, a mechanical stirrer was employed to attain uniformity in the TiC particles. This process was continued for 30 minutes for the complete reaction as well as to avoid any secondary intermittent formation. Here, the extra heat was generated due to the exothermic reaction between aluminum and titanium carbide minimizes the formation of any unwanted reaction particles. Thereafter, the slag was removed from the surface of the molten metals and the melts were poured into a preheated die. The composite sample was then machined and cut accordingly as per the ASTM E23 standard of cross-section 55mm × 10mm × 10mm with central v-notch of 45° for impact testing. Pendulum type Charpy impact testing machine was employed to estimate the impact strength of the aluminum composites. The Charpy impact tester (Tinius Olsen) used in this work is given in **Figure 2(a)** and the machined impact specimens are shown in **Figure 2(b)**. Finally, the morphological study of the fractured surface was also carried out with the help of a scanning electron microscope.

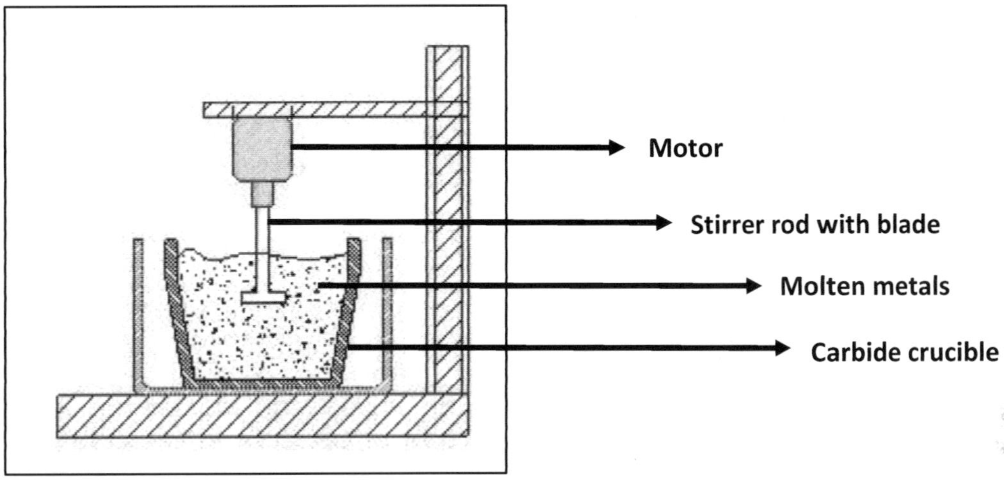

Figure 1. Schematic diagram of stir casting process.

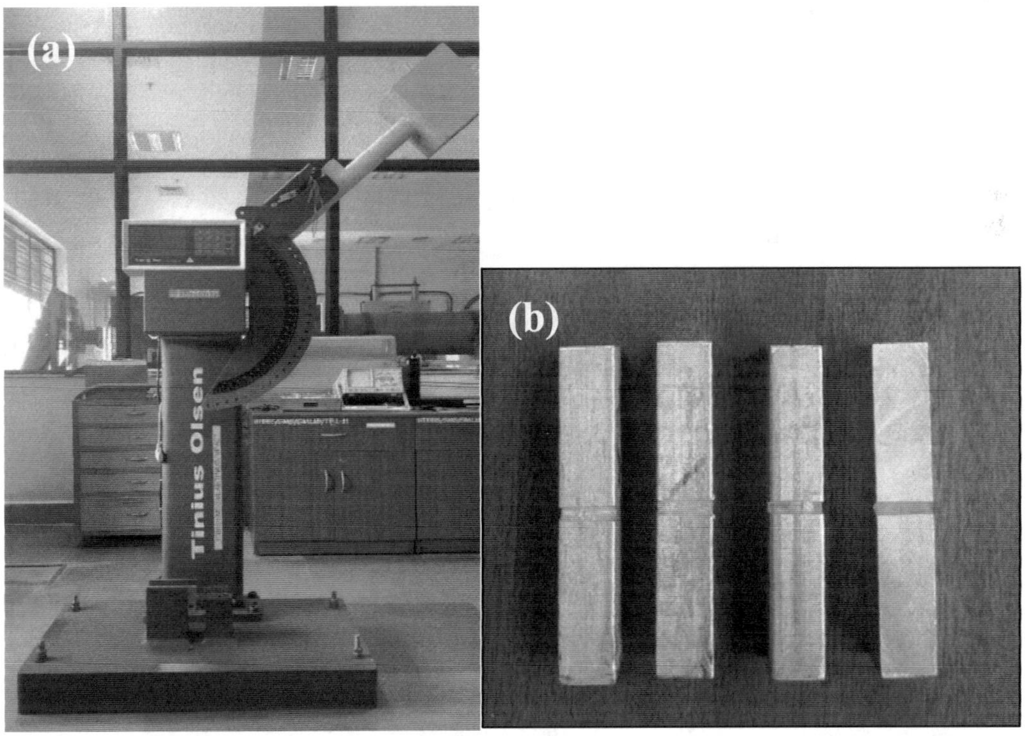

Figure 2 (a) Charpy impact tester, (b) Impact testing specimen according to ASME standard.

3. Results and Discussion

The effect of TiC on the impact strength of the composites is shown in **Figure 3** with appropriate error bars. Increase in the weight percentage of TiC in the composites reduces the impact strength. A maximum decrease of 42% was observed in case of the 5052-9wt% TiC composites as compared with the alloy. A material's impact strength normally calculated by the energy absorbed during its fracture. The addition of TiC in aluminum alloy increased its hardness, thus turned it into very brittle nature. As a result, the degree of plastic deformation energy for the composites reduced. This deformation energy increases the chances of debonding during the fracture which leads to the reduction in impact strength. Ravikumar et al. [8] described that the inclusion of TiC in the aluminum alloy improved the hardness, thus the composites are becoming more brittle. The aluminum alloy contains more plastic deformation energy. As the brittleness of the material decreases the plastic deformation energy thereby reducing the impact strength. When the aluminum alloys lose their ductility, the stress concentration areas increase. Therefore, it favors the crack formation and its propagation. These cracks create the debonding between aluminum and titanium carbide, thus decreasing its impact strength. Furthermore, it was also seen that the solid solubility of the composite is less than that of the alloy. Similar outcomes were also reported by Kumar et al. [9] by reinforcing fly ash in Al/3Cu/8Si alloy. Toughness of the aluminum composites was lowered due to the addition of the brittle fly ash particles. The toughness produced by the pure aluminum is generally better than carbide particles reinforced AMMCs. The aluminum alloy produced more toughness than the aluminum tungsten carbide reinforced composites [10]. The impact strength can be improved by employing secondary fabrication process such as hot forging.

The micrographs of the impact tested specimens are shown in **Figure 4**. The formation of dimples observed in the aluminum base alloy indicates the ductile fracture. As the content of TiC increases, the cleavages, crack formation and delamination could be visible in case of the composites. This shows the composites become brittle with the inclusion of titanium carbide in the base alloy. Presence of voids and subsequently the formation of cracks and propagation occurs because of the plastic deformation confirms brittleness of the composites. Furthermore, the loss of debonding energy due to the decrease in deformation energy of the composites can also be correlated to the reduction in impact strength.

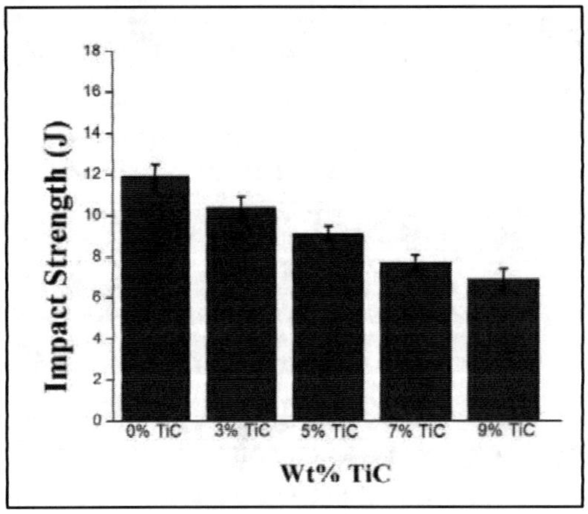

Figure 3. Variation of impact strength with wt% of TiC.

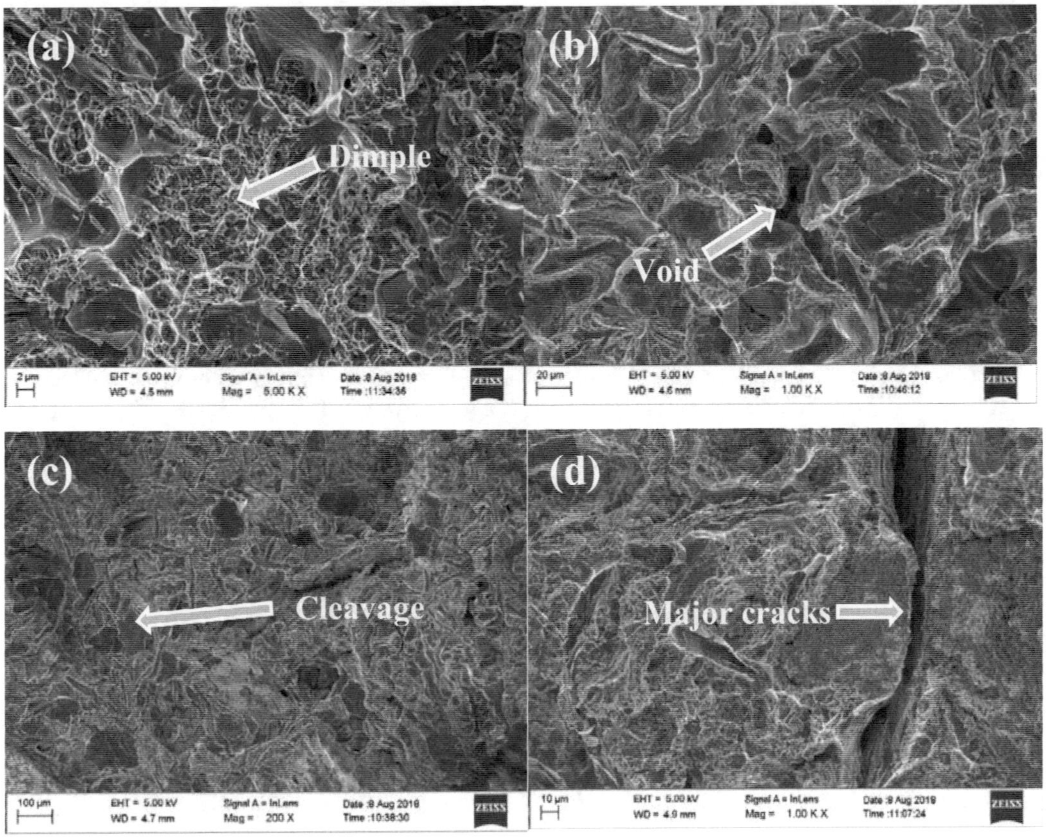

Figure 4. Micrographs of fractured impact specimen (a) 5052 alloy, (b) 5% TiC, (c) 7% TiC, (d) 9% TiC.

4. Conclusions

The aluminum MMCs reinforced with titanium carbide were successfully fabricated by stir casting in-situ liquid metallurgy process. As TiC content increases, a decrease of 45% in impact strength of the composites was observed. The high degree of plastic deformation produced by the addition of TiC particles resulted in brittleness of the composites thereby reducing the impact strength of the composites. The base alloy shows ductile fracture indicated by the dimples. The fracture impact test surface also characterized by cracks, delamination, cleavage facets indicating the brittle nature of the composites.

Acknowledgements

The authors appreciate the encouragement and constant support from the Director and Head of School of Mechanical Sciences, IIT Bhubaneswar to conduct this research work.

Reference

1. Chen, F., Chen, Z., Mao, F., Wang, T. & Cao, Z. TiB_2 reinforced aluminum based in situ composites fabricated by stir casting. *Mater. Sci. Eng. A*, 2015, **625:** 357–368.
2. Rajan, H. B. M., Ramabalan, S., Dinaharan, I. & Vijay, S. J. Synthesis and characterization of in

situ formed titanium diboride particulate reinforced AA7075 aluminum alloy cast composites. *Mater. Des.,* 2013, **44:** 438–445.

3. Yilmaz, O. and Buytoz, S. Abrasive wear of Al_2O_3 reinforced aluminium-based MMCs. *Compos. Sci. Technol.,* 2001, **61:** 2381–2392.

4. Samal, P. R., Vundavilli, P. R., Meher, A. & Mahapatra, M. M. Influence of TiC on dry sliding wear and mechanical properties of in situ synthesized AA5052 metal matrix composites. *J. Compos. Mater.* 2019, 1–14. ***Online First***

5. Gomez, L., Matrix, D. B., Amigo, V. & Salvador, M. D. Analysis of boron carbide aluminum matrix composites. *J. Compos. Mater.,* 2009, **43:** 987–995.

6. Zheng, J., Li, Q., Liu, W. & Shu, G. Microstructure evolution of 15 wt% boron carbide/ aluminum composites during liquid-stirring process. *J. Compos. Mater.,* 2016, **50:** 3843–3852.

7. Sundriyal P. and Singh V.K. Investigation of impact strength and hardness of aluminum based SiC particle reinforced composites. Journal of New Technology and Materials, 2012, **02:** 22-25.

8. Ravi Kumar, K., Kiran, K. & Sreebalaji, V. S. Micro structural characteristics and mechanical behaviour of aluminium matrix composites reinforced with titanium carbide. *J. Alloys Compd.,* 2017, **723:** 795–801.

9. Kumar, K. R., Mohanasundaram, K. M., Subramanian, R. & Anandavel, B. Influence of fly ash particles on tensile and impact behaviour of aluminium (Al/3Cu/8.5Si) metal matrix composites. *Sci. Eng. Compos. Mater.,* 2014, **21:** 181–189.

10. Narayan S. and Rajeshkannan A. Hardness , tensile and impact behaviour of hot forged aluminium metal matrix composites. *J. Mater Research Technol*, 2017, **6**: 213-219.

ANSYS-based Force Analysis on a New Horizontal Axis Wind Turbine Shield

Daniyar Seitenov[1], N.Mir-Nasiri [2*], and Md. Hazrat Ali[2]

[1] Ph.D. Student, Nazarbayev University, 010000 Astana, Kazakhstan
[2] School of Engineering, Nazarbayev University, 010000 Astana, Kazakhstan
* nazim.mir-nasiri@nu.edu.kz

Abstract. The importance of renewable energy sources cannot be denied in the era of increase of greenhouse gas emission and depletion of fossil fuels. The wind energy and design of new yet effective windmills become a necessity nowadays. This paper introduces a new configuration of wind turbine with a horizontal axis and ANSYS-based dynamic analysis of its high-speed protection shield that protects the turbine from overspending at extreme wind conditions. In comparison to traditional horizontal wind turbine structure, HAWT blades have a flat shape, and they are exposed to wind stream in such a way that the wind forces are strictly perpendicular to the blade surfaces. This structure extracts more wind energy and minimizes the useless forces affecting shaft joints in traditional wind turbines. The ANSYS-based force analysis presented in this paper proves the advantage of using of protection shield in such windmill structure.

Keywords: Protection Shield, Wind Turbine, Force Simulation, HAWT.

1 Introduction

One of the primary goals of the energy sector is to develop cheap alternative energy sources and reduce greenhouse gas emissions. Solar and wind power are the main sources of renewable energy in the energy sector. However, wind power contributes more than solar PV sources into the energy sector. The main focus of this paper is on the development of new wind power turbine system. Traditional wind turbines are classified depending on the axis of rotation, either horizontal or vertical as well as on location, either onshore or offshore [1]. Horizontal-axis wind turbines (HAWT), in fact, have higher efficiency than vertical-axis turbines (VAWT). One drawback of HAWT is that it requires aerodynamic modelling to identify the optimum height of the tower, control systems, and blade and rotor size [2]. Based on rotor location HAWTs are classified into upwind and downwind types. In case of upwind rotor type, it is faced to the wind, whereas for downwind type the wind passes tower and nacelle first only then reaches rotor for downwind. Cianetti et al. [3] include additional components such as yaw motor, gearbox, and foundation. The operation (mechanical rotation) of HAWT is based on aerodynamic lift force when wind reaches the rotor and blades. The rotor is connected with the low-speed shaft, which is, in turn, connected to the gearbox and creates a high-speed motion. The high-speed shaft has a

IOP Conf. Series: Materials Science and Engineering **653** (2019) 012048 doi:10.1088/1757-899X/653/1/012048

rigid connection with the generator. The design of blades is a major concern for such turbines. Mouhsine et al. [4] have analysed the effect of the blade on the performance of the wind turbine. It was concluded that blades should have a low angle of attack for HAWTs in order to reach a higher aerodynamic lift and thus higher efficiency for the turbine. Additionally, the design of HAWTs considers the protection system to eliminate possible damage to the rotor at high wind speeds [5]. Sorenson et al. [6] have identified the two major operational speed terms. The first is "cut-in speed" at which wind turbine starts to generate power and the second is "cut-off speed" at which wind turbine must stop to avoid possible damage of rotor and entire turbine. One of the methods to avoid this problem is based on keeping turbine rotation at a preferable operation region [7]. The operation is based on torque control, i.e. on the calculated reference torque as a function of rotational speed. Vertical-axis wind (VAWT) turbines, as opposed to HAWT, are widely used in an urban area because of its ineffectiveness in chaotic, less predictable and turbulence region [8]. The operation principles of VAWT are based on the difference of aerodynamic drag on each concave and convex blade [9]. One advantage of VAWTs is due to its ability to operate regardless of height and wind speed [10]. Additionally, it has a low cut-in speed, less structural support and does not require a yawing mechanism. Nevertheless, the major drawbacks of the turbine are because of its low efficiency and the absence of reliable performance prediction. The approximate point force analysis (when the result of wind forces are applied at the centre of the blade area) and preliminary proof of work efficiency of HAWT has been described in details in the paper [11]. This paper is focused on the shield design by using ANSYS simulation tools.

2 Design of a Horizontal Axis Wind Turbine (HAWT)

Fig. 1 shows the overall view of a new horizontal axis wind turbine. Platform 1 supports the turbine generator high enough to catch the streams of wind. The generator shaft 2 has horizontal orientation and is fitted with four flat blades 3 with slightly bent edges. The shaft has two sets of four blades symmetrically located along the shaft. The orientation of blades in two sets is selected to be offset in order to make the rotation of the shaft smooth and less jerky. For ANSYS simulation, only one set of blades have been used in this paper. Fig. 1 shows that only two blades are exposed to the wind at any time. Other two blades below the axis of the shaft are hidden by the sloped platform 1. Platform 1and vertical post 4 is connected by means of thrust bearings. It allows the platform to rotate about a vertical axis and select the best orientation for the blades with respect to changing wind direction. The orientation and rotation of the platform are controlled by the vertically oriented tail fin 5 that always keeps its plane parallel to the wind stream. The fin ensures that perpendicular forces of the wind stream always attack the blades. The distinctive and important feature of the designed system is the oscillating shield 6 that is connected to the main frame by means of horizontal hinge 7 and able to rotate about its axis. The angle of rotation is defined by the stiffness of the set of springs 8 that links shield 6 to platform 1. The stronger is wind, the larger is a deflection of springs and thus larger is the angle of the shield rotation.

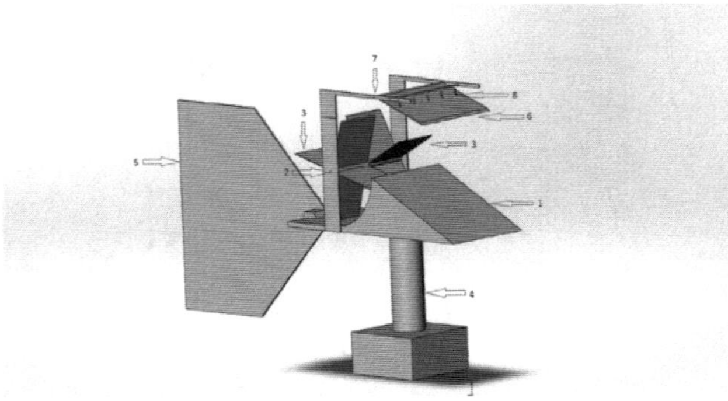

Fig. 1. Schematic view of the designed turbine system

3 Simulation and Force Analysis of the Shield

The simulation is conducted for the large scale wind turbine rotor that implies using a big size of turbine blades to generate sufficient power. For example, for the generator of 6 kW power, the selected one set of blades have 2 m width and 4 m height. For these dimensions of the blades, the shield width is determined to be W_S = 4 m wide to cover both visible to wind blade sets. The maximum wind speed that may cause full stretch of the shield is selected to be v_{max} = 25 m/sec. For the height of the selected blades H_b = 4m, the height of the shield is defined to be H_S =6 m.

3.1 Forces analysis of the shield under varying wind speed and selection of springs

This analysis first step is to analyse the deflection of the shield under various wind conditions and select the proper springs that will hold it from deflection for speeds less than 10 m/sec and start deflecting when speed exceeds 10 m/sec as well as able to hold the fully defected shield when speed reaches maximum assumed speed of 25 m/sec. This methodology can be used successfully for other possible boundary values of wind speeds, if necessary. The model of the shield was constructed using Solidworks and parameters given in paper [**100**,"Conceptual Design and Simulation of Horizontal Semi-exposed Wind Turbine (HSWT) Structure]. Computational domain has a size of 22.5x16x16 m, i.e., large enough to include the rotor blades later into the simulations (Fig. 2).

IOP Conf. Series: Materials Science and Engineering **653** (2019) 012048 doi:10.1088/1757-899X/653/1/012048

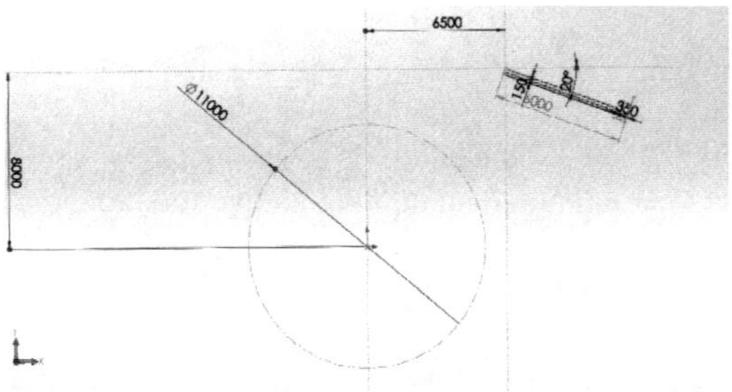

Fig. 2. Shield dimensions and location.

Patch Conforming method using tetrahedrons was chosen to save up time for simulations. The total number of elements is approximately 310 000. The set-up domain for an inlet wind speed of 10 m/sec and the selected initial shield angle of 20 degrees is shown in Fig. 3.

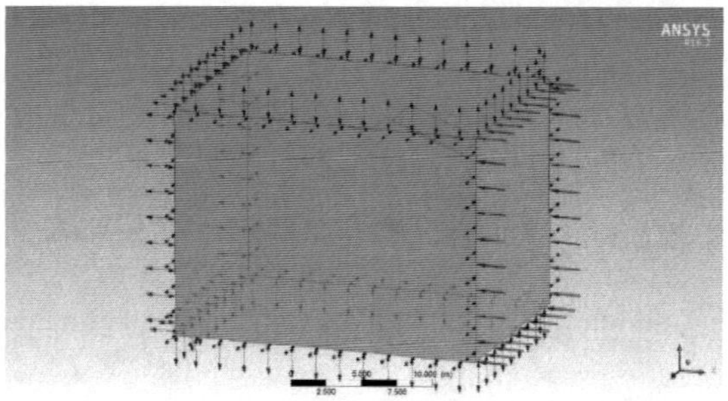

Fig. 3. Set-up of a domain with an inlet wind velocity of 10 m/s.

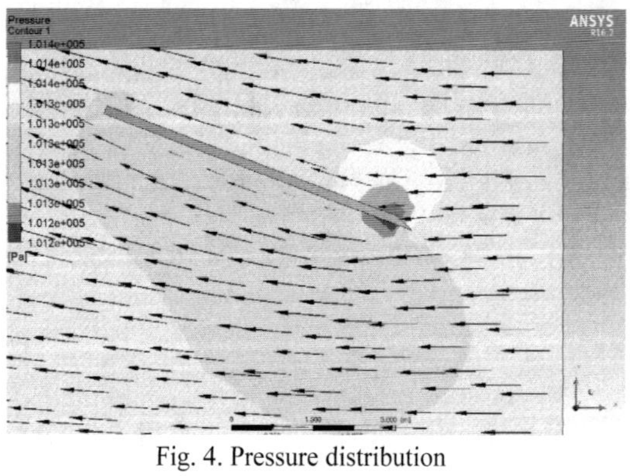

Fig. 4. Pressure distribution

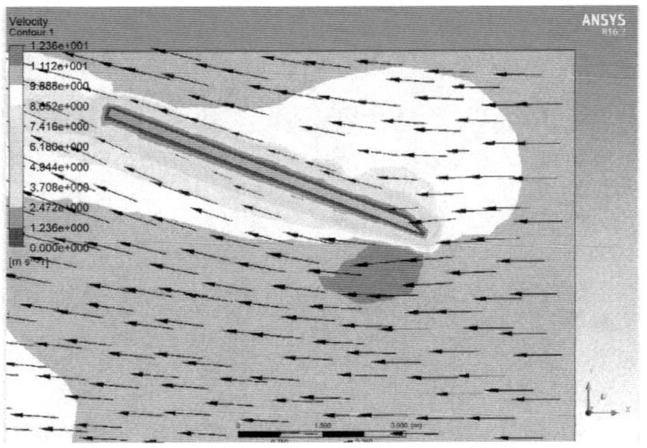

Fig. 5. Velocity distribution

Figures 4 and 5 show the pressure distribution due to wind flow and velocity contours at the center plane of the shield. In order to calculate torque on the axis of a shield, a new coordinate frame was specified to predict its value using a function calculator accurately. As a result, the torque acting on a shield around z-axis revealed to be -3401.4 N*m, which is apparently in a clockwise direction. Springs selection is based on the following methodology. The distance which is left open due to windshield position is dependent on the wind speed attacking the shield surface and selected spring parameters.

4 Conclusion

The developed HASWT system is able to fully utilize the wind energy due to the perpendicular orientation of the blades towards the wind flow. The wind forces applied on the blades generates only radially rotating forces on the shaft, and no or very minimum components of the force are generated in the axial direction of the shaft. Also, the vertically oriented tail of the platform enables the system to keep the rotor axis strictly perpendicular to the wind direction, similar to a rudder of the aeroplane. Its structure increases the efficiency of wind power absorption. The paper is focused on a small part of the project, i.e. pressure distribution on the protective shield of the wind. This distribution, as well as wind stream distribution, are generated by ANSYS software (Fig. 4 and Fig. 5) based on the given parameters and geometry of the shield in Fig. 1 and Fig. 2.

References

1. W. Tong, "Wind Power Generation and Wind Turbine Design," *WIT PRess*, 2010.
2. S. Thyagarajan, A. Bhalla, I. T. Masmanidis, C. N. Nayeri, T. P. Philippidis, and C. O. Paschereit, "HAWT for Power Generation," *International Journal of Engineering and Management Research*, vol. 3, no. 4, pp. 43–47, 2014.
3. F. Cianetti, A. Cetrini, M. Becchetti, F. Castellani, and C. Braccesi, "Dynamic modeling of wind turbines. Experimental tuning of a multibody model," *Procedia Structural Integrity*, vol. 8, pp. 56–66, 2018.

4. S. Mouhsine, K. Oukassou, M. Inchenical, B. Kharbouch, and A. Hajraoui, “Aerodynamics and structural analysis of wind turbine blade,” *Procedia Manufacturing*, pp. 747–756, 2017.
5. S. Thyagarajan, A. Bhalla, I. T. Masmanidis, C. N. Nayeri, T. P. Philippidis, and C. O. Paschereit, “HAWT for Power Generation,” *International Journal of Engineering and Management Research*, vol. 3, no. 4, pp. 43–47, 2014.
6. S. Sorensen, K. Brodbaek, M. Moller, A. Augustesen, and L. Ibsen, “Evaluation of the Load-Displacement Relationships for Large-Diameter Piles in Sand,” *Proceedings of the Twelfth International Conference on Civil, Structural and Environmental Engineering Computing*, 2009.
7. M. Rossander, A. Goude, and S. Eriksson, “Critical Speed Control for a Fixed Blade Variable Speed Wind Turbine,” *Energies*, vol. 10, no. 11, 2017.
8. P. Gulve and S. Barve, “Design and construction of vertical axis wind turbine,” *International Journal of Mechanical Engineering and Technology (IJMET)*, vol. 5, no. 10, pp. 148–155, Oct. 2014.
9. K. Pope, G. F. Naterer, and E. Tsang, “Canadian Society for Mechanical Engineers 2008 forum,” in *Canadian Society for Mechanical Engineers* , 2008.
10. S. Shah, R. Kumar, K. Raahemifar, and A. Fung , “Design, modeling and economic performance of a vertical axis wind turbine,” *Energy Reports*, pp. 619–623, 2018.
11. N. Mir-Nasiri, D. Seitenov, “Conceptual Design and Center-point Force Dynamic Simulation of a New Horizontal Axis Semi-exposed Wind Turbine”, *International Journal of Renewable Energy Research*, Vol. 6, no. 1, pp. 1-9, 2016 .

Study of controlling parameters during electro-discharge machining of titanium alloy using grey relation analysis coupled with Taguchi philosophy

Biswajit Debnath[1], Dipraj Banik[2], Rahul[3*], Hu Huang[4], B.C Routara[5]
[1,2,5]KIIT Deemed to be University, Bhubaneswar, India
[3,4]School of Mechanical and Aerospace Engineering,
Jilin University, Changchun, 130025, China
*Corresponding author: rahulkumar589@gmail.com

Abstract. Titanium grade 4 is a widely used material in the manufacturing industry due to its properties like high strength to weight ratio, low density, high-temperature resistivity, high corrosiveness. But due to its high strength, it is hard to machine in conventional machines. So here a different approach is shown to machine Titanium. Electro-discharge machine(EDM) is used for machining this metal. In this research work, an investigation is done to find the set of process parameters (peak current, gap voltage, pulse on time) for which responses like metal removal rate will be highest, but electrode wear rate and surface roughness is least. Optimization is done using Grey relation analysis coupled with Taguchi philosophy.

Keywords: Electro Discharge Machining, Titanium grade-4, copper tool, Grey Relation Analysis, Taguchi method.

1. Introduction

Electro-discharge machining (EDM) is a broadly used non-conventional machining method. The main benefit of using EDM is that it works on electric spark generation and metal erosion technology. That's why it can machine hard materials easily. Here EDM is used for machining of titanium metal. Titanium Grade-4 is a tough material with very less weight, high corrosion resistant, high-temperature resistant, and non-toxic. That's why it is used in different fields like aerospace industry, power plants, medical operations for making spacecraft, missiles, airplane parts, naval ships, armor plating, pipes of power plants, human body parts, etc. But it is quite hard to machine by conventional machining method. That's why EDM is used in this study. During EDM machining we have taken peak current, gap voltage and pulse on time as input parameters and an investigation are done to find the optimum values of input parameters for which material removal rate (MRR) will be high, but electrode wear rate (EWR) and surface roughness (SR) will be low.

Jatti [1] have studied the outcome of different parameters on responses like MRR and TWR during EDM machining of NiTi alloy and Ni Cu alloy. He used Utility method coupled with Taguchi philosophy for optimization. He observed that the peak current is the most significant factor among all the parameters. Rouniyar et al. [2] experimented the effects of controlling parameters on various responses during powder mixed EDM machining of Titanium alloy using grey relation analysis. He found that discharge current has the highest impact on responses (i.e., MRR and SR). Johan et.al.[3] has presented an experimental study to find the surface integrity of nickel-based alloy, Inconel 718 through Electro Discharge Machine(EDM) and also studied the fatigue performance of this hard material. Manohar et al. [4] studied to print metallic parts using laser powder bed fusion from hard alloy like Inconel 718 and also investigate the milling behavior of this hard material. Kumar et al. [5] have done an experimental study to find temperature profile, Material Removal Rate (MRR). Material Removal is also calculated on multi-discharge machining by calculating a number of a pulse. Kliuev et.al.[6] has presented an experimental study which describes a large setup for high-speed imaging of Electro Discharge Machine drilling process with an electrode and also with multi-hole flushing. Tripathi et.al.[7] has done a research work which studies the output responses of machining nickel-based alloy, Inconel 718 like Material Removal Rate(MRR), Tool wear rate(TWR) and Surface Roughness.

Klocke et.al.[8] has presented an experimental study that studied technological evaluation and machining technologies for the production of fir tree slots by EDM process by using hard material alloy. Devendranath et.al.[9] has done an experimental study to determine the optimal process parameters of CO_2 laser welding after iteration with different velocity for Nickel-based superalloy, Inconel 718. Yong et al. [10] have presented an experimental study to investigate the output performance like Material Removal rate(MRR), low electrode wear and small working gap by Grey-Taguchi method of a nickel-based alloy, Inconel 718. Geethapriyan et.al.[11] has presented an experimental study to drill the nickel-based superalloy, Inconel 718 through Electro Discharge machining processes. Qu et al. [12] studied in which the brazed diamond wheel was employed on the alloy to substitute an electrodeposited diamond wheel. Rahul et al. [13] also experimented the optimization of parameters on Inconel 601, 625, 718 and 825 super alloy through electro-discharge machine based on 5 factors 4 levels L16 orthogonal array and calculated the optimum results using the Topsis and Taguchi method for it. Rahul et al. [14] experimentally investigated machinability aspects of machining of Inconel 718 with a copper tool. They conducted the research work based on L25 OA by taking various parameters such as Vg, Ip, Ton, τ and Flush pressure and taken various responses such as Tool Wear Rate (TWR), Surface Roughness (Ra), Surface Crack Density (SCD) and White Layer Thickness (WLT).

The primary aim of this research work is to study the output responses like MRR, TWR, SR on a Titanium grade-4 using Electro Discharge Machine (EDM) by giving input process parameters like Peak Current(IP), Open Voltage (Vg), Pulse-On-Time (Ton).

2. Experimentation

Titanium grade 4 (3.7065, R50700) round workpiece is used as work material, and the diameter of the work material is 15mm. The chemical formation (wt%) of titanium grade-4 is shown in Table.1. Copper rod of the round cross-area with a diameter as 10 mm has been utilized as an electrode (Fig. 2). The investigations have been completed on die-sinking EDM (Make: Smart ZNC, Electronica Machine Tools Ltd, Pune, India) setup (Fig. 3). 125cc SAE-40 grade oil is utilized as a dielectric liquid. The viscosity of EDM oil is 2.16 cs at 38°C. The workpiece is kept as positive. Analysis has been done utilizing three controllable process parameters (gap voltage, peak current, pulse-on time) each fluctuated at three distinct levels (Table. 2). The structure of analysis has been arranged according to 3-factor-3-level L_9 Orthogonal Array (OA). The machining term for each test run has been kept fixed (30 minutes). The machining reactions viz. Average roughness(Ra), Metal Removal Rate(MRR), and Tool Wear Rate(TWR) have been measured for each trial run. The depiction of EDMed titanium grade-4 has been furnished in (Fig. 1). For all the experiments flushing fluid pressure is taken as 8 kg/cm^2.

Table 1. Chemical composition of Titanium Grade-4 alloy

Elements	Carbon	Iron	Hydrogen	Nitrogen	Oxygen	Titanium
Weight %	0.1	0.5	0.015	0.05	0.335	99

IOP Conf. Series: Materials Science and Engineering **653** (2019) 012049 doi:10.1088/1757-899X/653/1/012049

Fig.1. EDMed Titanium

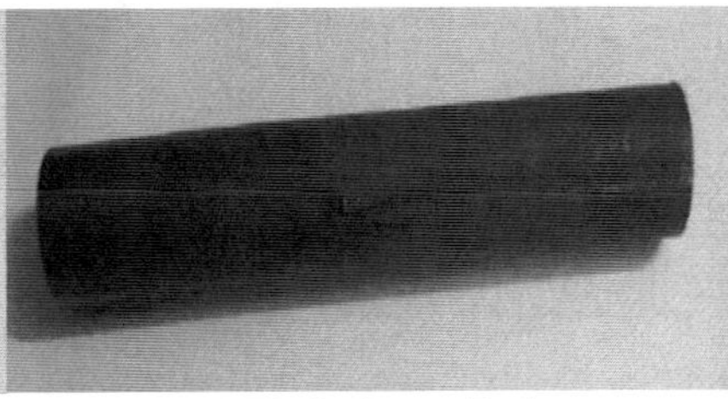

Fig.2. Copper Tool

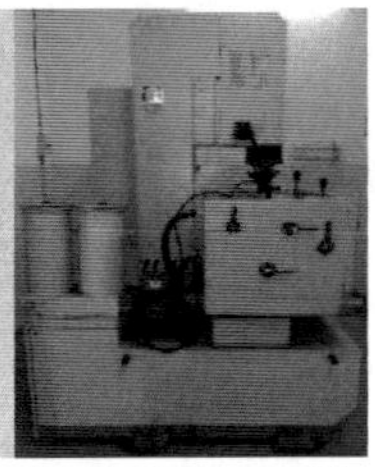

Fig.3. EDM machine

Table 2. Machining control parameters: domain of experiments

Parameters	Unit	Notation	Levels of variation		
			1	2	3
Peak current (I_p)	Amp	A	20	22	24
Pulse on time (T_{on})	µs	B	205	305	405
Gap voltage (V_g)	V	C	57	67	77

3. Methodology

3.1 Grey Relation Analysis

Grey relational analysis (GRA) was developed by Julong Deng. It is one of the most widely used methods of Grey system theory. In this investigation, the input parameters are converted into standardized value in a scale range of 0 and 1. The response parameter MRR is Larger the better type, and TWR and SR is Smaller the better type. The Normalized value is x_i^*.

Step1: Calculation of normalized value.

The Normalized value for MRR is written as $x_i^* = \frac{x_i - x_{min}}{x_{max} - x_{min}}$

and for TWR and SR the normalized value is written as $x_i^* = \frac{x_{max} - x_i}{x_{max} - x_{min}}$

Here, i = Experiment number i.e. 1,2,3.......9; x_i=standardized value of MRR, TWR, and SR for ith experiment number. x_{min} =Minimum value of MRR, TWR, and SR among the 9 experiments. X_{max} =Maximum value among the 9 experiments.

Step. 2: Evaluate Deviation Sequences (Δ) for the normalized data.

After the standardization method, all the performance value like MRR, TW, and SR will be scaled into [0,1]. The main objective of deviation arrangement is to find out the alternative whose standardized values is closest to the reference value. We have to calculate the deviation sequence with the help of normalized values of performance parameters for each experiment.

Deviation sequence Δ_i =1- the Normalized value of performance parameter for up to ith experiment.

Here I = experiment number, i.e. 1, 2...9.

Step. 3: Calculation for Grey Relation coefficient (GRC) and Grey Relational Grade (GRG).

The GRC is expressed as GRC = $\frac{\Delta_{min} - \Gamma\Delta_{max}}{\Delta_{ij} + \Gamma\Delta_{max}}$

Here i = experiment number and j = response parameter. Δ_{min}= minimum values of deviation for j^{th} response parameter among all the 9 experiments, Δ_{max} =maximum values of deviation for j^{th} Response parameter among all the 9 experiments. Γ = distinguishing factor and the value is taken as 0.5. After finding the value of GRC for all 9 experiments for all performance parameter, GRG was calculated for every experiment.

The GRG is expressed as,

$$GRG_i = (W_g * GRG_{MRRi}) + (W_g * GRG_{TWRi}) + (W_g * GRG_{Rai})$$

Here, i = Experiment number, W_g = weightage of response parameter = 0.33

4. Data analysis

At first, from experimental data's of MRR, SR, and TWR, we calculated the derivation sequence of each response. Then we calculated the grey relation coefficient for each response. Then the grey relation grade (GRG) is found out and used it for S/N ratio calculation (Table.3). Then response table for means is found out on Minitab software (Table.4). Figure.4 shows the graph of mean effects for S/N ratios.

Table 3. GRG and S/N ratio Calculation

Exp. No.	Δ_{MRR}	Δ_{TWR}	Δ_{Ra}	GRC$_{MRR}$	GRC$_{TWR}$	GRC$_{SR}$	GRG	S/N Ratio	Predicted Value
1	0.044	0.991	1.000	0.917	0.335	0.333	0.523	0.523	
2	0.063	1.000	0.828	0.887	0.333	0.376	0.527	0.527	
3	0.000	0.967	0.698	1.000	0.340	0.417	0.580	0.580	
4	0.404	0.946	0.696	0.552	0.345	0.417	0.434	0.434	
5	0.550	0.000	0.589	0.475	1.000	0.458	0.638	0.608	0.61653
6	0.603	0.838	0.311	0.453	0.373	0.615	0.476	0.476	
7	0.529	0.949	0.565	0.485	0.344	0.469	0.428	0.428	
8	0.813	0.937	0.060	0.380	0.347	0.892	0.534	0.534	
9	1.000	0.997	0.000	0.333	0.333	1.000	0.550	0.550	

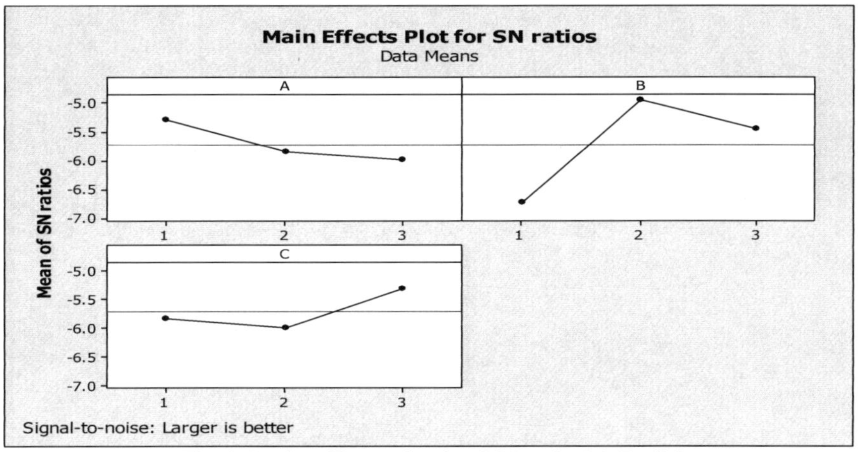

Fig.4. Main effects plot for S/N ratios[$A_1B_2C_3$]

Table 4. Response table for means of GRG

Level	A	B	C
1	0.5435	0.4623	0.5114
2	0.5163	0.5668	0.5039
3	0.5047	0.5354	0.5492
Delta	0.0388	0.1046	0.0453
Rank	3	1	2

5. Conclusion

From the above research work, we concluded the points that are given below.

- The Grey relation analysis integrated with the Taguchi approach utilized here to find a set of controlling parameters such as gap voltage, peak current, pulse-on-time to get high MRR but with low TWR and SR simultaneously.
- The optimum controlling parameters found out as ($A_1B_2C_3$), i.e. I_P=20A, T_{on}=305μs, and V_g=77V.
- Mean response table (mean S/N ratio of GRG) shows that pulse on time (T_{on}) has the highest significance on influencing the machining performances.

6. References

[1] Jatti V S 2018 Multi-characteristics optimization in EDM of NiTi alloy, NiCu alloy, and BeCu alloy using Taguchi's approach and utility concept. Alexandria Eng. J. 57 (3) 1173–2198

[2] Rouniyar A K and Shandilya P 2018 Multi-Objective Optimization using Taguchi and Grey Relational Analysis on Machining of Ti-6Al-4V Alloy by Powder Mixed EDM Process. Mater. Today: Proc. **5** 23779–23788

[3] Johan Z C, Lin M R and Johansson P S 2016 Surface Integrity and Fatigue Performance of Inconel 718 in Wire Electrical Discharge Machining. Procedia CIRP. **45** 307-310

[4] Manohar M, Selvaraj T, Sivakumar D, Gopinath S and George K M 2014 Experimental Study to Assess the Effect of Electrode Bottom Profiles while Machining Inconel 718 through EDM Process. Procedia Mater. Sci. **6** 92-104

[5] Kumar A, Bagal D K and Maity K P 2014 Numerical Modeling of Wire Electrical Discharge Machining of Super Alloy Inconel 718. Procedia Eng. **97** 1512-1523

[6] Kliuev M, Baumgart C, Büttner H and Wegener K 2018 Flushing Velocity Observations and Analysis during EDM Drilling. Procedia CIRP **77** 590-593

[7] Tripathi V, Armstrong A, Gong X, Manogharan G, Simpson T and Meter E D 2018 Milling of Inconel 718 block supports fabricated using laser powder bed fusion. J. Manuf. Processes **34** 740-749

[8] Klocke F, Welling D, Klink A, Veselovac D, Nöthe T and Perez R 2014 Evaluation of Advanced Wire-EDM Capabilities for the Manufacture of Fir Tree Slots in Inconel 718. Procedia CIRP **14** 430-436

[9] Ramkumar K D, Dev S, Prabhakar K V P, Rajendran R, Mugunda K G and Narayanan S 2019 Microstructure and properties of Inconel 718 and AISI 416 laser welded joints. J. Mater. Process. Technol. **266** 52-62

[10] Lin M, Tsao C C, Hsu C, Chiou A, Huang P and Lin Y 2013 Optimization of micro milling electrical discharge machining of Inconel 718 by Grey-Taguchi method. Trans. Nonferrous Met. Soc. China **23** 661-666

[11] Geethapriyan T, Kalaichelvan K and Muthuramalingam T 2016 Influence of Coated Tool Electrode on Drilling Inconel Alloy 718 in Electrochemical Micro Machining. Procedia CIRP **46** 127-130
[12] Qu N S, Zhang Q L, Fang X L, Ye E K and Zhu D 2015 Experimental Investigation on Electrochemical Grinding of Inconel 718. Procedia CIRP **35** 16-19
[13] Rahul, Abhishek K, Datta S, Biswal B B and Mahapatra S S 2017 Machining performance optimization for electro-discharge machining of Inconel 601, 625, 718 and 825: an integrated optimization route combining satisfaction function, fuzzy inference system and Taguchi approach. J. Braz. Soc. Mech. Sci. Eng. 39 (9) 3499–3527
[14] Rahul, Datta S, Biswal B B and Mahapatra S S 2018 Optimization of Electro-Discharge Machining Responses of Super Alloy Inconel 718: Use of Satisfaction Function Approach Combined with Taguchi Philosophy. Mater. Today: Proc. **5** 4376-4383

AUTHOR INDEX